D0854964

Community Participation and Geographic Information Systems

Community Participation and Geographic Information Systems

Edited by

William J. Craig,
Trevor M. Harris
and Daniel Weiner

London and New York

First published 2002 by Taylor & Francis
11 New Fetter Lane, London EC4P 4EE

Simultaneously published in the USA and Canada
by Taylor & Francis Inc,
29 West 35th Street, New York, NY 10001

Taylor & Francis is an imprint of the Taylor & Francis Group

Typeset in Sabon by
Integra Software Services Pvt. Ltd, Pondicherry, India
Printed and bound in Great Britain by
The Cromwell Press, Trowbridge, Wiltshire

British Library Cataloguing in Publication Data
A catalogue record for this book is available from the British Library

Library of Congress Cataloging in Publication Data
Community participation and geographic information systems/[edited by] William J. Craig, Trevor M. Harris and Daniel Weiner.
p. cm.
Includes bibliographical references (p.).
1. Geographic information systems–Social aspects. 2. Geographic information systems–Citizen participation. I. Craig, William J. II. Harris, Trevor M. III. Weiner, Daniel.

G70.212.C65 2002
910′.285–dc21 2001053008

ISBN 0415–23752–1

Contents

Figures

Tables

Contributors

Kheir Al-Kodmany is Assistant Professor in the Urban Planning and Policy Program at the University of Illinois at Chicago.

Stuart C. Aitken is Professor in the Department of Geography at San Diego State University in San Diego, California.

Michael Barndt is Coordinator of the Data Center Program at the Nonprofit Center of Milwaukee. At the time of this research, Michael was also Associate Professor of Urban Studies Programs at the University of Wisconsin-Milwaukee.

Crystal Bond is Cartographer at the Cherokee Nation GeoData Center, Tahlequah, Oklahoma.

Mark Bosworth is GIS Program Supervisor at Metro in Portland, Oregon.

Patrick Burke is Director of Planning at the Board of Education in Fulton County, Georgia. He is a Ph.D. candidate in City and Regional Planning at the Georgia Institute of Technology and was, at the time of this research, Project Manager with Georgia Tech's Data and Policy Analysis Group (DAPA).

Liza Casey is Senior Consultant of ESRI Inc. in Washington DC. At the time of this research, Liza was Director of Enterprise GIS, City of Philadelphia.

Richard E. Chenoweth is Professor in the Department of Urban and Regional Planning at the University of Wisconsin-Madison.

Alison Cottrell is Lecturer in the Department of Tropical, Environmental Studies and Geography at James Cook University in Townsville, Australia.

Paul Couey is GIS Communications Specialist at Metro in Portland, Oregon.

William J. Craig is Associate Director at the Center for Urban and Regional Affairs and Co-Director of the Master of Geographic Information Science Program at the University of Minnesota in Minneapolis.

Jack Dangermond is President of ESRI Inc. in Redlands, California.

Raymond de Lai is Centre Manager of the Herbert Resource Information Centre in Ingham, Australia.

John Donovan is Public Involvement Specialist at Metro in Portland, Oregon.

Sarah Elwood is Assistant Professor in the Department of Geography at DePaul University in Chicago, Illinois. At the time of this research, Sarah was a Ph.D. candidate in Geography at the University of Minnesota.

Gisela Frias is a Ph.D. candidate in the Department of Geography at McGill University in Montréal, Canada.

Trevor M. Harris is Eberly Professor of Geography and Chair of the Department of Geology and Geography at West Virginia University in Morgantown, West Virginia.

Andrew K. L. Johnson is Principle Research Scientist at CSIRO Sustainable Ecosystems in Brisbane, Australia.

Gavin Jordan is Senior Lecturer in the National School of Forestry at the University of Central Lancashire in Carlisle, England.

Richard Kingston is Research Officer in the School of Geography at the University of Leeds, England.

John B. Krygier is Assistant Professor of Geography in the Department of Geology and Geography at Ohio Wesleyan University in Delaware, Ohio.

Peter A. Kwaku Kyem is Assistant Professor in the Department of Geography at Central Connecticut State University in New Britain, Connecticut.

Melinda Laituri is Associate Professor in the Department of Earth Resources at Colorado State University in Fort Collins, Colorado.

Anne M. Leitch is Journalist at the CSIRO Sustainable Ecosystems in Brisbane, Australia.

Helga Leitner is Professor in the Department of Geography at the University of Minnesota in Minneapolis.

Robert B. McMaster is Professor in the Department of Geography at the University of Minnesota in Minneapolis.

Susanna McMaster is Associate Program Director of Master of Geographic Information Science Program at the University of Minnesota in Minneapolis.

Paul Macnab is Oceans Policy Officer at Fisheries and Oceans Canada in Halifax, Nova Scotia, Canada. At the time of this research, Paul was a graduate student in the Department of Geography at the University of Waterloo.

Thomas C. Meredith is Associate Professor in the Department of Geography at McGill University in Montréal, Canada. Thomas is also a principal of McGill University's Community-Based Environmental Decision Support (CBED) project.

Bernard J. Niemann, Jr. is Professor in the Department of Urban and Regional Planning at the University of Wisconsin-Madison.

Cheryl Parker is Principal at Urban Explorer in Berkeley, California (www.theurbanexplorer.com). At the time of this writing, Cheryl was Economic Development Specialist at the South of Market Foundation.

Amelita Pascual is Manager of the Department of Cellular and Molecular Pharmacology at the University of California at San Francisco. At the time of this writing, Amelita was Executive Director of the South of Market Foundation.

Tom Pederson, ESRI Inc. in New York City. At the time of the research, Tom was finishing his Ph.D. in City Planning and serving as Director of Research and Development at the Cartographic Modeling Lab, University of Pennsylvania.

David Randall Peterman is Transportation Analyst at the Congressional Research Service, Library of Congress in Washington, DC. He is a Ph.D. Candidate in City and Regional Planning at the Georgia Institute of Technology.

David Pullar is Lecturer in the Department of Geographical Sciences & Planning at the University of Queensland in Brisbane, Australia.

Eric Sheppard is Professor in the Department of Geography at the University of Minnesota in Minneapolis.

David S. Sawicki is Professor in the City and Regional Planning Program and in the School of Public Policy at the Georgia Institute of Technology in Atlanta. Sawicki is also Director of Georgia Tech's Data and Policy Analysis group (DAPA).

Renée E. Sieber is Assistant Professor in the Department of Geography and in the School of Environment at McGill University in Montréal, Canada. At the time most of this research was conducted, Renée was a Ph.D. candidate in Urban Planning at Rutgers University.

Michael J. Shiffer is Associate Professor in the Urban Planning & Policy Program and Director of the Digital Cities Lab at the University of Illinois at Chicago.

Susan C. Stonich is Professor in the Department of Anthropology and in the Interdepartmental Graduate Program in Marine Science; she is also chair of the Environmental Studies Program; University of California, Santa Barbara.

Todd L. Sutphin is Outreach Specialist in the Land Information and Computer Graphics Facility at the University of Wisconsin-Madison.

David L. Tulloch is Assistant Professor in the Department of Landscape Architecture and Associate Director of the Center for Remote Sensing and Spatial Analysis at Rutgers University in New Brunswick, New Jersey.

Stephen J. Ventura is Professor at the Institute for Environmental Studies and in the Department of Soil Science at the University of Wisconsin-Madison.

Daniel H. Walker is Principal Research Scientist at CSIRO Sustainable Ecosystems in Townsville, Australia.

Daniel Weiner is Professor of Geography in the Department of Geology and Geography and Director of the Office of International Programs at West Virginia University in Morgantown, West Virginia.

Gregory G. Yetman is Geographic Information Specialist in the Center for International Earth Science Information Network (CIESIN) at Columbia University in New York City. At the time of this research, Gregory was a graduate student in the Geography Department, McGill University.

Foreword

It is now almost 40 years since Roger Tomlinson coined the term *geographic information system* (GIS), and led the development of the world's first, the Canada Geographic Information System (CGIS), in the mid-1960s (for a history of GIS see Foresman 1998). Today's technology would be almost unrecognizable to the pioneers of the 1960s, not only because of the almost unbelievable advances in information technology (IT) that have occurred since then, but also because of dramatic changes in the functionality, appearance, use, and societal context of GIS. This book addresses one of the most recent manifestations of those changes, the developing use of GIS by grassroots community organizations, and participation in its use by ordinary citizens.

Early GIS was massively expensive. CGIS required a large, dedicated mainframe computer costing several millions of 1965 dollars; the development of hundreds of thousands of lines of computer code in very primitive programming language; and the invention of novel devices for converting maps to digital form. Although the project was based on sound cost-benefit analysis, the technical problems of building CGIS were such that by the early 1970s, and despite the expenditure of tens of millions of dollars, CGIS was essentially unable to deliver the results that had been promised to its sponsors, and several more years of effort were required to bring it to operational status.

CGIS was a child of its time. Only senior governments were able to afford the cost of early GIS, and only skilled experts were able to do successful battle with its primitive interfaces. As with other early computer applications, early GIS was designed to augment the limited and fallible skills of humans, by performing tasks that humans found too difficult, tedious, or inaccurate when done by hand; in the case of CGIS, these tasks included measuring areas from maps, and overlaying maps, both on a massive scale. In essence, CGIS was performing the geographic equivalent of other applications driving early computer development – the massive numerical simulations of nuclear explosions being performed by Los Alamos National Laboratory, or the massive cryptographic computations of the National Security Agency.

Early computers quickly gained a popular image of mechanical efficiency, lightning speed, and perfect accuracy that was in sharp contrast to supposed human characteristics of clumsiness, sluggishness, and vagueness. As such, both they and GIS fed the human appetite for enlightenment. Computerized maps would replace the stained, creased, and tattered maps of the glove compartment. Instead of inaccurate maps recording someone's impression of land-use at some undetermined point in the past, the civilian remote sensing satellites that began to appear in the early 1970s would continuously monitor the Earth's surface and ensure a constant, up-to-date, and precise digital record.

Early GIS was also firmly grounded in science, and its associated ideas of objectivity and replicability. We knew, of course, that some of the data being entered into CGIS had been invented by poorly-paid undergraduates idling in coffee shops, but once in the computer and stripped of this awkward human lineage the data appeared to all intents and purposes as if they had been measured by the most precise of scientific instruments. The scientific measurement model dominated early thinking in GIS, and may have reached its apogee in the Digital Earth speech of Vice President Al Gore in 1998, describing a future in which it would be possible to enter and explore a virtual world based on a perfect digital replica of the planet that included measurements of practically everything.

Early GIS was not surprisingly much more attractive to users whose applications lay in the physical and natural sciences, than in the social sciences. Although GIS made useful inroads into marketing and site selection (Martin 1996), by and large it was the physical aspects of the planet that dominated early GIS use. GIS was adopted by forest management agencies and lumber companies; by engineering consultants and utility companies; by Earth system scientists, landscape ecologists, and agronomists. But only recently has there been substantial interest among sociologists, economists, and political scientists in the potential of GIS to elucidate social processes (for more on the social science applications of GIS see CSISS.org).

Many factors have contributed to the evolution of GIS over the past 40 years, and brought it to the state in which we find it today. First and perhaps foremost is the cost of hardware. The power of the multimillion-dollar computer used by CGIS is now vastly exceeded by the average laptop, and the most advanced GIS applications now run on computers costing less than $2,000. At this level, GIS is affordable by many libraries, schools, households, and community organizations, although it is still far beyond the reach of others, particularly in developing countries. The cost of software has also dropped substantially, in tandem with the cost of hardware, as demand for both has grown.

Second, developers of GIS software have made great progress in facilitating use through improved user interfaces. Early GIS required its users to learn its specialized language, and by the late 1980s command languages had grown to include thousands of terms, to be used in precise and unforgiving syntax. But the early 1990s brought WIMP (windows, icons, menus, pointers) interfaces into the computing mainstream. Learning to use GIS is still a challenge, but it is now at least possible for children in elementary school to use it effectively. We are still a long way from the kind of intuitive interface that would be readily usable by a child of ten, but GIS users no longer require skills comparable in complexity and sophistication to those of concert pianists.

This trend towards more intuitive interfaces is part of a deeper, third trend, towards a more human-centric vision of GIS. Researchers in the early 1990s noted how GIS interfaces were essentially *intrusive*, requiring their users to learn the system's language, rather than adopting the intuitive language of humans. We humans work every day with geographic information, as we share driving directions, describe distant places to each other, or reason about the information we acquire through our senses. Much is known about how children acquire spatial skills, and how they build mental models of their surroundings. If the conceptual structures of GIS were similar, it was argued, then GIS would be necessarily easier to use, and accessible to a much larger proportion of the general public, including children. GIS researchers began to discover cognitive science, and those parts of linguistics that deal with concepts about the geographic world (Frank and Mark 1991).

Human discourse is inherently vague, and science has long been concerned with providing an alternative to vague subjectivity. Instead of describing things as hot or cold, scientists measure temperature on standard scales in order to ensure replicability and shared meaning, and early GIS similarly imposed requirements of precision on its users, forcing them to replace vague terms like *near* with precise measurements of distance. So, while on the one hand this ensured objectivity and meaningfulness, it also acted as a filter. Human discourse is vague, but it is at the same time semantically rich, with nuances that allow one word to have many context-specific shades of meaning. By comparison, the scientific GIS is precise, but also crude in its simplicity.

The final twist in this transition was brought about by the IT revolution of the 1990s, which not only put computers into the hands of millions, but demanded that they address everyday needs. No longer would users be required to learn the language of computers – the new software interfaces of the 1990s were designed to do something useful for the average person almost immediately upon installation. The spreadsheet software of the 1980s probably did the most to precipitate this trend, but by the late 1990s, even GIS was starting to enter the application mainstream. Computers are now seen not as calculating wizards but as connections to the Internet, providing essential channels of communication between humans (Goodchild 2000). By extension, a GIS was no longer a way of doing things that humans found tedious, time-consuming, or clumsy to do by hand, but the means by which humans exchanged information about the world around them. A GIS ought to be able to accommodate all ways of describing the world, from maps and images to stories, pictures, and sketches. When such descriptions are sufficiently precise, it should be possible to reason and analyse them automatically by selecting from a battery of standard techniques; but precision should not be a requirement for entry into the GIS world.

Any community attempts to maximize its well-being and potential within the constraints imposed on it by its technology and its environment. Communication and sharing of geographic information are essential to the successful functioning of communities, because they are the means by which each individual extends his or her knowledge beyond the limits of the human senses – we learn what is beyond the mountains, or across the river, or what used to be in the past, by communicating with others who live there or have been there, or by sharing the products of satellites that can see from space. New technologies are adopted if they loosen the constraints of older ones, allowing the community to reach new levels of well-being. In that sense, the IT revolution of the late twentieth century has greatly loosened the constraints on human communication by providing new channels that are capable of transmitting virtually any form of information at high speed and minimal cost.

This book is about some of the potential that new IT offers to communities. It could not have been written in the 1960s, and perhaps not as recently as ten years ago, so it is very timely. If GIS is indeed ubiquitous and easy to use – and there are many reasons why we have not yet reached that point – then it has the potential to revolutionize the ways in which communities develop consensus about their surroundings, resolve disagreements and difficulties, and plan for the future. But there are few guidelines on how to take advantage of the undoubted power of modern GIS, and community participation will clearly occur in settings that are very different from those faced by the GIS pioneers and the users of early GIS: indeed, in many cases communities will find themselves using GIS to oppose and disarm the agencies that adopted GIS much earlier, under the older paradigm. An early instance of this occurred in the 1980s on the North Slope of Alaska, when the local burough government found itself with no alternative but to adopt the then-expensive technology already being used by the oil companies in Prudhoe Bay to build arguments in support of land-use permits, and paid for it with oilfield royalties.

This book is a collection of reports from pioneers who are developing the guidebooks, and creating the roadmaps. They are experimenting with new forms of visualization that are more readily understood by non-experts; with new forms of representation that recognize multiple perspectives on the same reality; and with new forms of community interaction through the Internet and its communication technologies. Many of these experiments will ultimately enrich GIS technology by driving a new generation of technological developments. But the ultimate test will occur some time in the future, when it will be possible to ask the fundamental question: how does GIS affect the ways in which communities are able to build awareness of their surroundings, develop consensus, and argue persuasively for a better future?

As Chair of the Executive Committee of the National Center for Geographic Information and Analysis, I am honoured to have been asked to write the foreword to this book, which stems in part from an NCGIA-sponsored workshop on public participation GIS (PPGIS), funded under the Varenius project, NCGIA's effort to advance geographic information science. The National Science Foundation provided funding for the workshop and the Varenius project under Cooperative Agreement SBR 9600465. The leaders of the workshop and the editors of this book are to be congratulated on stimulating a new level of interest in the use of GIS in community development and planning, and I hope this book will make the work in this area accessible to a larger audience, including practitioners as well as researchers in the many disciplines that overlap PPGIS.

Michael F. Goodchild
Professor, Department of Geography
University of California, Santa Barbara
Director, Center for Spatially Integrated Social Science
Chair of the Executive Committee
National Center for Geographic Information and Analysis

REFERENCES

Foresman, T. W. (1998) *The History of GIS: Perspectives from the pioneers*, Upper Saddle River, NJ: Prentice Hall PTR.

Frank, A. U. and Mark, D. M. (eds) (1991) *Cognitive and Linguistic Aspects of Geographic Space*, Dordrecht: Kluwer Academic Publishers.

Goodchild, M. F. (2000) 'Communicating geographic information in a digital age', *Annals of the Association of American Geographers* 90(2): 344–355.

Martin, D. (1996) *Geographic Information Systems: Socioeconomic applications*, second edition, London: Routledge.

Acknowledgements

This book grew out of a 1998 Specialist Meeting held at the University of California, Santa Barbara and sponsored by Project Varenius of the National Center for Geographic Information and Analysis (NCGIA). We owe thanks to the many who participated in the meeting. In addition, we would like to thank Harlan Onsrud for encouraging us to propose this initiative. We also owe thanks to the staff of the NCGIA, especially LaNell Lucius and Karen Kemp, for providing excellent support for the meeting.

This resulting book contains the work of nearly four-dozen authors who wrote and revised their chapters to fit into a coherent collection of case studies and reflections on *Community Participation and Geographic Information Systems*. We wish to thank them for their willingness to share their work and for their patience with us as we went through the process of moving the book from concept to reality.

Equally important, we wish to acknowledge the communities where our authors undertook their work. Many of these communities are struggling to survive, and yet they were willing to invest some of their time to see whether this new GIS technology could be of any assistance to them. These communities are the real pioneers and we hope that their daring has paid adequate dividends to them. We know that other communities will benefit greatly from their documented experiences.

Finally, we wish to thank Mike Greco, communications director for the Center for Urban and Regional Affairs at the University of Minnesota. When we were short on time and needed professional help in editing the manuscripts, Mike stepped forward and provided excellent service for us.

Part I

Introduction

Chapter 1

Community participation and geographic information systems

Daniel Weiner, Trevor M. Harris and William J. Craig

It is not enough for a handful of experts to attempt the solution of a problem, to solve it, and then apply it. The restriction of knowledge to an elite group destroys the spirit of society and leads to its intellectual impoverishment.

Attributed to an address by Albert Einstein at Caltech, 1931
Source: The expanded quotable Einstein, Alice Calaprice (ed.), Princeton University Press, Princeton, 2000

1.1 INTRODUCTION

Geographic information systems (GIS) and geographic information technologies (GIT) are increasingly employed in research and development projects that incorporate community participation. For example, there are now applications involving indigenous natural resource mapping in arctic and tropical regions within the Americas (Marozas 1993; *Cultural Survival Quarterly* 1995; Bond, this volume). There is also a rapidly growing network of planning professionals interested in how GIS can merge with community participation in the context of neighbourhood revitalization and urban planning (Aitken and Michel 1995; Craig and Elwood 1998; Leitner *et al.*, this volume; Sawicki and Peterman, this volume; Talen 1999; 2000). Environmental groups are experimenting with community GIS applications to promote environmental equity and address environmental racism (Sieber 2000; Kellogg 1999). Furthermore, NGOs, aid organizations, and governmental agencies are linking communities with GIS as they seek to promote more popular and sustainable development projects (Dunn *et al.* 1997; Elwood and Leitner 1998; Gonzales 1995; Harris *et al.* 1995; Hutchinson and Toledano 1993; Jordan and Shrestha 1998; Kwaku-Kyem 1999; Mitchell 1997; Obermeyer and Pinto 1994; Rambaldi and Callosa 2000; Weiner *et al.* 1995; Weiner and Harris 1999).

Importantly, these applications have in common the linking of community participation and GIS in a diversity of social and environmental contexts

(Abbot *et al.* 1998; Harris and Weiner 1998). They also demonstrate a variety of methodological approaches. In October 1998, an NCGIA (National Center for Geographic Information and Analysis) sponsored Varenius initiative (Craig *et al.* 1999) brought together academics and practitioners experimenting with public participation GIS (PPGIS) (see Goodchild *et al.* 1999 for an overview of the Varenius project). Case studies were presented that were drawn from many world regions and included applications in urban and community development, environmental management, and development planning.

This volume on *Community Participation and Geographic Information Systems* draws upon Varenius project case studies and conceptual contributions. The book situates PPGIS within broader GIS and Society debates, and addresses six core concerns:

1. differential access to geographic information and technology,
2. integration and representation of multiple realities of landscape within a GIS,
3. identification of the potential beneficiaries of participatory GIS projects,
4. development of place-based methodologies and methods for more inclusive community participation in spatial decision-making,
5. situating of PPGIS production and implementation in its local political context, and
6. identification of community GIS contributions to geography and GIScience.

A key assumption of the Varenius initiative was that community-based GIS projects simultaneously promote the empowerment and marginalization of socially differentiated communities. As a result, the nature of the participatory process itself is critical for understanding who benefits from access to GIS and why. PPGIS explicitly situates GIS within participatory research and planning and, as a result, local knowledge is incorporated into GIS production and use. There are formidable social and technical challenges involved in the successful design and implementation of PPGIS. The enthusiasm for undertaking PPGIS is thereby complicated by the difficulties encountered in its implementation (Barndt 1998).

Community Participation and Geographic Information Systems is intended for a broad audience of students, academics, planners, policy-makers, and GIS practitioners. When reading the book, we caution that substantive GIS and Society concerns should not be ignored because of the growing fascination for developing more inclusive GIS. Johnston (1999: 45) argues that 'GIS usages have been subject to substantial critiques...and the role of GIS in creating new images of the world is increasingly appreciated...but the technology's positive potential has been submerged under the weight of this (usually valid) assessment of likely negative impacts.' This book and its 46 contributors suggest an alternative interpretation whereby the critique of GIS has helped to launch a flood of alternative community-based GIS applications. Indeed, we are concerned that the rapid growth of PPGIS might have the opposite effect of submerging a critical theory of GIS. PPGIS is not a panacea, and must not undermine the robust debate on the political economy of GIS, its epistemology, and the philosophy and practice of GIScience. Pickles (1999) and Sheppard *et al.* (1999) provide valuable overviews of these issues.

1.2 GIS AND THE COMMUNITY

Community can be defined by physical proximity to others and the sharing of common experiences and perspectives. The word has become synonymous with neighbourhood, village or town, although communities can also exist in other forms – e.g. through professional, social, or spiritual relationships. Communities can thus be virtual (Kitchin 1998; Graham 1998). *Public participation* in this book refers to grassroots community engagement. Jane Jacobs (1961) has eloquently documented how neighbourhoods attain vitality through the collective efforts of individuals who care about their common place. Castells (1983) has provided evidence that community-based action has occurred in a wide variety of cultures and is universal.

For several reasons, communities formalize themselves and create official organizations with which the state can negotiate. Participants in such organizations see opportunities to achieve individual goals through collective action (Olson 1965). Politicians are responsive to community organizations when they represent sufficient numbers of committed voters (Grant and Omdahl 1993). Planners, in particular, pay attention to public participation and community organizations (Jones 1990) because community input is critical for defining local issues. Planners accept that community-developed solutions are feasible because they tend to be reasonable, realistic, and sustainable. Public participation is important in community planning, but has been practiced in ways that range from evasion to full empowerment. This range may be seen as a ladder of increasing participation (see Figure 1.1). On the lowest rung, citizens are (sometimes) provided with requested information. At the top rung, the public has a full voice in the final decision, usually through a community organization.

Geographic information systems can assist community organizations regardless of the rung they are placed on, and assist them to climb the ladder further. Better information will help develop appropriate responses, and the technology will support the creation of map products and analysis. GIS can also help a community organization climb the participation ladder, and the state may be willing to share more power with a credible partner. Similar community organizations see one organization's status grow, and are more

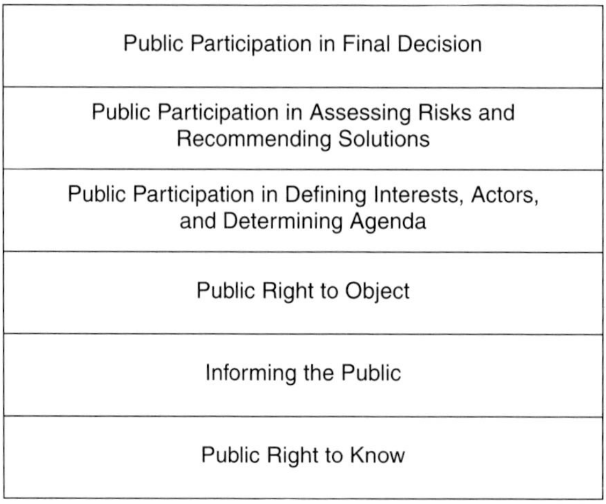

Public Participation in Final Decision
Public Participation in Assessing Risks and Recommending Solutions
Public Participation in Defining Interests, Actors, and Determining Agenda
Public Right to Object
Informing the Public
Public Right to Know

Figure 1.1 The citizen participation ladder adapted from Weidemann and Femers 1993.

likely to enter into collaborative efforts with them. However, even the most homogeneous community contains individuals whose goals differ from those of the group, and who may be marginalized by this process.

1.3 THE CONCEPTUAL ORIGINS OF PPGIS

Although PPGIS projects are being implemented within the context of an academic debate over GIS and Society, there is also a spontaneous fusion of participatory forms of development planning with new ITs. As a result, PPGIS has a rich and diverse conceptual history that draws upon several intellectual traditions including political economy and critical theory, participatory planning and community development, democracy and social justice, anthropology and ethnography, political ecology, and philosophies of science.

Although the GIS and Society debates emerged in the 1990s, Tomlinson had earlier recognized the importance of non-technical institutional and managerial issues in the success or failure of a GIS effort (Chorley Report 1987), and Chrisman had provided valuable insight into the social, political, and ethical implications of GIS use (Chrisman 1987). In the early 1990s, however, several researchers entered into a social-theoretical critique of the perceived positivism and hegemonic power relations embedded within GIS (Curry 1995; Goss 1995; Lake 1993; Pickles 1991; 1995; Smith 1992; Taylor 1990; 1991; Taylor and Overton 1991; 1992). Much of this concern focused on the claimed objectivity and value-neutral nature of GIS.

Taylor (1990) argued that with the increasing popularity of GIS within the field of geography, 'facts' had risen to the top of the geographical agenda, accompanied by a concomitant retreat from knowledge to data. As a result, GIS was viewed as a return to empiricism and positivism (Taylor 1990: 212). Pickles (1991) and Edney (1991) also questioned the potential anti-democratic nature of GIS brought about by differential access to data and technology, as well as the surveillance capabilities of GIS that reinforced both particular knowledge-power configurations and the technologies of normalization, knowledge engineering, and control of populations (see also Rundstrom 1995; Yapa 1991). Openshaw's (1991) response captured the surprise, frustration, and anger of the GIS community to the scale and intensity of such critiques. Goodchild (1995) and Sheppard (1995), on the other hand, acknowledged the validity of some of these critiques and offered a valuable prospective for GIS and Society research.

Building on this literature and conference discussions of these themes, a workshop sponsored by the NCGIA on 'Geographic Information and Society' was organized in 1993 at Friday Harbor by Poiker, Sheppard, Chrisman, and others. Some 23 prepared papers were discussed, and several were subsequently published in a special issue of *Cartography and Geographic Information Systems* (Sheppard and Poiker 1995). The workshop exchanges were surprisingly positive, and laid the foundation for an ongoing dialogue and research agenda that identified issues of access, ethics and values, representation, democratic practice, privacy, and confidentiality as particularly significant (Sheppard 1995). Contemporaneously, the influential book *Ground Truth: The Social Implications of Geographic Information Systems* (Pickles 1995) sought to capture the essence of the critique of GIS, and to build on what Pickles perceived as the 'creative tensions' between the social theory and GIS communities.

Building on the enthusiasm of the Friday Harbor meeting, the NCGIA sponsored Initiative #19: 'GIS and Society – The Social Implications of How People, Space, and Environment are Represented in GIS.' The first specialist meeting of this initiative was held in March 1996 in Minnesota (Harris and Weiner 1996). Three broad conceptual issues were identified: the epistemologies of GIS; GIS, spatial data institutions, and access to information; and developing alternative GIS. Participants at the meeting questioned whether a 'bottom-up' GIS could be successfully developed, and discussed what forms this system might take. A number of other probing questions were raised, including how community participation could be incorporated into a GIS, and to what extent such participation would serve only to legitimize conventional top-down decision-making. It was at this meeting that a further question was posed regarding what an alternative GIS – what became known as GIS2 – might look like. It was from these reflections that the concept of public participation GIS arose. This theme was developed and the term defined at a subsequent meeting held in Orono, Maine (Shroeder

1996). The discussion about 'alternative' types of GIS production, use, access, and representation is based on an understanding of the social impacts of existing applications of GIS. Thus, it is unwise to detach the PPGIS discussion from its broader conceptual base in GIS and Society issues.

At the 1997 University Consortium of GIS summer meeting in Bar Harbor, Maine, it was proposed that PPGIS be incorporated into a new Varenius initiative. A core planning group was established, and a proposal was submitted to the NCGIA. From the beginning, it was presumed that the initiative would focus on field experiences and alternative GIS implementations reflecting the existence of PPGIS in many socio-geographic contexts. The workshop reviewed a variety of PPGIS initiatives, considered critical social and technical issues associated with their implementation, and discussed the successes and failures of existing PPGIS projects.

The formal presentations and the discussions that ensued, generated a number of perspectives about community uses of GIS and GIT. The chapters that follow are a result of this workshop, and are valuable not only for shedding light on the conceptual core of PPGIS, but also for providing case studies of how PPGIS are presently constructed and implemented. The chapters also point to the importance of the social, historical, and political contexts in which PPGIS initiatives are pursued.

1.4 EMERGING PPGIS THEMES

Community Participation and Geographic Information Systems identifies PPGIS as a broad tent with multiple meanings and a global reach. The introductory chapters in Part I confirm that there are many emerging forms of community interaction with GIS that are linked to the social and geographic context of PPGIS production and implementation. Sawicki and Peterman report on the already extensive PPGIS suppliers in the United States. Although their survey generated low response rates, and the broad definitions of PPGIS created difficulties when compiling the database, they identified 67 organizations in 40 cities that claimed to have some form of PPGIS. Four types of institutional location for PPGIS delivery in the United States are identified: nonprofit organizations (31), universities (18), government agencies (16), and private companies (2).

Leitner *et al.* draw on experiences in Minneapolis and St Paul to identify six models of PPGIS delivery for community and grassroots organizations:

1 community-based (in house) GIS,
2 university–community partnerships,
3 publicly accessible GIS facilities at universities and libraries,
4 map rooms,

5 Internet Map Servers, and
6 neighbourhood GIS centres.

Based on a review of these six models, they conclude that 'community organizations do not just choose one model, but draw on different ways of gaining access to GIS, changing their strategies over time and perhaps developing novel ways of accessing and utilizing GIS.'

Part II of this volume contains 18 case studies that highlight the diversity of contexts in which PPGIS has been applied. The Inner City examples offer a fascinating view of the complexities of PPGIS production and implementation in established urban neighbourhoods. Parker and Pascual, for example, report on a project that is empowering to participants because the PPGIS helps them express their views and aspirations in ways that were previously unavailable, even though the particular gentrification struggle detailed in the case study was not successful. Casey and Pederson are working with the City of Philadelphia in a project that incorporates local community knowledge of historically marginalized neighbourhoods. The project illustrates how neighbourhood mapping by local residents can contribute to the development of a 'neighbourhood planning GIS' that goes well beyond data provided by the city by adding place-based knowledge and the capacity of local data manipulation. In so doing, the project also contributes to building local capacity for neighbourhood improvement. Elwood is working with the Powderhorn Park Neighbourhood Association in Minneapolis in a project focused on GIS and community housing improvement. While noting considerable progress in incorporating neighbourhood input to address critical housing issues, she also observes that the power relationships within the community organization were altered. Specifically, a neighbourhood discourse about the local landscape was replaced, in part, by an official housing discourse associated with technical planning methods. As a result, the residents most affected by this shift in language and expertise were those who traditionally have been marginalized from neighbourhood organizations – people of colour, renters, senior citizens, and non-native English speakers. Sawicki and Burke, in their chapter on the 'Atlanta Project' PPGIS effort, are more optimistic about the empowering capabilities of GIS technology: 'We illustrate that there is no fundamental incompatibility between the use of technology and community empowerment. In the code enforcement case, citizen mobilization was the determining factor in the successful change in the city's approach to enforcement.'

These inner-city PPGIS case studies begin to identify the differing, and sometimes contradictory, nature of PPGIS applications because they empower and marginalize simultaneously and are locally dependent. The chapters also indicate the growing use of the Internet to connect community members with GIS, and point to the Internet as a central component of PPGIS delivery. For example, Kingston provides an example from the United Kingdom of a 'virtual Slaithwaite' planning experiment. He suggests that a PPGIS is more robust because of the interactivity and connectivity provided by the Internet. He raises concerns, however, about the implications for planners when seeking to incorporate 'fuzzy information' that is not easily mapped or verified. Ventura *et al.* give a case study of a land information system that performs a number of functions in support of land-use planning. The system also integrates conventional planning methods with innovative web-based planning tools, including the solicitation of community perspectives through chat rooms and the equivalent of an electronic town hall meeting. Using the Internet in this way broadens community participation in land-use planning, and is augmented by a citizenry that is, in this case, highly computer literate. As a result, the planners simultaneously train community members and gain valuable local input into the planning process. Bosworth and his colleagues tell a similar story from Portland based on public engagement in growth management and transportation planning. A PPGIS has been operationalized for 'real-time' urban planning using the Internet. In this way, they suggest that planners can reach a much wider audience. 'A public workshop is considered a success if 60 people attend, while a website on the topic can reach 6,000 people a week.' In rural Australia, Walker and Pullar involve communities in a watershed GIS in which the catchment is dominated by industrial sugar production. They establish a participatory planning methodology using GIS in the context of community resource information centres.

The next set of case studies revolves around environmental management and activism. Sieber discusses five GIS applications in the California environmental movement, and finds that the availability of technological expertise within the groups is not much of a constraint. Access to digital data is, however, a problem because it tends to 'favour groups engaged in proactive and non-confrontational agendas.' Activist groups encounter much greater difficulty in gaining access to digital spatial information. Macnab's case study of participatory GIS in a Newfoundland fishing community is an innovative demonstration of the integration of local and 'expert' knowledge. Tulloch is working with a New Jersey umbrella NGO that oversees PPGIS projects and finds that 'identifying the extent of participation may become increasingly difficult as citizens learn to support and rely upon these groups for the employment of sophisticated technologies on their behalf.' In a different arena, Meredith and colleagues are building local capacity for PPGIS applications for biodiversity conservation, and argue that community GIS applications can contribute to ecosystem sustainability.

The final group of case studies is concerned with development planning in underdeveloped regions. Kyem's study of forest management in Ghana is an excellent example of established participatory development methods being merged with GIS. The case study highlights important political aspects of PPGIS projects: 'We soon realized that some rich and powerful people in the community objected to the open and participatory uses of GIS.' This suggests that PPGIS methods need to be politically integrated into the local development infrastructure for them to be empowering. Jordan's work in Nepal and Harris and Weiner's field work in South Africa supports this conclusion. Jordan also reminds us that a critical aspect of PPGIS projects is the actual form of participation and not the hardware/software configuration; PPGIS is as much about participation as it is about GIS. These three case studies are also a reminder that PPGIS projects can be exploitative as advocates and researchers 'capture local knowledge.'

Stonich employs PPGIS in a global NGO coalition project to fight the hegemony of industrial forms of shrimp production. The coalition uses the Internet to politically 'scale up' from local ethnographic cases of struggle to link regional and global resistance movements. She finds that NGOs are enthusiastic about using advanced information technology, but that the challenges they face are magnified with a global coalition that includes communities with significant differences in power, language, culture, and wealth. Despite such obstacles, the Internet-enabled global resistance coalition supports a common opposition to industrial shrimp production. The final two case studies also focus on ways to represent alternative knowledge systems and resist the hegemony of a Western, scientific, Cartesian understanding of space and territory. Laituri's work is with a Maori community in New Zealand, while Bond is working with the Cherokee Nation in Oklahoma. Both studies map culturally relevant information that is important for local resource management decisions, and challenge the epistemological limits of conventional GIS.

These case studies demonstrate how the socio-geographic context of PPGIS production and implementation impacts community access and use of GIS and technologies. Furthermore, the context of PPGIS is intricately linked to the nature of the participation process itself. In core industrial regions, community GIS applications are rapidly incorporating Internet capacity for connectivity, and multimedia forms of representation with virtual – and sometimes shifting – communities. In underdeveloped regions, PPGIS is comprised mainly of participatory development research and planning methods with a GIS–GIT interface. In such cases, the type of participation remains field-based within established communities. In all regions, however, there is evidence of the simultaneous empowerment and marginalization of people and communities. PPGIS does impose a technological layer to complex political struggles that are locally based, and this can alter existing community power relations. Issues of data cost and access also remain a concern, and can actually be compounded due to the high costs and time involved in collecting, maintaining, and updating local knowledge databases. Another interesting PPGIS characteristic is its contribution to computerized 'countermapping' and spatial story telling.

Significantly, most current PPGIS projects do not utilize GIS functionality for advanced spatial analysis. In PPGIS applications with an Internet GIS backbone, the Internet and its multimedia capabilities form the core of the application, with the GIS providing the digital maps. In this respect, the evolving generation of Internet mapping systems will probably play a significant role in future PPGIS projects. The final section of the book gazes into these possible PPGIS futures. Dangermond of ESRI offers a very optimistic view of Internet mapping systems and how they will service communities while also educating the lay public about geography: 'By combining a range of spatially referenced data, information media, and analytic tools, GIS technology enables citizens to prioritize issues, understand them, consider alternatives, and reach viable conclusions.' This, he suggests, will act to reinforce and promote democracy. Dangermond also reiterates that 'One key element that has affected the growth of public involvement in GIS is the Internet.' Shiffer focuses on the potential of Internet PPGIS for virtual communication and public access, but recognizes the problems that might arise due to the necessity of communicating with non-technical people, the technical problems of implementation, and differential understanding of information presented through virtual images and representations. Al-Kodmany develops this latter point and demonstrates how environmental design and visual representations of community perceptions and desires can be empowering in a Chicago community. His study concludes that 'The GIS helped highlight the importance of cultural values in history in the future design of the neighbourhood.' Krygier provides a similar story of a PPVisualization demonstration project in a Buffalo neighbourhood. Interestingly, his research suggests that 'the most vital issues for PPGIS and PPVis are not technical issues…but funding and [the] complexities within communities.… Unfortunately in most cases it will be those communities that are more stable, wealthy, and less vulnerable that can support the development of PPGIS and PPVis sites on the WWW.'

The issue of who has access to PPGIS and who benefits from such systems is a recurring theme in the book. Although PPGIS is intended to broaden access to GIS and GIT, Barndt rightly questions the criteria to be used for the evaluation of such implementations. PPGIS projects are, at their core, political because they attempt to broaden access to digital spatial information and empower historically disempowered people and communities. PPGIS projects are also political because they involve community participation, which is again essentially a political process. This suggests that understanding the politics and associated power relationships of PPGIS are critical for unpacking their impacts, wherever and however implemented. Community GIS is a reflection of the politics of the builders and users of such systems, although these politics extend beyond the local impacts on participating and non-participating communities.

In an insightful chapter, Aitken responds to the common assumption that community activism is spatially fixed and asks: 'Is it possible that PPGIS enables a breakthrough of local practices and community concerns from what John Agnew (1993: 252) calls the "hidden geographies" of scale?' The

Cartesian logic of GIS assumes a human agency bound by scale coordinates, but people operate at many scales simultaneously. As a result, Aitken questions the assumption that scale arises simply out of some simplistic notion of cartographic hierarchy and representation of space that enables political struggle to shape political discourse. He provocatively contends that 'PPGIS can be part of creating strong multiple publics that augment democracy. They do so by enabling people to become involved at a level that does not obfuscate their daily lives through maps and language drawn from instrumental, strategic logic. Rather, to be effective, the maps and language of PPGIS must communicate spatial stories that clarify and ultimately politicize the issues about which people feel concern.'

Community Participation and Geographic Information Systems is an eclectic collection of conceptual essays and case studies that demonstrate the social, political, epistemological, and methodological possibilities and boundaries of PPGIS. We have genuine concerns, however, that academics engaged with PPGIS will tire and fall back to their familiar role as researcher. In such cases, PPGIS has the potential to become another form of community exploitation. But the evidence from this volume suggests a more optimistic scenario as a growing coalition of professional planners, community activists, NGOs, government agencies, private sector groups, and academics find innovative and progressive ways that enable ordinary people and historically marginalized communities to benefit from the technologies of the digital age.

ACKNOWLEDGEMENT

We are indebted to one of our authors, Richard Kingston, for bringing the ladder of participation to our attention.

REFERENCES

Abbot, J., Chambers, R., Dunn, C., Harris, T., de Merode, E., Porter, G., Townsend, J. and Weiner, D. (1998) 'Participatory GIS: opportunity or oxymoron', *PLA Notes* 33: 27–34.

Agnew, J. (1993) 'Representing space: space, scale and culture in social science', in James Duncan and David Ley (eds) *Place/Culture/Rrepresentation*, London and New York: Routledge, pp. 251–271.

Aitken, S. and Michel, S. (1995) 'Who contrives the 'Real' in GIS? Geographic information, planning, and critical theory', *Cartography and Geographic Information Systems* 22(1): 17–29.

Barndt, M. (1998) 'Public participation GIS – barriers to implementation', *Cartography and Geographic Information Systems* 25(2): 105–112.

Castells, M. (1983) *The City and the Grassroots: A Cross-Cultural Theory of Urban Social Movements*, Berkeley: University of California Press.

Chorley Report (1987) *Handling Geographic Information*, London: HMSO.

Chrisman, N. R. (1987) 'Design of geographic information systems based on social and cultural goals', *Photogrammetric Engineering & Remote Sensing* 53(10): 1367–1370.

Craig, W., Harris, T. and Weiner, D. (1999) 'Empowerment, marginalization and public participation GIS', Specialist Meeting Report compiled for Varenius: NCGIA's Project to Advance Geographic Information Science. NCGIA, University of California at Santa Barbara, February.

Craig, W. and Elwood, S. (1998) 'How and why community groups use maps and geographic information', *Cartography and Geographic Information Systems* 25(2): 95–104.

Cultural Survival Quarterly (1995) Special issue on Geomatics, 18(4).

Curry, M. R. (1995) 'Rethinking rights and responsibilities in geographic information systems: beyond the power of image', *Cartography and Geographic Information Systems* 22(1): 58–69.

Dunn, C., Atkins, P. and Townsend, J. (1997) 'GIS for development: a contradiction in terms?' *Area* 29(2): 151–159.

Edney, M. H. (1991) 'Strategies for maintaining the democratic nature of geographic information systems', *Papers and Proceedings of the Applied Geography Conferences* 14: 100–108.

Elwood, S. and Leitner, H. (1998) 'GIS and community-based planning: exploring the diversity of neighbourhood perspectives and needs', *Cartography and Geographic Information Systems* 25(2): 77–88.

Gonzales, R. M. (1995) 'KBS, GIS and documenting indigenous knowledge', *Indigenous Knowledge and Development Monitor* 3(1): http://www.nuffic.nl/ciran/ikdm/3-1/contents.html.

Goodchild, M. F. (1995) 'GIS and geographic research', in Pickles, J. (ed.) *Ground Truth: The Social Implications of Geographic Information Systems*, New York: Guilford.

Goodchild, M., Egenhofer, M., Kemp, K., Mark, D. and Sheppard, E. (1999) 'Introduction to the Varenius project', *International Journal of Geographical Information Science* 13(8): 731–745.

Goss, J. (1995) 'We know who you are and we know where you live: the instrumental rationality of geodemographic information systems', *Economic Geography* 71: 171–198.

Graham, S. (1998) 'The end of geography or the explosion of place? Conceptualizing time and information technology', *Progress in Human Geography* 22: 165–185.

Grant, D. R. and Omdahl, L. B. (1993) *State and Local Government in America*, Madison WI: Brown and Benchmark.

Harris, T. M., Weiner, D., Warner, T. and Levin, R. (1995) 'Pursuing social goals through participatory GIS: redressing South Africa's historical political ecology', in Pickles, J. (ed.) *Ground Truth: The social implications of geographic information systems*, New York: Guilford, pp. 196–222.

Harris, T. and Weiner, D. (1996) 'GIS and society: the social implications of how people, space and environment are represented in GIS', Scientific Report for NCGIA Initiative # 19 Specialist Meeting, University of California at Santa Barbara, November.

Harris, T. and Weiner, D. (1998) 'Empowerment, marginalization and community-integrated GIS', *Cartography and Geographic Information Systems* 25(2): 67–76.

Hutchinson, C. F. and Toledano, J. (1993) 'Guidelines for demonstrating geographical information systems based on participatory development', *International Journal of Geographical Information Systems* 7(5): 453–461.

Jacobs, J. (1961) *The Death and Life of Great American Cities*, New York: Vintage Books.

Johnston, R. (1999) 'Geography and GIS', in P. Longley, M. Goodchild, D. Maguire and D. Rhind (eds) *Geographical Information Systems: Principles, Techniques, Management, and Applications*, New York: John Wiley, pp. 39–47.

Jones, B. (1990) *Neighbourhood Planning: A Guide for Citizens and Planners*, Chicago and Washington, DC: Planners Press, American Planning Association.

Jordan, G. and Shrestha, B. (1998) 'Integrating geomatics and participatory techniques for community forest management: case studies from Yarsha Khola Watershed, Dolakha District, Nepal', International Centre for Integrated Development, Kathmandu, Nepal.

Kellogg, W. (1999) 'From the field: observations on using GIS to develop a neighbourhood environmental information system for community-based organizations', *URISA Journal* 11(1): 15–32.

Kitchin, R. (1998) 'Towards geographies of cyberspace', *Progress in Human Geography* 22(3): 385–406.

Kwaku-Kyem, P. (1999) 'Examining the discourse about the transfer of GIS technology to traditionally non-western societies', *Social Science Computer Review* 17(1): 69–73.

Lake, R. W. (1993) 'Planning and applied geography: positivism, ethics, and geographic information systems', *Progress in Human Geography* 17(3): 404–413.

Marozas, B. A. (1993) 'A culturally relevant solution for the implementation of geographic information systems in Indian country', *Proceedings of the Thirteenth Annual ESRI User Conference* 1: 365–381.

Mitchell, A. (1997) *Zeroing In: Geographic Information Systems at Work in the Community*, Redlands, CA: ESRI.

Obermeyer, N. and Pinto, J. (1994) *Managing Geographic Information Systems*, New York: Guilford.

Olson, M. (1965) *The Logic of Collective Action; Public Goods and The Theory of Groups*, Cambridge, MA: Harvard University Press.

Openshaw, S. (1991) 'A view on the GIS crisis in geography, or, using GIS to put Humpty-Dumpty back together again', *Environment and Planning A* 23: 621–628.

Pickles, J. (1991) 'Geography, GIS, and the surveillant society', *Papers and Proceedings of Applied Geography Conferences* 14: 80–91.

Pickles, J. (1995) 'Ground Truth: The social implications of geographic information systems', New York: The Guilford Press.

Pickles, J. (1999) 'Arguments, Debates, and Dialogues: The GIS–social theory debate and concerns for alternatives', in P. Longley, M. Goodchild, D. Maguire and D. Rhind (eds). *Geographical Information Systems: Principles, Techniques, Management, and Applications*, New York: John Wiley, pp. 49–60.

Rambaldi, G. and Callosa, J. (2000) *Manual on Participatory 3-Dimensional Modeling for Natural Resource Management (Vol. 7). NIPAP, PAWB-DENR*: Philippines Department of Environment and Natural Resources.

Rundstrom, R. (1995) 'GIS, indigenous peoples, and epistemological diversity', *Cartography and Geographic Information Systems* 22(1): 45–57.

Schroeder, P. (1996) Report on public participation GIS workshop, in T. Harris and D. Weiner (eds) GIS and Society: The Social Implications of How People, Space and Environment are Represented in GIS. *NCGIA Technical Report 96–97, Scientific Report for Initiative 19 Specialist Meeting*, South Haven, MN, 2–5 March 1996.

Sheppard, E. (1995) 'GIS and society: towards a research agenda', *Cartography and Geographic Information Systems* 22(1): 5–16.

Sheppard, E. and Poiker, T. (1995) GIS and society special issue. *Cartography and Geographic Information Systems* 22(1) (January).

Sheppard, E., Couclelis, H., Graham, S., Harrington, J. and Onsrud, H. (1999) 'Geographies of the information society', *International Journal of Geographical Information Science* 13(8): 797–823.

Sieber, R. (2000) 'Conforming (to) the opposition: the social construction of geographical information systems in social movements', *International Journal of Geographical Information Science* 14(8): 775–793.

Smith, N. (1992) 'History and philosophy of geography: real wars, theory wars', *Progress in Human Geography* 16(2): 257–271.

Taylor, P. J. (1990) 'GKS', *Political Geography Quarterly* 3: 211–212.

Taylor, P. J. (1991) 'A distorted world of knowledge', *Journal of Geography in Higher Education* 15: 85–90.

Taylor, P. J. and Overton, M. (1991) 'Further thoughts on geography and GIS', *Environment and Planning A* 23: 1087–1094.

Taylor, P. J. and Overton, M. (1992) 'Further thoughts on geography and GIS: a reply.' *Environment and Planning A* 24: 463–466.

Talen, E. (1999) 'Constructing neighbourhoods from the bottom up: the case for resident generated GIS', *Environment and Planning B* 26: 533–554.

Talen, E. (2000) 'Bottom-up GIS. A new tool for individual and group expression in Participatory Planning', *APA Journal* 66(3): 279–294.

Weidemann, I. and Femers, S. (1993) 'Public participation in waste management decision making: analysis and management of conflicts', *Journal of Hazardous Materials* 33: 355–368.

Weiner, D., Warner, T., Harris, T. M. and Levin, R. M. (1995) 'Apartheid representations in a digital landscape: GIS, remote sensing, and local knowledge in Kiepersol, South Africa', *Cartography and Geographic Information Systems* 22(1): 30–44.

Weiner, D. and Harris, T. (1999) 'Community-integrated GIS for land reform in South Africa', WVU Regional Research Institute Research Paper # 9907, Morgantown W. V. (http://www.rri.wvu.edu/wpapers/1999wp.htm).

Yapa, L. S. (1991) 'Is GIS appropriate technology?' *International Journal of Geographical Information Systems* 5: 41–58.

Chapter 2

Surveying the extent of PPGIS practice in the United States

David S. Sawicki and David Randall Peterman

2.1 INTRODUCTION

Many of the key themes in PPGIS research revolve around knowing who produces and who consumes small area GIS products. Examples include the multiple ways in which PPGIS are being designed and implemented, and identifying community information needs and how PPGIS might contribute to those needs. However, many of the questions posed are difficult to answer because of a lack of comprehensive inventories of either PPGIS providers or consumers. Two exceptions are Craig's inventory of consumers (community groups) in the Twin Cities (Sawicki and Craig 1996) and the Urban Institute's list of 30 citywide neighbourhood data providers (Urban Institute 1996). Neither is comprehensive, nor were they meant to be. They do provide a start, however.

In this chapter, we provide a definition of PPGIS and report on results of a search for PPGIS providers (see Table 2.1). Research interest on the use of GIS as a tool for enhancing public policy activities by community groups has been in evidence for a number of years. As part of the National Center for Geographic Information and Analysis's Project Varenius, a suggestion was made to undertake an inventory of PPGIS activities. The group decided that its primary concern was to learn from those using GIS and Information Technology (IT) to support community initiatives. Our original intent was to produce a comprehensive inventory of PPGIS groups throughout the United States. We quickly realized that it was not a reasonable goal, in part because advancing technology is making PPGIS activity ever more widespread, and in part because delineating PPGIS activity within the universe of GIS was not a simple task. The concepts we used to generate an inventory of PPGIS organizations are reflected in the following introductory statement to the survey instrument:

> The Public Participation GIS effort of the National Center for Geographic Information and Analysis (NCGIA) requests your assistance in identifying significant information technology projects providing

Table 2.1 PPGIS suppliers contacted in fall 1998 survey

Organization	*Host organization*	*City*	*State*	*Phone*	*URL*	*e-mail*
Office of Community Development	City of Birmingham	Birmingham	AL	205-254-2309		
Morrison Institute	Arizona State University	Phoenix	AZ	480-965-4525	www.asu.edu/copp/morrison	
Urban Data Centre	Arizona State University	Phoenix	AZ	602-965-3046	www.asu.edu/xed/urbandata	
Los Angeles Neighbourhood Early warning System (LANEWS)	Neighbourhood Knowledge Los Angeles	Los Angeles	CA		nkla.sppsr.ucla.edu	
Bay Area Shared Information Consortium (BASIC)		San Franscisco	CA		www.basic.org	
San Diego Association of Governments (SANDAG)		San Diego	CA	619-595-5300	www.sandag.cog.ca.us	webmaster@sandag.cog.ca.us
Urban Strategies Council		Oakland	CA	510-893-2404		
GreenInfo Network		San Francisco	CA	415-979-0343	www.greeninfo.org	info@greeninfo.org
ESRI Conservation Programme-Conservation GIS Consortium	Environmental Systems Research Institute (ESRI)	Redlands	CA	909-793-2853	www.esri.com	ecp@lists.desktop.org
Piton Foundation		Denver	CO	303-825-6246	www.piton.org	eberman@piton.org
City Room (aka New Haven On-Line)	Institution for Social and Policy Studies, Yale University	New Haven	CT			

Public Access Network Directory	The Council on Library and Information Resources (CLIR)	Washington	DC	202-939-4750	www.clir.org/pand/pandhome.htm	info@clir.org
The Right-to-Know Network (RTK NET)		Washington	DC	202-234-8494	rtk.net	webmaster@rtk.net
Food Research and Action Center		Washington	DC	202-986-2200		
DC Agenda Project	Federal City Council	Washington	DC	202-223-2598	www.dcagenda.org/index.htm	
Poverty and Race Research Action Council		Washington	DC	202-387-9887	www.prrac.org www.povertyandrace.org	info@prrac.org
GIS and Software Development Team, Development and Implementation Branch, Engineering Division, Information Technology Services Directorate	Federal Emergency Management Agency	Washington	DC	202-646-3071		
Environmental Projection Agency		Washington	DC		www.epa.gov/enviro	
Community Networks	Office of Minority Health; Federal Dept. of Health and Human Services	Washington	DC		www.hhs.gov/progorg/ophs/omh/community/htm	
The Center for Civic Networking		Washington	DC	202-362-3831	www.civicnet.org/index.html	webmaster@civicnet.org
Community Planning and Development	Department of Housing and Urban Development	Washington	DC		www.HUD.gov/cpd/2020soft.html	

Table 2.1 (Continued)

Organization	*Host organization*	*City*	*State*	*Phone*	*URL*	*e-mail*
Jacksonville Community Council Incorporated		Jacksonville	FL	904-396-3052	www.jcci.org	
Overtown Neighbourhood Partnerships	Miami-Dade Community College	Miami	FL			
Center for Economic Development Research	University of South Florida	Tampa	FL	813-905-5854	www.coba.usf.edu/centers/cedr	
Office of Research and Planning	Florida Department of Children and Families, District 11	Miami	FL		www.state.fl.us/cf_web/district11	
Office of Data and Policy Analysis (DAPA)	The Atlanta Project (TAP)	Atlanta	GA	404-206-5015		
Hawaii Community Services Council		Honolulu	HI	808-521-3861	www2.hawaii.edu/~cssdata	
Center for Neighbourhood Technology		Chicago	IL	773-278-4800	www.cnt.org	info@cnt.org
Chicago Area Geographic Information Study (CAGIS)	University of Illinois at Chicago	Chicago	IL	312-996-5274	www.cagis.uic.edu	httpadm@cagis.uic.edu
East St. Louis Action Research Project (ESLARP)		Champaign	IL	217-265-0202	www.imlab.uiuc.edu/eslarp	

Polis Centre	Indiana University-Purdue University Indianapolis (IUPUI)	Indianapolis	IN	317-274-12455	www.savi.org	polis@iupui.edu
Boston Persistent Poverty Project	Boston Foundation	Boston	MA	617-723-7415		
Coalition for Low-Income Community Development		Baltimore	MD	410-752-7222	www.clicd.org	
Michigan Metropolitan Information Center (MIMIC)	Center for Urban Studies, Wayne State University	Detroit	MI	313-577-8996	www.cus.wayne.edu/mimic/mimhome.htm	
Automated Cartographic Information Center	John R. Borchert Map Library, University of Minnesota	Minneapolis	MN	612-625-9024	www.map.lib.umn.edu/acic.html	
The Urban Coalition		St. Paul	MN	612-348-8550	http://www.urbancoalition.org	gen@urbancoalition.org
Neighbourhood Planning for Community Revitalization	Center for Urban and Regional Affairs, University of Minnesota	Minneapolis	MN	612-625-1551	http://www.npcr.org	nelso193@tc.umn.edu
Neighbourhood Network	Kansas City Neighbourhood Alliance	Kansas City	MO	816-753-8600	www.kcneighbornet.org	
Greater Kansas City Community Foundation		Kansas City	MO	816-842-0944	www.gkccf.org	
Community Information Network	St Louis Enterprise Community Programme, City of St Louis	St Louis	MO	314-622-3400	stlouis.missouri.org	cin@stlouis.missouri.org

Table 2.1 (Continued)

Organization	*Host organization*	*City*	*State*	*Phone*	*URL*	*e-mail*
Community Research Center	Asheville Chamber of Commerce	Asheville	NC	828-258-6128	www.ashevillechamber.org/CRC.htm	
Center for Urban Policy Research	Rutgers University	New Brunswick	NJ	732-932-3133	policy.rutgers.edu/cupr	
Community Mapping Assistance Project	New York Public Interest Research Group	New York	NY	212-349-6460	www.cmap.nypirg.org	
Urban Technical Assistance Programme	Graduate Programme in Urban Planning,Columbia University	New York	NY		www.arch.columbia.edu/UTAP	UTAP@columbia.edu
Green Mapping	Modern World Design	New York	NY	212-674-1631	www.greenmap.com	web@greenmap.com
United Neighbourhood Houses of New York		New York	NY	212-967-0322	www.unhny.org	
Center on Urban Poverty and Social Change	Mandel School of Applied Social Sciences, Case Western Reserve University	Cleveland	OH	216-368-6946	povertycenter.cwru.edu/cando2.htm	info@poverty.cwru.edu
NeighbourhoodLink	Center for Neighbourhood Development	Cleveland	OH	216-687-2134	little.nhlink.net/nhlink	
Northern Ohio Data and Information Service (NODIS)	The Urban Center, Levin College of Public Affairs, Cleveland State U	Cleveland	OH	216-687-2134	urban.csuohio.edu/~ucweb/index.htm	rosemary@wolf.csuohio.edu

Economic Development Information Center	Tulsa City-County Library	Tulsa	OK	918-596-7991		mgregor@tccl.lib.ok.us
PSU@Home, Institute of Portland Metropolitan Studies	College of Urban and Public Affairs, Portland State University	Portland	OR	503-725-5170	www.upa.pdx.edu/IMS/About IMS	
Oregon Progress Board		Salem	OR	503-986-0039	http://www.econ.state.or.us/opb	
Interrain Pacific	Ecotrust	Portland	OR	503-226-8108	www.interrain.org (also www.inforain.org)	info@interrain.org
Portland Multnomah Progress Board	Multnomah County Auditor's Office	Portland	OR	503-823-3504	www.p-m-benchmarks.org	
Philadelphia Neighbourhoods On-line	Institute for Study of Civic Values	Philadelphia	PA		www.libertynet.org/nol	
The Alliance for Aquatic Resource Monitoring	Dickinson College Environmental Studies Department	Carlisle	PA	717-245-1135	www.dickinson.edu/allarm	
Housing Association of Delaware Valley		Philadelphia	PA	215-545-6010	www.libertynet.org/hadv	hadv@libertynet.org
Rhode Island Geographic Information System (RIGIS)	Rhode Island Dept of Administration-Planning	Providence	RI	401-222-6483	www.edc.uri.edu/gis	
The Providence Plan		Providence	RI	401-455-8880	www.providenceplan.org	
Department of Ecology	State of Washington	Olympia	WA	360-407-7128	www.wa.gov/ecology/gis	
Nonprofit Center of Milwaukee	Nonprofit Center of Milwaukee	Milwaukee	WI	414-344-3933	http://www.execpc.com/~npcm	npcm@execpc.com

community information to community groups around the world.... Your response to this request can be either a full reply to the questionnaire below or a brief note or call to us, which we can follow up with you in more detail.

There are other surveys underway that look at the use of information technology by nonprofit organizations. This survey is broader than just nonprofit organizations, and narrower than the entire range of information technology. Our goal is to assemble an inventory of organizations that contribute to public participation in community decision-making by providing local-area data to community groups.

We are looking for organizations that:

(a) collect demographic, administrative, environmental or other local-area databases,
(b) do something to the data to make it more useful locally (e.g., address matching of individual records; creating customized tables), and
(c) provide this information to local nonprofit community-based groups at low or no cost. This can include local non-profit community groups that are collecting and processing data in-house, or data 'intermediaries' that process and analyze data for others (data intermediaries might be government offices, nonprofit groups, university-based centers, etc.).

This working definition of PPGIS generated valuable discussion and is explored in more detail below.

2.2 GIS AND INFORMATION TECHNOLOGY, OR JUST GIS

Information technology is a broad term. There are many sites on the World Wide Web that offer advice to non-profit/public service organizations on making use of IT. A glance at representative sites suggests the primary uses (so far) of IT by non-profits are:

1 word-processing programs for report writing, newsletters,
2 database programs for accounting, fund-raising, volunteer management, project management, training, mailing lists,
3 e-mail for communication, and
4 Internet access to create websites, disseminate information, and for research.

We decided that we were not attempting to inventory all IT activities (though that could be an interesting, tough, and rewarding task), but rather were searching for organizations with a significant spatial analysis component. Many groups use GIS to simply display spatial information, a task that might be done as well or better by hand. This is not necessarily a trivial activity. Displaying spatial information on a map can enable viewers to see patterns that would otherwise not be apparent. But in our view, the power of GIS is in analysing information, not merely displaying it; using a GIS system just to draw maps ignores most of its functionality. It was decided, therefore, that the GIS component of the PPGIS activity must include some analytical capability to be included in the survey.

Spatial analysis need not be expressed in maps. Just as the real power of GIS is analysing information, rather than just drawing maps, the presentation of data in a table or a report is still representative of spatial analysis. Nor is it critical, in our definition, that the organization even is using a GIS software program; after all, for some time people did spatial analyses by hand. The important thing is that some sort of analysis is being carried out.

But by whom? The Census Bureau has long been a great source of spatial information for community organizations. Now with its use of IT to make its data more accessible to users (e.g. the 1990 Census Lookup site), the Census Bureau's role as a provider as well as generator of spatial data has been expanded. It also offers analysis of data via reports, though usually at the national level. Thus as a tremendous source of local-level spatial information, the Census Bureau would have to be on any list of organizations promoting public participation through providing geographic information. The Bureau collects demographic data, does something to the data to make it more useful, then distributes the data to community groups at low or no cost.

Of course, the Census Bureau is a special case. The primary reason for our requiring that a PPGIS organization do something to the data to make it more useful was to exclude many organizations that merely redistribute local-area Census data without any further analysis. Nevertheless, an argument could be made that even this sort of activity may assist community groups, by making local-area Census information even more widely available.

2.2.1 Geographic scale

At what geographic scale would a community GIS activity operate? 'Community' has many possible meanings. We take community in this context to be a spatial as well as a social term: a relatively small, roughly defined area, populated with people who feel themselves to have something in common. We were thinking of it interchangeably with neighbourhood and perhaps small town. We exclude virtual communities, though we include organizations comprised of members with non-contiguous residence whose object of analysis might be a particular small place.

It is difficult to limit the scale for other reasons. For example, regional planning agencies tend to work with large land areas, often metropolitan areas. So, on the face of it, their work would be excluded. But some of their work may have important implications for small areas. Thus, were they to provide residents of neighbourhoods with spatial data to be used by residents (say in a planning process) we would want to include them. Most obviously, though, we are trying to find examples of organizations providing spatial analysis to persons who share the fate of their small place.

There is a definite urban bias to this definition of community scale. The clearest shortcoming of the definition is the use of GIS by environmental groups. Environmental concerns centre around natural systems rather than social systems, and many of these systems operate on a large scale, e.g. air pollution, watersheds. To address this issue, we divided our PPGIS survey into two conceptual parts. For organizations that dealt primarily with social issues, we looked for the use of social data at the local-area scale. For organizations that dealt primarily with environmental issues, we looked for the use of environmental data at a regional or smaller scale. Demographic data is readily available at the level of standard political divisions (nation, states, counties, cities). Thus we decided that organizations that provided demographic data at other levels, whether sub-city (as in our local-area focus) or super-county (not just aggregating counties, but crossing county or even state lines), were adding something to the database.

2.2.2 Whose data?

The US Census Bureau provides data for small areas at their Lookup website. Anyone with access to http://venus.census.gov/doc/lookup_doc.html on the Internet can get information at the tract or block group level. The user can even see the information displayed on a map. The United States Department of Housing and Urban Development provides GIS software and data to hundreds, maybe thousands, of local communities, allowing community groups to display local Census information on maps.[1] Are those activities community GIS? It seems that if 'community GIS activity' is to mean anything, it must go beyond simply redistributing the work of these organizations.

We defined 'major community GIS activity' as one in which some organization collects data for small areas. Local, state or federal governments might first collect the data, or residents themselves might collect it. By 'collect' we mean only acquiring the use of, not necessarily generating primary data. However, we hoped to find organizations that did engage in primary data collection.

2.2.3 Whose analysis?

An important distinction can be made between organizations that take a supply side approach (e.g. post data on the web but have little or no contact with data users) and organizations that are demand-driven (provide data to individual clients in response to specific requests). The existence of supply side organizations is too extensive to ignore, suggesting that analysis should be viewed as a continuum, with organizations that disseminate data with little analysis at one end, and organizations that perform custom queries for individual clients at the other. We are more interested in organizations that engage clients.

An ideal PPGIS could be where neighbourhood residents collect their own spatial data and process it themselves using GIS software. We have found a number of organizations that have as part of their mission the training of community citizens in uses of GIS. Thus, the producers are also the consumers. However, this endeavour is challenging and the successes, as far as we can say, have been few. An additional dimension is whether the organization has a single client or multiple clients. Clearly, a community-based organization (CBO) could provide GIS services to just itself, or it could provide services for free or a fee to others. And organizations without a direct involvement in community building or neighbourhood development could provide GIS services to CBOs that do. We call such organizations *data intermediaries*.

2.2.4 From data to information to action

We see a continuum from data to action. Spatial data gets gathered and processed using a variety of analytical techniques. With the right mind-set and experience, analysts can turn spatial data into spatial information that can be insightful for local communities. But beyond insight is the notion that such analytical products as tables, graphs, charts and maps can be useful in a public policy context. It can be employed in an action agenda. We favour locating organizations that contribute to an action agenda as opposed to those which simply shovel data out the door or provide reports that describe situations that are not fertile ground for action by citizens at the local level. But this is a stringent criterion indeed. And thus we asked our survey respondents to reflect on their work and share examples with us of major successful and unsuccessful actions taken as a result of data and information generated by a community GIS. Our results thus far indicate that more action successes are borne out of the pairing of an action expert (a community organizer) with an educated GIS/policy analyst. Working collaboratively, the right questions seemed to get framed and the appropriate GIS products produced.

2.2.5 PPGIS data intermediaries

Our assumption is that not many people with GIS skills volunteer their time to work with grassroots groups. We are prepared to believe otherwise. However, this brings in a definitional problem. If the definition of community GIS activity is too inclusive, it is of little use, and the same is true if it is too exclusive. In one of our classes, we had a student who used newfound GIS skills to help her church select an alternative location. This work was unpaid, and might be the only time the church will make use of her skills. That may be an interesting example of the use of GIS by a community group, but it is so ephemeral that it would be hard to capture in a general survey. And it does not represent a major community GIS activity.

It is possible for a community organization to develop GIS capacity in-house. However, that is likely to be rare, for several reasons. Most community organizations have small staffs and small budgets, surviving from year to year on annual receipt of grants. Although the user friendliness of GIS systems – and the power of desktop GIS – is increasing, expertise in GIS work still requires a significant investment of time on the part of a user. Once the users have made that investment, they find themselves with a valuable skill for which organizations with bigger budgets are willing to bid. Also, even with desktop mapping programs, and free Census data in digital format, the cost of setting up a system is not inconsequential. The ability to make use of information is a skill in short supply as well. And few local community organizations would need GIS work often enough to justify the investment in training a staff person to use it. For these reasons, organizations that do have the resources and inclination to train or hire GIS users are likely to be valuable sources of expertise for community groups. These data intermediaries could allow community groups to focus on what questions to ask, rather than spending lots of time learning how to use the tool to answer a question.

Data intermediaries can be divided into four general classifications: (1) *government agencies*; (2) *university centres*; (3) *quasi-autonomous non-governmental organizations*; and (4) *non-profit organizations*.

1 *Government agencies* – These can be federal, state or local government agencies. At the state and local level, these are most likely to be planning offices. They would typically have an in-house GIS capacity for their own work, and might provide information to CBOs on request. However, these offices are most likely to limit their community-oriented work to sharing simple information or the results of their own projects and tend to be reluctant to undertake extensive work on behalf of a community organization.
2 *Quasi-autonomous non-governmental organizations* – These are mainly planning commissions and are likely to be functionally similar to those in classification #1.
3 *University research centres* – These are most likely to be associated with political science, sociology, geography, urban planning, or public policy departments. Their focus is typically on the work of professors and the staff are likely to be students. Such centres tend to have lots of turnover

in the student staff, limiting the development of expertise in local community work. Moreover, the culture of higher education does not reward professors or students for community service. Rewards come from publishing research of interest to other university researchers, and increasingly from getting large grants for projects. More formal organizations with full-time staff may have a more professional–client orientation, but may also do work only on a fee basis, which may put their services out of reach of many small community organizations.

4 *Non-profit organizations* – CBOs that have their own in-house GIS capacity tend to be better funded and may develop their own GIS expertise, particularly, if they choose a mission of providing such expertise to other community groups. These organizations are attractive to foundations, which increases the stability of their funding and hence staffing.

In addition to these four, a fifth type of organization, community learning centres, can be identified. These are locations where the public can use computers with Internet access and, in this context, GIS software. However, there doesn't appear to be any expectation that these organizations would provide much specialized assistance to users. Rather, they appear to be largely passive in approach, providing a place where members of the public can access computers and perhaps receive limited training in computer programs, though probably not enough training to make use of GIS software.

2.3 THE SURVEY INSTRUMENT

The survey was implemented in three ways: by telephone, by e-mail, and by web search. The e-mail survey consisted of three parts: an introductory section describing the origin and purpose of the survey; the survey questionnaire (see Table 2.2); and a sample response, based on our own organization.[2] We had written it out as a test of the questions, and included it as a guide to help others interpret the questions. The telephone survey also used the survey questionnaire. The web search collected information about organizations that had put sufficient data on the Internet to be able to answer our survey questionnaire. It was a way of collecting information in non-office hours and from organizations we were not able to contact directly.

We understood that the primary purpose of the survey was to create an inventory of PPGIS efforts going on currently. This inventory would presumably help researchers understand the extent of the activity, and give them a universe of organizations to contact for further study. Since we were trying to compile an inventory, most of the questions were chosen to get basic information about the operation of these organizations: their structure, when they were founded, their funding sources and level of support, how many people work there, what data they collect, and what services

Table 2.2 Survey questions

1	What is the administrative structure of this organization (e.g. connected to a college or university, part of a municipal government agency, free-standing NGO, etc.)?
2	When was the organization established?
3	What is the financial base of this organization (e.g. supported by annual grants, line-item in government budget, fees charged, etc.)?
4	What is the annual budget, and number of staff (please indicate full-time and part-time staff)?
5	What types of data do you collect (e.g. Census, local administrative data, environmental, transportation, etc.)? What are your major databases?
6	Does the organization provide services to nonprofit community groups? What charges, if any, are made for these services? Or does the organization provide data or information but no direct services?
7	Who are your major clients?
8	What have been some of your major projects? Notable successes or notable failures? Explain.
9	Other comments:

they offer. We also asked about major clients in an attempt to better identify those organizations that do most of their work for grassroots organizations.

In the e-mail message and in our telephone calls, we asked that respondents share any compelling stories about the use of their work by community groups. However, we heard very few stories. We guess that people are more likely to tell stories in person than to write them down or think of them in the course of a telephone survey, perhaps because the stories seem vague, would take too long to write, and do not flow naturally in the slight pressure of a telephone interview.

Our starting universe was two lists: (1) the respondents to the Urban Institute's first survey with the Neighborhood Indicators Projects, and (2) a group of nine organizations identified by Craig and invited to a PPGIS meeting at the 1995 URISA conference (see appendix in Sawicki and Craig 1996). The Urban Institute's National Neighborhood Indicators Project list, contained in their first year report (Urban Institute 1996), consisted of a paragraph describing the response they had received to a survey of US cities, looking for organizations compiling neighbourhood indicators. They surveyed many cities and got responses from 30. Many of those responses indicated that no organization in the community was doing anything similar to the neighbourhood indicators project, or that some organization was thinking of doing something similar.

PPGIS survey implementation began by e-mailing targeted organizations for which we could find e-mail addresses, and following up with a phone call if we did not get a reply. When we could not find an e-mail address for an organization, we telephoned directly. In the e-mail message, and in each phone conversation, we asked each respondent if they knew of other organizations engaged in PPGIS activities.

We also searched the Internet for references to the organizations mentioned in these two lists, primarily to find contact information about them, but also to see what additional information might be accessed. Finding references to these organizations on websites identified other relevant organizations for surveying. Given that the target group was loosely defined, searching the web for information was like trying to drink from a firehose; the quantity of potentially relevant sites was overwhelming. Fortunately, the web also provided a means for quick and easy communication, and the survey was e-mailed to many new organizations discovered by the search. Sometimes much of the requested information was available on the organization's website. The survey was also e-mailed to several mail lists that seemed pertinent.

In the e-mailed questionnaire, and in the telephone interviews, the last question was 'what other organizations do you know of that are doing similar work?' We got very few responses to this question. Interestingly, sometimes organizations in the same city would not even mention each other. One explanation might be that the respondents had a narrower interpretation what the target organizations might be. Another possible explanation is that, respondents feel that to admit (to themselves, or to us) there were other organizations in their community doing the same thing they were doing, might call into question the value of their own work. That would obviously be more true of an organization that was taking a supply side approach (just putting information out) than of an organization that was acting as a consultant to specific neighbourhood groups. Finally, it might have just been ignorance of the existence of other organizations doing a similar kind of community work.

2.4 SURVEY RESULTS

At the conclusion of the survey (1 November 1998), there were 65 organizations in the database sponsoring some PPGIS-related project. This included 30 non-profits, 18 affiliated with universities, 15 government offices, and 2 private companies. They came from 40 cities. Washington, DC led the list, followed by New York City and several other cities with multiple organizations. Their budgets ranged from $1 million and lower, though most did not report a budget figure and in the cases that did, it was not always clear how much of the budget went to PPGIS activities; the same was true of staffing, which ranged from 35 downward with many blank responses.

Two types of lessons were learned from the survey results: (1) lessons about the process of surveying these groups; and (2) lessons shared with us by the respondents about their work.

2.4.1 Lessons about the process

It cannot be said too often that pilot surveys are essential. We piloted the survey questions on our own organization, and sent that response along with the survey as a guide. But not until the survey was sent to others did we see the failure to clearly indicate to the respondent organizations that questions about dates and budget pertained only to the PPGIS activity and not to the entire organization. There was a very low response rate to the e-mail survey. It was sent to several maillists, and two responses were immediate, but then only a handful more were received over the next month. The lesson seems to be that if people are going to respond to it, they will respond immediately.

Another lesson is that maillists are not always open to surveys. Some maillists are unmoderated, meaning that messages sent to the list are automatically posted to it. Other maillists are moderated, meaning a person reads each message sent to the list and decides whether to post it or not. The list administrator of one moderated list refused to post the survey. In response to a query as to whether the survey had been posted, she wrote that she routinely discards most posts that have to do with GIS, and feels that posting surveys to the list is of no benefit to the list members (the topic of that list is the provision of state and local government information on-line!). This response was surprising, but also amusing, since one of the ideas underlying the Varenius PPGIS initiative is the value of providing information to people and letting them determine if it is of interest to them, versus the old model of having someone else decide for them what information they should have access to.

The questionnaire was also emailed to specific organizations for whom we had an e-mail address, to give them the opportunity to fill the form out at their own convenience. We got some responses from this, but ended up having to call most of the organizations for a response. When we called, we asked to talk to someone. If the person hesitated, we offered to e-mail or fax the questionnaire. Some people asked for that, others responded to us right then.

2.4.2 Lessons learned about the organizations

One important lesson learned was that there are a variety of PPGIS activities going on around the country, some throwing data over the wall to the public, others working actively with neighbourhood groups to respond to their expressed needs, with a variety of databases. One of the most frequent comments we heard was that community organizations don't know how to make effective use of data. Another frequent comment, which may simply be a different perspective on the first one, was that community groups don't attach much significance to the data that social scientists find interesting. We were told that neighbourhood groups tend to be uninterested in demographics, except when filling out a grant application. Several respondents noted that the information most frequently requested by neighbourhood groups was ownership records for buildings and property in their neighbourhoods.

Another respondent challenged one of the fundamental assumptions of PPGIS – the idea that information is power. Knowledge is power, no doubt, but knowledge of the demographics of a neighbourhood is not necessarily very powerful knowledge. As planners have often had occasion to learn, the facts about a situation that are important to social scientists do not necessarily influence the political decision-making process. To paraphrase a political scientist, facts count but resources decide,[3] and the kinds of resources that make a difference in the local decision-making process are not directly supplied by PPGIS activities.

However, facts do count for something. The federal Home Mortgage Disclosure Act of 1975 required that lending institutions disclose the location of their loans, by Census tract. This information enabled neighbourhood groups to demonstrate that banks were systematically not making loans in certain neighbourhoods. That analysis contributed to the passage of the Community Reinvestment Act (CRA) of 1977, which gave community groups some leverage with which to negotiate with banks for increased lending activity in their neighbourhoods. Without the power of the CRA, the information provided by the Home Mortgage Disclosure Act would mean little; but with that information, the CRA was passed, and the combination of the mortgage location information and the power of the CRA has led to more than $60 billion in community reinvestment agreements (Cincotta 1996).

Another observation concerned the activities of grassroots organizations themselves. Ryan (1998) argued that most grassroots organizations spend 90 per cent of their time seeking grants to keep themselves alive, leaving only 10 per cent of their time to actually do anything to make a difference in the community. Rather than pushing community organizations to use GIS to make marginally better use of that 10 per cent, helping them become more efficient at getting grants would free up much more time for them to focus on community work.

This point is supported by the findings of a 1995 survey of neighbourhood organizations in Ohio. The authors mailed surveys to 613 organizations; they received 183 responses. They attributed the low response rate to having included many very small organizations in the population surveyed. Even in the 183 respondents, approximately half of the organizations had annual budgets of $100,000 or less, and about the same percentage had two or fewer staff members. In response to a question about the types of information most important to them, the two most frequent answers were information about their service area and information about funding opportunities (Stoecker nd).

No one in our survey mentioned any privacy issues about the information they were collecting and distributing, though this was mentioned by Charles Kindleberger in the 1998 URISA conference session on 'Community Information Networks'. This issue has several facets. Often the local-area information, especially administrative records, are confidential, though organizations can sometimes get access to records that have been edited (e.g. by deleting individual names) or sign confidentiality agreements limiting their distribution of the data. But Kindleberger noted that people can, and do, object to having public information about themselves being made more easily available. In one example, it was property tax records that could be searched by name; the police in one community objected, because their names were in the database, and thus their addresses could be located by criminals who'd had encounters with them (the police had unlisted telephone numbers for the same reason). This same point was raised in 1998 when a group of doctors sued to shut down an Internet site called 'The Nuremberg Files', which listed the names, addresses, phone numbers, and sometimes the names of the children of doctors who performed abortions (and which crossed out the names of doctors who were killed); they won a $107 million verdict against the site and its supporters. Similarly, criminals convicted of child molestation while in prison have collected information about families from newspaper articles (e.g. the names of parents and children, parents' occupations and schedules, etc.) and made it available over computer networks.[4] After all, one of the points often made by IT proponents is that just being able to collect a variety of information in one place can greatly increase the uses that people can make out of that information.

There is also an issue of the impact on a neighbourhood of providing information about it. Kretzmann and McKnight (1996) noted that the kind of information social scientists collect and make use of is often information about social pathologies. What is the impact on an inner-city minority neighbourhood, e.g. of putting data on the web that makes it easier for anyone to see that the neighbourhood has high rates of poverty, teen mothers, welfare recipients, and criminal activity? Even if the same information is available about other neighbourhoods in the community, will people bother to make comparisons, and find, e.g. that some types of crime may be more common in more 'desirable' neighbourhoods than in 'underclass' ones? Or will people just use the information to justify decisions to abandon these neighbourhoods?

One of the most basic issues in PPGIS is whether to charge for services. There is a practical aspect: how will a data intermediary organization maintain itself if it doesn't charge for its services? On the other hand, how useful will it be to grassroots groups surviving on a shoestring if they can't afford to make use of its services? There is another aspect – rationing service: without any charge for its services, a group may be overwhelmed by requests beyond its capacity to respond in a timely fashion. It then has to decide how to ration its services. If it charges something, even a nominal amount, the flow of requests is likely to be smaller. Our inventory did not address this issue directly, because our focus was on organizations that made no charge, or only a nominal charge for their information, and there proved to be many.

One issue rarely mentioned was whether there was any real value to grassroots groups internalizing GIS capacity, or whether it is a better use of their resources to focus on other things, such as learning to ask the right questions, or making good use of the answers. The concept of democratization of data carries the implication that eventually grassroots organizations would have their own GIS capacity and do their own analysis. But short of the time when GIS systems become as simple to use as word processing software (if ever), becoming skilled at GIS takes time and effort, and in the end is a marketable skill. Larger, more established neighbourhood groups, with stable funding and relatively large staff, may have the resources to internalize GIS capacity, but is it worth it? A graduate student, quite taken with GIS upon first exposure to it years ago, expressed an interest in majoring in GIS to a veteran planner. The planner offered the following advice: 'You don't want to be the GIS expert. You want to be the person who poses the questions for the GIS expert to answer.' The student decided that was good advice and we think it might apply equally well to neighbourhood groups.

2.5 CONCLUSION

We set out to compile a comprehensive inventory of PPGIS providers, a task that quickly became more complicated and larger in scope than originally thought. The first step in this process was to create a definition of PPGIS providers. We see them on a continuum, from those organizations that work closely with community groups to collect and produce local-area data and analysis, to those organizations that repackage local-area Census data and make it available to whoever wants it. We began by defining local-area as sub-city (neighbourhood level). But we soon realized that this definition excluded much of the work of community environmental groups, so we tried to include them by defining local-area as areas other than standard political areas, e.g. other than states, counties, and cities whether smaller than these, or cutting across them.

The inventory reflects this definition, although everyone who responded was included. Some organizations did not seem to be PPGIS organizations by our definition, but they identified themselves in this way. We do not claim that the inventory is comprehensive; it includes organizations that do not quite meet our criteria, and certainly omits many that do. And given the progress of desktop GIS technology, more organizations undertake PPGIS work every month. This list does however provide a good starting point for further research into PPGIS activity in the United States.

NOTES

1. HUD's Community 2020 software: http://www.hud.gov:80/cpd/c2020/2020soft.html.
2. The Office of Data and Policy Analysis and Evaluation (DAPA). This office was created by Sawicki in 1992, staffed by Master's and Ph.D. students from Georgia Tech's City Planning programme, and administratively located within The Atlanta Project, a community development project of the Carter Presidential Center which has since been transferred to Georgia State University.
3. 'Votes count but resources decide', Stein Rokkan, quoted in Stone, C. (1989) *Regime Politics: Governing Atlanta 1946–1988*, Lawrence: University of Kansas, p. 239.
4. 'On prison computer, files to make a parent shiver', Nina Bernstein, *New York Times*, 18 November 1996, p. 1.

REFERENCES

Cincotta, G. (1996) 'From redlining to reinvestment – the need for eternal vigilance', Keynote speech at 4th International Conference on Financial Service: European Monetary Union and the Regional Responsibility of Financial Institutions Toward the Customer, 27 September, Strasbourg, France, http://www.iff-hamburg.de/Strasburg_virtuell/Reden/cincotta.htm.

Kindleberger, C. (1998) 'Community networks', presented at Urban and Regional Information Science Association Conference, Charlotte, North Carolina.

Kretzmann, J. P. and McKnight, J. L. (1996) *Building Communities from the Inside Out*, Chicago: ACTA Publications.

NCGIA Workshop, Public Participation GIS, http://ncgia.spatial.maine.edu/ppgis/ppgishom.html.

Ryan, D. (1998) Personal communication. Ryan was the founder and administrator of New Haven On-Line.

Sawicki, D. S. and Craig, W. J. (1996) 'The Democratization of data: bridging the gap for community groups', *Journal of the American Planning Association* 4: 512–523.

Stoecker, R. (nd) 'Putting neighborhoods on-line; putting academics in touch: the urban university and neighborhood network', http://131.183.70.50/docs/UUNN/CDSPPR.htm.

Urban Institute (1996) 'Survey findings by city', in *Democratizing Information: First Year Report of the National Neighborhood Indicators Project*, pp. 118–136.

Chapter 3

Models for making GIS available to community organizations: dimensions of difference and appropriateness

Helga Leitner, Robert B. McMaster, Sarah Elwood, Susanna McMaster and Eric Sheppard

3.1 INTRODUCTION

The research agenda addressing public participation GIS is, broadly speaking, evolving in two different directions. First, there is research examining the conventional use of standard GIS technologies by organizations with strong traditions of direct democracy; addressing issues of access; and whether or not this GIS can empower such groups, particularly those already occupying a marginalized social or geographical location (cf. Allen 1999; Jordan 1999; Kyem 1999). Second, some researchers, concerned that such GISs are not necessarily empowering, are beginning to examine alternatives to conventional use of GIS (cf. Krygier 1996; Harris and Weiner 1998; Shiffer 1998). These alternatives extend from the integration of narratives and local knowledge within current GIS software, to multimedia GIS, the design of collaborative decision support systems, and the use of non-hierarchical systems of information flow.

While the latter body of work was the inspiration for theorizing GIS2 and then PPGIS, and began in discussions at the NCGIA Initiative 19 specialist meeting (Harris and Weiner 1996), this chapter is within the former tradition. We seek to investigate the appropriateness of current GIS technologies for neighbourhood and grassroots organizations (henceforth 'community organizations'), in their tasks of articulating and pursuing the interests of those whom they are supposed to represent. The work reported here is based on a variety of experiences with models designed to make GIS available to community organizations in Minneapolis and St Paul (cf. Elwood and Leitner 1998). Rather than report in detail on these experiences, we seek to abstract from them and to position our experiences within a conceptual framework. This chapter is organized as follows. First, a discussion is provided, in general terms, of the different ways in which the appropriateness of GIS for community organizations can be assessed. Second, different models for making GIS available to community organizations are conceptualized and described. Third, a discussion of the putative advantages and disadvantages of these models for empowering community organizations seeking to use GIS is provided.

3.2 THE APPROPRIATENESS OF GIS FOR COMMUNITY ORGANIZATIONS[1]

The appropriateness of GIS for advancing the interests and concerns of communities can be assessed at three levels (Leitner *et al.* 1996). First, is the question of how GIS is made available to community organizations (Yapa 1991; Hutchinson and Toledano 1993; Barndt and Craig 1994; Sawicki and Craig 1996; Barndt 1998; Clark 1998; Elwood and Leitner 1998; Harris and Weiner 1998). GIS availability will be governed by financial considerations and the ability to purchase and maintain the appropriate hardware and software; by the expertise available locally and the geographical and technical skills necessary to make use of GIS; and by the availability of data, often depending on the openness of government agencies and freedom of information regulations. Some of these barriers are falling as computing costs decline and expertise spreads, although this is mitigated by an increased tendency of local governments to charge for the use of their databases.

Second, is the question of how successful implementation of the technology affects democratic processes in the community. The literature on organizations is full of cases where a new technology or body of expertise creates divisions within the organizations adopting them (cf. Bikston and Eveland 1990; March and Sproul 1990). The adoption of GIS may reduce the cohesion of the community organization as rifts develop between the new experts and often longer-term members of the organization. These rifts can be particularly acute in community organizations where goals are often negotiated through communicative action rather than being given by bottom-line imperatives.

Third, apparently successful and democratic implementation of GIS within an organization need not advance the participation of all of those that the organization is supposed to represent. Indeed, increased use of GIS may alter the priorities of the community organization such that it becomes less representative of the community at large. This is more likely to happen when the community is heterogeneous, and when diverse local concerns and understandings cannot easily be made consistent with the technology. Rundstrom (1995), e.g. expresses the fear that use of GIS by Native American organizations is inconsistent with Indian understandings of space and place (see also Brown *et al.* 1995; Jarvis and Spearman 1995; Kemp and Brooke 1995; Nietschmann 1995).

Since this research is taking place at a time when community organizations in Minneapolis and St Paul are just beginning to use GIS (Craig and Elwood 1998; Elwood and Leitner 1998), it is too soon to make any judgements about the second and third aspects of appropriateness sketched above.

Instead, in the sections that follow, issues affecting the first dimension – that of the availability of GIS to community organizations – will be examined. It is recognized that, in practice, community organizations will use a variety of ways to assemble the expertise they believe to be advantageous to their goals. This mix of expertise may constantly shift as circumstances change. Furthermore, the efficacy of different ways of making GIS available will vary with the context of the organization concerned, for no single way of providing GIS to community organizations is necessarily superior. In this sense, the evolution of GIS-related practices within community organizations would be characterized by the path-dependent dynamics associated with the development of any social technology in use, and as conditioned by the particular context of those using it. This evolutionary aspect is addressed more generally by those examining the intellectual history of GIS (Chrisman 1988; Sheppard 1995; Harvey and Chrisman 1998). Nonetheless, in order to gain insight into why certain ways of making GIS available may be favoured in certain circumstances, abstractions are drawn from these complexities to compare and contrast different models for making GIS available to community organizations. In the following section, a conceptual framework is proposed for distinguishing between different models of GIS access; a framework that can be applied to categorize models already in use and to think about other possibilities. This framework is then applied to six models, drawn largely from our experiences to date in the Twin Cities.

3.3 CONCEPTUALIZING MODELS OF AVAILABILITY

Models for making GIS available to community organizations can be differentiated along five important inter-related dimensions: The communication structures connecting community organizations with GIS facilities; the nature of the interaction with GIS; the physical (geographical) accessibility of the GIS to the community organization; the stakeholders involved in making the technology available; and legal and ethical ramifications (see Table 3.1).

Communication structures include: (1) independent nodes, whereby each community organization operates its own GIS in relative isolation from one another; (2) radial structures in which community organizations' use of GIS centres on separate use of a common facility; and (3) network structures, in which community organizations communicate directly with one another as they use GIS. The nature of interaction with the GIS can include: (1) no direct use at all; (2) passive use by individuals, where use is dictated by available databases and maps and by standardized GIS procedures; (3) active use, whereby users are free to develop their own operations and classifications of given databases; and finally (4) proactive use, where users can enter their own data and benefit from a variety of information technologies best suited to

Table 3.1 Differentiating models of availability

Dimensions	*Attributes*
Communication structures	Independent nodes Radial connectivity Network connectivity
Nature of interaction with the GIS	No direct use Passive use Active use Proactive use
Location of a GIS	In-house GIS Virtual (web-based GIS) Remote GIS (outside the community)
GIS stakeholders	Local & non-local state agencies Non-governmental organizations (NGOs) Private industry Educational institutions Within community stakeholders
Legal and ethical issues surrounding GIS use	Ownership of/responsibility for spatial databases Access to publicly held information Issues of privacy and surveillance Checks and balances governing appropriate GIS use

those data (cf. Harris *et al.* 1995; Weiner *et al.* 1995). The interaction can also vary from individual to collaborative user interfaces, with the latter facilitating collective negotiation and decision-making (Couclelis and Monmonier 1995; Nyerges *et al.* 1997). Models also differ in the geographical location of the GIS, ranging from local access in the community (in-house GIS), to virtual access over the information networks (e.g. Web-based GIS), to remote access, where physical travel to a location outside the community is necessary in order to use GIS. These three dimensions relate directly to questions raised within research into public participation in GIS (Brown *et al.* 1995; NCGIA 1996; Barndt 1998; Dangermond 1988; Obermeyer 1998).

Yet another dimension involves the stakeholders. Stakeholders include individuals and institutions external to the community, such as local and non-local state agencies, NGOs, private industry and educational institutions. These actors and institutions have their own priorities and interests that can affect the responsiveness of the GIS system to community organization needs. This dimension relates directly to research addressing the institutional perspective on GIS and society (cf. Onsrud and Rushton 1995; Ventura 1995; Tulloch and Niemann 1996). In addition, community organizations often represent diverse communities within which there are local stakeholders with conflicting understandings and priorities.

A final dimension involves legal and ethical issues. Legal and ethical issues, a separate area of research in GIS and society (Onsrud and Rushton

1995), refers in general terms to questions of intellectual property rights in spatial databases, access rights of citizens to publicly held information, privacy rights and principles, liability in the use and distribution of GIS data and products, and ethical issues in the use of geographic information (Onsrud 1992a; 1992b; 1995; Sheppard *et al.* 1999). Models for making GIS available to community organizations will differ in terms of several factors. Some of these include: (1) who has legal ownership and responsibility for the accuracy of the spatial databases used or created by these organizations in GIS analysis; (2) whether the communities represented by community organizations have access to publicly held information; (3) the potential for abuse of the privacy of those in the community; and (4) the checks and balances that can guard against this and other unethical activities related to the use of the GIS.

Models for making GIS available to community organizations will differ from one another along one or more of these dimensions, which represent a means for differentiating and classifying models that are currently in use (as we seek to demonstrate in the subsequent section). They can also aid in both normative reasoning and in conceptualizing the desired attributes of other models not net developed. One might speculate, e.g. that a model might be particularly advantageous for community organizations if it were characterized by: (1) a network communication structure; (2) a collaborative proactive use of GIS; (3) no other stakeholders with conflicting interests or goals; (4) local accessibility; and (5) where community organizations carefully regulate legal and ethical responsibilities.

3.4 SIX MODELS OF GIS AVAILABILITY

In the first column of Table 3.2, we list six models for making GIS available to community organizations. In this section, the nature of these models is discussed, and the differences among them based on the conceptual framework of the previous section are laid out.

Table 3.2 Six models for making GIS available to community organizations

Community-based (in-house) GIS	(e.g. Powderhorn Park, Prospect Park)
University–community partnerships	(e.g. Urban GIS class, Macalester Action Research, University Neighborhood Network)
GIS facilities in universities and public libraries	(e.g. ACIC, St Louis Public Library)
'Map Rooms'	(e.g. City of Minneapolis Map Room)
Internet Map Servers	(e.g. Phillips Neighborhood Environmental Inventory)
Neighbourhood GIS centre	(e.g. Milwaukee Data Centre)

3.4.1 Community-based (in-house) GIS

The establishment of an in-house GIS capability and database by community organizations for community-based planning is still a rare and recent phenomenon in the Twin Cities. In the city of Minneapolis, very few neighbourhood organizations have an in-house GIS. A community-based GIS is usually designed as an independent node located within the community organization, usually at its office. Neighbourhood organizers and residents do not have to physically travel outside the neighbourhood, but are able to gain direct and immediate access to information as needed for neighbourhood planning and organizing purposes. Furthermore, an in-house system can be tailored to the specific needs of community organizations because it allows them to create and interactively manipulate their own databases and maps, rather than relying only on pre-defined data sets or maps.

The responsiveness of an in-house GIS to neighbourhood needs is potentially enhanced by the fact that neighbourhood organizations are the primary stakeholders in an in-house system. This does not imply, however, that there exists a consensus among neighbourhood residents regarding neighbourhood priorities, or that the community organization will represent all of these priorities. Rather, the diversity of neighbourhood residents usually means that there are a variety of stakeholders with often differing agendas. Thus the responsiveness of an in-house GIS to meeting community needs must also be evaluated in the context of diverse internal stakeholders.

Neighbourhood/community organizations do not assume primary legal responsibility regarding the ownership, control, and accuracy of public data, but do have to face legal issues regarding community-generated databases. For example, local government or the media might try to gain access to sensitive community-generated databases that the community organization, or stakeholders in the community, might not want to release.

3.4.2 University–community partnerships

Increasingly universities, through a variety of mechanisms, are attempting to assist community organizations with their spatial information and mapping needs. One common approach is to provide assistance through community service learning requirements in urban GIS courses, whereby students provide a service to community organizations, such as developing a GIS application based on a community request, and then reflect on and share the lessons learned with the class. The service provided to the community organizations is generally limited to the duration of the class. Action research is an alternative, fully collaborative, inclusive, and longer-term approach to community–university partnerships that emphasizes the importance of full participation by community members in both research and the generation of knowledge. A key aspect of this approach is to actively involve community members in defining and examining community issues and problems and deriving solutions. Active involvement of the community in this way potentially empowers the community to employ GIS to generate social change and affect public policy.

Another approach to university–community partnerships occurs when faculty research projects are linked to community-based problems. Here, faculty and student research teams work with a community organization over a longer period of time, not only assisting with basic mapping problems, but also with the analysis and interpretation of data.

University–community partnerships operate within a radial communication structure, because community organizations separately use university facilities. Community organizations rely strongly on university GIS expertise, rarely maintaining their own systems or making direct use of the technology. Other stakeholders are rarely involved in such partnerships. In our experience, many community organizations seeking such partnerships begin with no existing experience or expertise with GIS, but envision using the relationship to acquire in-house GIS capability. Since the university provides data for community organizations, there are no significant concerns with ownership of data.

3.4.3 Publicly accessible GIS facilities at universities and libraries

A further means by which community organizations gain access to GIS and spatial data is through publicly accessible GIS facilities at universities and libraries. Typically, the facility creates and maintains certain basemaps and spatial data and makes these available for use with GIS software. Community organization staff or volunteers must travel to the facility to create or print out maps, or work with the database. Such publicly accessible GIS facilities rely on a radial type communication structure in which community organizations separately make use of the same facility. While such an arrangement may mean that the GIS facility staff develop and disseminate expertise about how to solve standard problems, it also may mean that community organizations do not communicate directly with one another about solving common problems.

The nature of interaction that users have with a GIS at public GIS facilities can vary from passive to proactive use, depending on whether users can manipulate the database or enter individualized data for their community. At the St. Louis Public Library GIS users select from predefined data sets and maps (Krofton 1993). At other facilities, like the University of Minnesota's Automated Cartographic Information Center (ACIC), users can manipulate existing data or bring in their own data for mapping and analysis.

Publicly accessible GIS facilities in universities and libraries involve at least two major stakeholders in addition to the community organization and the community it represents. These comprise the organization managing the facility, and the agency providing the funding. These stakeholders influence a GIS and its use by community organizations in several ways. The organization managing the facility determines which data and maps are made available to users and what kinds of support services the facility staff provides. At most facilities, staff are available to help with technical problems concerning the operation of hardware or software. However, these facilities differ with respect to the amount of guidance or advice provided by staff in analysis of data or maps. The involvement of multiple stakeholders in community organizations' use of public GIS facilities presents a complex set of legal issues about who has responsibility for the data and maps prepared in such facilities and their subsequent use.

3.4.4 Map rooms

There are several examples of 'map rooms' used by community organizations to acquire spatial information. The city planning office, e.g. may provide citizens with land use/land cover, taxation, and other maps relevant to their planning mission. Many other city and state offices, such as departments of natural resources and pollution control agencies, create and distribute maps. In Minneapolis, the city maintains the map room, operated by the City Engineering Department (which is responsible for maintaining the city's spatial database). For a fee, this facility will create custom maps on demand for community organizations and citizens. Many of the community organizations in Minneapolis make extensive use of this facility as a surrogate for the lack of in-house mapping capability.

The map room represents radial connectivity because information, in the form of maps, flows from the map room to citizens. In most cases, those who use map rooms are unlikely to even know others using the facility. The users of this facility have no direct interaction with the system and thus gain no expertise with GIS. Local and non-local state agencies represent active stakeholders in that the facility is owned and maintained by the city and extensively utilize city-generated data. The ownership of data for map rooms lies with the agency, and is not a significant concern to community organizations. Although the city maintains confidential information, it is normally not released unless strict confidentially agreements are signed, as in the case of public housing, certain public health variables, or data based on economic measures.

3.4.5 Internet map servers

Internet map servers make pre-defined maps available to community organizations over the Internet, most often residing at websites. This model requires that some existing institution, such as city-government, colleges and universities, private companies, NGOs or even another community organization, establish a website with a series of already designed and symbolized maps. Although most sites are still oriented towards cartographic display, increasing numbers of sites now allow users to make simple spatial queries or perform analyses, thereby lessening the potential requirement for a fully functional in-house GIS. One example of such a site is a neighbourhood environmental inventory developed by the author for the Phillips Neighbourhood in Minneapolis (http://www.geog2.umn.edu/mapserve/pneiweb/Pneinet/PNEI.html). As richer websites are created, Internet map servers have the potential to become a major source of spatial information for community organizations.

The communication structure for Internet map servers theoretically represents one of the most egalitarian methods for distributing spatial information to neighbourhoods, a method of ubiquitous access both in terms of space and time. With relatively low-level equipment, neighbourhoods can access this rich source of information, albeit filtered by those who designed the site. Interaction with the server can vary from passive use, where spatial information is 'served' in a typical server–client model, to a more active model where users can remotely query and manipulate data. The former is far more common than the latter. The related stakeholders most often involve local or non-local public or quasi-public agencies, which are often responsible for designing the website. Since GIS software and hardware is not needed in house, this represents a virtual interaction with the 'system.'

3.4.6 The neighbourhood GIS centre

A neighbourhood GIS centre, a model with which we have little empirical experience, is created when neighbourhoods pool their expertise and resources to provide a central facility that all affiliated community organizations can use. The funding to maintain such a centre could come from the community organizations themselves, but continuity of funding is best provided by a non-profit foundation, by the private sector, or by the state. The governing principles of a neighbourhood GIS centre would be that its goals are set by the community organization(s) that it serves, and that it provides those organizations with the capacity not only to gain access to pre-existing databases but also to input information gathered by the communities themselves.

Such a centre, if successfully implemented, would be characterized by a network communication structure, since the collaboration of communities in development and use of the centre will encourage both the sharing of knowledge and expertise, and joint action to address emerging problems. A neighbourhood GIS centre would have the capacity for proactive use. Communities could enter their own information into the GIS, and the shared infrastructure and expertise might create the capacity for innovative integration of other kinds of information with GIS. It could also be an ideal environment for collaborative learning and decision-making, as representatives of different communities can gather around a single computer terminal. This would probably require community representatives to travel to the centre, thereby limiting the degree to which neighbourhood residents can participate in this process.

The role of other stakeholders will depend on how the centre is funded and equipped. If the centre is based on a 'block grant' and contractual agreements giving neighbourhoods wide-ranging access to local databases, free choice of GIS software, and suitably equipped physical space, then the influences of external stakeholders will be less. At the other extreme, a local government may make only limited data available; external funding may dictate the software that can be used and thereby the GIS capabilities and the types of data that can be entered; and the centre may have to be housed in space owned by external stakeholders who thereby exert control over how that space is used. There is also a sense in which different community organizations become stakeholders in, and thus may attempt to exert influence over, each other's GIS-related activities as a result of collaboration in the centre. This can occur when community organizations are compelled to use GIS in ways that go against their best judgement, as a result of majority decisions about how the centre should operate.

A neighbourhood GIS centre faces complex legal and ethical issues. Legal responsibility for the accuracy and reliability of databases and software acquired from external stakeholders lies outside the centre. In addition, active collaboration between community organizations in the centre can facilitate dialogue about the development of ethical and legal standards for data acquisition, and use and analysis, which are appropriate to the needs and responsibilities of such organizations. At the same time, however, less attention may be paid to the ethical or legal implications of those standards for individuals or organizations outside the centre. Furthermore, the sharing of expertise and data between community organizations creates the possibility that information about individuals from one organization is inappropriately made available to other organizations.

3.5 ASSESSING THE APPROPRIATENESS OF THE SIX MODELS FOR COMMUNITY ORGANIZATIONS

Table 3.3 lists a variety of potential possible advantages and disadvantages associated with each of the six models of making GIS available to community organizations. This list is based on our own informal assessments, and thus is necessarily tentative. Broadly speaking, these advantages and disadvantages are of two types: first, those that address the question of model flexibility and responsiveness to the needs of community organizations; and

Table 3.3 Advantages and disadvantages of the six models

	Advantages	*Disadvantages*
Community-based (in-house) GIS (e.g. Powderhorn Park, Prospect Park)	Can be tailored to local needs Can be made directly available to community organizers and residents Allows direct monitoring of community/neighbourhood change by community organization Allows for quick and flexible response to community issues Potential for community-based employment and related skill-building	Difficulties in raising funds to purchase hardware and software Difficulties in long-term maintenance of GIS due to monetary costs and personnel turnover Requires technologically skilled community organizers and/or residents Unnecessary duplication of effort across communities Independent data access reliant on political connectedness (differs between organizations)
University – community partnerships (e.g. Urban GIS class, Mac Action Research, University Neighbourhood Network)	Easier access to GIS expertise and data Can be made responsive to specific data and application needs of community organizations Costs to community are lower (both monetary costs and time necessary to learn and maintain the system) Possibility of improved communication and interaction between partners	University has limited capacity, and can provide services to only a few communities University or research project agenda may not fit with that of community Faculty/students may not fully understand community needs Timing – university help may not be available when needed Lack of long-term commitment of university to communities Unnecessary duplication of efforts across communities Hard to directly involve community members in analysis
GIS facilities in universities and public libraries (e.g. ACIC, St Louis Public Library)	Community members may have direct access to GIS and databases, and to expert advice in how to use these Lower costs for community organizations Reduces duplication of effort (expertise about how to solve standard problems is developed and hopefully retained within the facility) Longer-term resource	Analysis is limited to publicly available data sets, unless community organizations have capacity to enter their own data Can be intimidating to use such facilities (particularly university) Use requires travel outside the neighbourhood and is limited to times when facilities are open

Table 3.3 (Continued)

	Advantages	*Disadvantages*
'Map Rooms' (e.g. City of Minneapolis Map Room)	Relatively easy availability of basic mapped spatial data (little expertise needed) Pay as you go; no up-front investments required	Limited to the databases maintained by the government agency Use requires travel outside the neighbourhood and is limited to times when facilities are open Services available reflect priorities of institution maintaining the map room Limited advice from map room staff regarding which maps to use or what these maps mean with regard specific neighbourhood context, concern or area of interest
Internet Map Servers	Allows direct access to spatial data Potential for 2-way interactions	Dependent on computer capacity of community organization (people and hardware/software) Limited ability to manipulate data according to specific needs No access to external expertise to interpret maps and data
Neighbourhood GIS centre	Provides economies of scale, in terms of expertise and resources Continued operation does not depend on the fortunes of individual community organizations Responsive to community needs Potentially promotes collaboration among community organizations, including problem-solving	Difficult to realize because it requires collaboration in advance of GIS implementation Significant external funding must be secured Conflicts between neighbourhood organizations about priorities could reduce effectiveness GIS facility is not located in community/neighbourhood

second, those that address the difficulties of implementing and maintaining a GIS.

Flexibility and responsiveness attempts to capture the various ways in which a particular model can be flexible and sensitive to the particular context and needs of different community organizations. Community organizations are highly heterogeneous, making flexibility an important feature enhancing the appropriateness of GIS for them. Flexibility and responsiveness will depend on ease of use; on any geographical or social barriers reducing access to GIS; and on the degree to which the user can directly interact with and control the GIS. These aspects all influence the capacity for a broadly participatory and inclusive use of GIS that in turn can contribute to empowerment and democratic decision-making within the community itself.

Difficulties of implementation and maintenance refers to the monetary and non-monetary costs of developing and using GIS for community organizations. First, it is important to distinguish between the individual costs to a community organization, and the collective costs that result when a number of organizations simultaneously seek to use GIS. Because of economies of scale, the cost-minimizing solution for a single organization is not necessarily the best for the organizations as a group, as it may result in the unnecessary duplication of costs in different community organizations. Second, it is important to distinguish between set-up and maintenance costs. The failure to plan for maintenance, resulting in technologies not being used after they have been made available, is a classic barrier that reduces the appropriateness of technologies. This is particularly challenging for community organizations that generally face limited and rapidly changing financial and human resources. Human capital costs are important to the successful operation and maintenance of a GIS, because of the rapid turnover of community organization personnel. It may be time-consuming to build up the relevant skills and expertise within the community, and there is always the danger that those who have gained such skills will leave, thereby compromising the maintenance of the GIS.

3.6 CONCLUSION

In this chapter, a tentative conceptualization of different models for making GIS available to, and for assessing their efficacy for, community organizations is provided. We note that this is just one of the issues that must be addressed in discussing how GIS can empower and/or marginalize community organizations and the communities that they represent. Yet, even on this issue, many questions remain unanswered. First, we need to determine whether the conceptualization, models and provisional assessments reported here are robust. To this end, a survey is currently underway of community organization representatives working with a variety of these models in the Twin Cities and Milwaukee to cross-check their evaluations against ours.

Second, it is necessary to empirically examine the relationship between these models and the evolving practices of community organizations. Based on our experience thus far, community organizations do not choose just one model, but draw on different ways of gaining access to GIS, changing their strategies over time and perhaps developing novel ways of accessing and utilizing GIS. One test of the utility of an exercise such as this is whether it helps make sense of such strategies and the implications for the empowerment and/or marginalization of community organizations.

NOTE

1. Note that we are assuming, for the purposes of this discussion, that the technology is more-or-less given, something with which community organizations must come to terms. This rather static view of the technology neglects the ways in which users can change the nature of the technology itself, to their advantage or disadvantage. Considerations of this aspect – the influence of society on GIS – suggest that there may be limitations to the 'appropriateness of technology' way of thinking about this issue (cf. Harvey and Chrisman 1998).

REFERENCES

Allen, M. (1999) 'A participatory model of information system design: a case study of the economic human rights documentation infomanagement system of the Kensington welfare rights union', paper presented at the First International Conference on Geographic Information and Society, 20–22 June, Minneapolis, MN.

Barndt, M. (1998) 'Public participation GIS – Barriers to implementation', *Cartography and Geographic Information Systems* 25(2): 105–112.

Barndt, M. G. and Craig, W. J. (1994) 'Data providers empower community GIS efforts', *GIS World* 7(7): 49–51.

Bikson, T. and Eveland, J. (1990) *Technology Transfer as a Framework for Understanding Social Impacts of Computerization*, Santa Monica, CA: Rand.

Brown, I. F., Alechandre, A. S., Sassagawa, H. S. Y. and Aquino, M. A. (1995) 'Empowering local communities in land-use management: the Chico Mendes Extractive Reserve, Acre, Brazil', *Cultural Survival Quarterly* 18(4): 54–57.

Chrisman, N. R. (1988) 'The risks of software innovation: a case study of the Harvard lab', *The American Cartographer* 15(3): 291–300.

Clark, M. J. (1998) 'GIS – democracy or delusion?' *Environment and Planning A* 30(2): 303–316.

Couclelis, H. and Mark Monmonier (1995) 'Using SUSS to resolve NIMBY: how spatial understanding support systems can help with the 'Not In My Back Yard' Syndrome', *Geographical Systems* 2(2): 83–101.

Craig, W. J. and Elwood, S. (1998) 'How and why community groups use maps and geographic information', *Cartography and Geographic Information Systems* 25(2): 95–104.

Dangermond, J. (1988) 'Who is designing geographic information systems for the public?' *Urban and Regional Information Systems Association*, pp. 37–45.

Elwood, S. and Leitner, H. (1998) 'GIS and community-based planning: exploring the diversity of neighborhood perspectives and needs', *Cartography and Geographic Information Systems* 25(2): 77–88.

Harris, T. and Weiner, D. (1996) 'GIS and society: the social implications of how people, space, and environment are represented in GIS', Scientific Report for the Initiative 19 Specialist Meeting, 96–97, NCGIA.

Harris, T. and Weiner, D. (1998) 'Empowerment, marginalization and "community-integrated" GIS', *Cartography and Geographic Information Systems* 25(2): 67–76.

Harris, T. M., Weiner, D., Warner, T. and Levin, R. (1995) 'Pursuing Social Goals Through Participatory GIS: Redressing South Africa's Historical Political Ecology', in J. Pickles (ed.) *Ground Truth: The Social Implications of Geographic Information Systems*, New York: Guilford Press, pp. 196–222.

Harvey, F. and Chrisman, N. R. (1998) 'Boundary objects and the social construction of GIS technology', *Environment and Planning A* 30(9): 1683–1694.

Hutchinson, C. F. and Toledano, J. (1993) 'Guidelines for demonstrating geographical information systems based on participatory development', *International Journal of Geographical Information Systems* (7): 5.

Jarvis, K. A. and Spearman, A. M. (1995) 'Geomatics and political empowerment: the Yaqui', *Cultural Survival Quarterly* 18(4): 58–61.

Jordan, G. (1999) 'Maximizing the benefits of participatory GIS in technology-poor countries: case studies from Zambia and Nepal', paper presented at the First International Conference on Geographic Information and Society, 20–22 June, Minneapolis, MN.

Kemp, W. B. and Brooke, L. F. (1995) 'Towards information self-sufficiency: the Nunavik Inuit gather information on ecology and land use', *Cultural Survival Quarterly* 18(4): 25–28.

Krofton, C. P. (1993) 'St. Louis library's GIS disseminates public information', *Geo Info Systems* (July/August): 46–50.

Krygier, J. (1996) 'Geographic visualization and the Making of a Marginal Landscape', in T. Harris and D. Weiner (eds) *GIS and Society: The social implications of how people, space, and environment are represented in GIS*. NCGIA Technical Report 96–97, Scientific Report for Initiative 19 Specialist Meeting, South Haven, MN, 2–5 March.

Kyem, P. (1999) 'Embedding GIS applications into resource management and planning activities of local community groups in sub-Saharan Africa: a desirable innovation or a disabling undertaking?' paper presented at the First International Conference on Geographic Information and Society, 20–22 June, Minneapolis, MN.

Leitner, H., McMaster, R. B., Miller, R. and Sheppard, E. (1996) 'I-19 initiative proposal: position paper', in T. Harris and D. Weiner (eds) *GIS & Society*, Santa Barbara: NCGIA, pp. D44-D45.

March, J. and Sproul, L. (1990) 'Technology, management, and competitive advantage', in P. Goodman, L. Sproul and Associates (eds) *Technology and Organizations*, San Francisco: Jossey-Bass, pp. 143–173.

NCGIA (1996) 'Public Participation GIS Workshop', http://ncgia.spatial.maine.edu/ppgis/ppgishom.html.

Nietschmann, B. (1995) 'Defending the Miskito reefs with maps and GPS: mapping with sail, scuba, and satellite', *Cultural Survival Quarterly* 18(4): 34–37.

Nyerges, T., Barndt, M. and Brooks, K. (1997) 'Public participation geographic information systems', *ACSM/ASPRS Annual Convention and Exposition Technical Papers*, Bethesda, MD, pp. 224–233.

Obermeyer, N. J. (1998) 'The evolution of public participation GIS', *Cartography and Geographic Information Systems* 25(2): 65–66.

Onsrud, H. J. (1992a) 'In support of cost recovery for publicly held geographic information', *GIS Law* 1(2): 1–7.

Onsrud, H. J. (1992b) 'In support of open access for publicly held geographic information', *GIS Law* 1(1): 3–6.

Onsrud, H. J. (1995) 'The role of law in impeding and facilitating the sharing of geographic information', in H. J. Onsrud and G. Rushton (eds) *Sharing Geographic Information Systems*, New Brunswick, New Jersey: Centre for Urban Policy Research, pp. 292–306.

Onsrud, H. J. and Rushton, G. (eds) (1995) *Sharing Geographic Information Systems*, New Brunswick, New Jersey: Centre for Urban Policy Research.

Rundstrom, R. A. (1995) 'GIS, indigenous peoples, and epistemological diversity', *Cartography and Geographic Information Systems* 22(1): 45–57.

Sawicki, D. and Craig, W. (1996) 'Democratization of data: bridging the gap for community groups', *Journal of the American Planning Association* 62(4): 512–523.

Sheppard, E. (1995) 'GIS and society: towards a research agenda', *Cartography and Geographic Information Systems* 22(1): 5–16.

Sheppard, E., Couclelis, H., Graham, S., Harrington, J. W. and Onsrud, H. (1999) 'Geographies of the information society', *International Journal of Geographic Information Science* 13: 797–823.

Shiffer, M. (1998) 'Multimedia GIS for planning support and public discourse', *Cartography and Geographic Information Systems* 25(2): 89–94.

Tulloch, D. and Niemann, B. J. Jr. (1996) 'Evaluating innovation: The Wisconsin Land Information Program', *Geo Info Systems* 6(10): 40–44.

Ventura, S. J. (1995) 'The use of geographic information systems in local government', *Public Administration Review* 55(5): 461–467.

Weiner, D., Warner, T., Harris, T. M. and Levin, R. M. (1995) 'Apartheid representations in a digital landscape: GIS, remote sensing, and local knowledge in Kiepersol, South Africa', *Cartography and Geographic Information Systems* 22(1): 30–44.

Yapa, L. (1991) 'Is GIS appropriate technology?', *International Journal of Geographical Information Systems* 5(1): 41–58.

Part II

PPGIS case studies

Chapter 4

A voice that could not be ignored: community GIS and gentrification battles in San Francisco

Cheryl Parker and Amelita Pascual

4.1 INTRODUCTION

In May 1998, San Francisco residents, night club owners and workers held a town-hall meeting to protest the 'Soho-ization' of their South of Market neighbourhood, affectionately known as 'SoMa'. The focus of their protests were the new developments being made next to or in place of diverse, mixed-use buildings that currently housed immigrant families, artists, start-up companies, and manufacturers. Geographic data about SoMa showed that it was fast transforming from a blue-collar neighbourhood into a chic residential and retail district. One central source of data was the SoMa community's GIS-based 'living neighbourhood map'. Used to start a conversation with policy-makers, the map sought to illustrate changes in development that portended zoning changes at a city-wide level. The May 1998 meeting came to represent the beginnings of a complex deliberative and democratic process focused around parcel politics and the politics of space.

Parcel politics is the politics of space at the smallest and most complex level. It is grounded in the idea that a great urban place is composed of a complex mix of spaces and places that can accommodate a wide variety of interdependent users. Until recently, few planning tools have enabled the kind of deliberation or debate characteristic of parcel politics. Instead, for the latter half of the twentieth century, planning tools and the political arena have focused on area-wide planning and development. For example, zoning practices have segregated land-uses and created vast districts of single use, while redevelopment and urban renewal practices have often razed neighbourhoods, thereby erasing intricate webs of streets and mixed uses, and replacing them with mega-block developments of just one or a few uses. SoMa's zoning, on the other hand, was an experiment created to foster a highly mixed-use district. This zoning, combined with SoMa's stock of highly flexible warehouse structures, presented the potential for the district to evolve in a myriad of ways.

Market-driven competition among builders and consumers at the parcel level poses an interesting challenge to the ideal of democratic planning.

Under such competitive development pressure, only informed and sophisticated voices will be heard. Politics at such a small scale, however, risks losing sight of a larger urban design and economic development vision. The story of SoMa illustrates the paramount role that PPGIS can play as a democratic planning tool by addressing both the complexity of development competition at the parcel scale while maintaining the larger economic and physical vision of an evolving neighbourhood within the context of a city. The story shows that information-based maps helped to educate a diverse neighbourhood about powerful and rapid forces of change and empowered its community to act. After mapping their neighbourhood and seeing abstract statistics portrayed as maps, SoMa's community of hundreds of very different, angry, reactive voices united into just one informed and very sophisticated collective voice. This collective then backed an alternative vision that embraced the existing character and flavour of the neighbourhood, while also accommodating its growth.

4.2 CONTEXT AND HISTORY

As the name implies, the South of Market area is just south of Market Street, one of San Francisco's most prominent streets (see Figure 4.1). SoMa is next to the financial district and near the bayfront, yet is not an area often associated with San Francisco tourism. Historically, it hosted the city's manufacturing and light industry and provided infrastructure serving the port. Long blocks, wide streets, and large areas of flat terrain unhindered by San Francisco's rolling hills made the area ideal for industry.

Historically, SoMa also served as a transient zone by acting as a portal where immigrants and urban poor could establish themselves before moving to a higher standard of living. As such, it hosted one of the last remaining affordable housing stocks in San Francisco, with over 30 Single Room Occupancy Residential Hotels (SROs) and several hundred units of low-rent family housing.

By 1995, a network of non-profit neighbourhood service providers were catering to the needs of SoMa's small businesses and low-income residents. This network provided job training and placement, built affordable housing, served small businesses, provided healthcare and childcare, and provided recreation and education programmes for inner-city youth. One of these non-profit service providers, the South of Market Foundation (SOMF), developed a GIS living neighbourhood map of SoMa, linking information about buildings, businesses and residents to produce a dynamic physical map.

After nearly 30 years of fighting development pressures in the city, this non-profit network established a strong political neighbourhood voice. This voice influenced the 1985 San Francisco Planning Department South of Market re-zoning study. The study revealed that SoMa was evolving into an

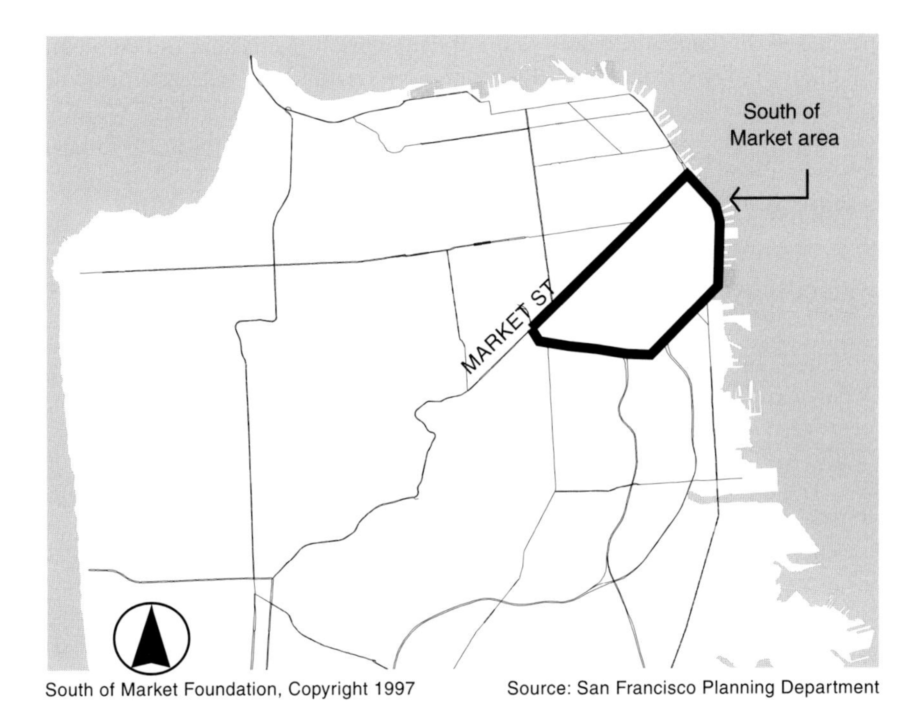

Figure 4.1 South of Market area.

important service district for the financial core and it was a haven for start-up businesses. Recognizing the district's increasing importance to the financial district, the Planning Department proposed zoning controls on a large area. Inside of the boundaries, a large mix of uses were permitted, from residential to light industry. In an effort to control gentrification and limit the displacement of industrial businesses that provided blue collar jobs and served San Francisco's Financial and Retail districts, office development was restricted to a few small pockets.

'Live–Work' was created at this time as a special type of mixed-use development that allowed people to legally have studios and workshops in the same space where they lived. It was created with the intent of legalizing a practice found all over SoMa, where artists, photographers, graphic designers and various other light-industrial artisans lived in old warehouses where they also had their studios and workshops.

The new zoning plan worked very well for a short time. SoMa evolved into a vibrant mix of service businesses, light industry, low-cost housing and artisans. Its mix of artists and abundant low-cost space was paramount in hosting San Francisco's so-called 'Multimedia Gulch' where hundreds of high-tech companies were incubating.

Up until 1996, there were few requests to construct new live–work housing. Based on the few requests to build live–work housing that did come in, the Planning Department quickly became aware that the code was difficult to enforce. It was almost impossible to verify whether people were actually working in the same spaces they were living in. Despite these problems, however, Planning Department staff did not attempt to correct the situation, because they did not anticipate much demand for live–work housing.

4.3 WINDS OF CHANGE

Intrigued by SoMa's evolution, in 1996 the San Francisco Planning Department collaborated with the SOMF in a series of GIS studies documenting traditional manufacturing and emerging high-tech industries (see Figure 4.2). In addition to documenting the complex side-by-side growth of two very different industry clusters, these studies revealed that construction of new live–work units was a serious threat to SoMa's economy due to business displacement caused by incompatible uses or evictions. The explosive growth of Silicon Valley high-tech firms led computer workers from throughout California and the United States to move to the Bay Area. Bay Area cities were not able to provide an adequate housing supply for this new demand, and consequently requests for housing permits, especially live–work permits, sky-rocketed in SoMa. The demand for housing among young, newly re-located high-tech workers, their inflated salaries and the stylish allure of loft-living in an industrial and night club district, all combined to cause an unforeseen demand for industrial land and, consequently, in conflicts between existing inhabitants and newcomers. In a short time, complaints about noise and truck traffic were commonplace, causing industrial and night club businesses to be cited, fined and eventually to relocate or close.

For builders, live–work was a popular type of development. Considered by planners to be commercial rather than residential development, the building codes for live–work did not require creation of open space, payment of school tax fees, notification of neighbours, conditional review, or parking development. In addition, live–work development could be built higher than could any other type of development, aesthetic concerns were minimal, and such development was permitted in industrial areas of the city with few existing residents and thus little risk of organized resistance. Finally, banks allowed developers to finance live–work construction with residential loans rather than commercial ones, and it is much easier to secure a residential construction loan.

Shortly after the first Planning Commission public hearing between live–work builders and the community regarding live–work-related land-use displacement, the GIS-based traditional manufacturing and multimedia industry studies were shelved by the Planning Department. This was a major blow to the community because the studies documented live–work as a

Garment Manufacturing in San Francisco

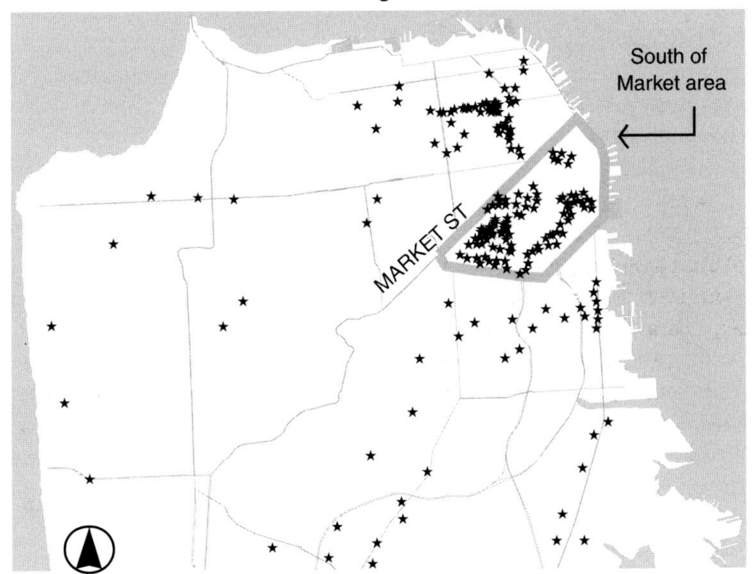

Multimedia Businesses in San Francisco

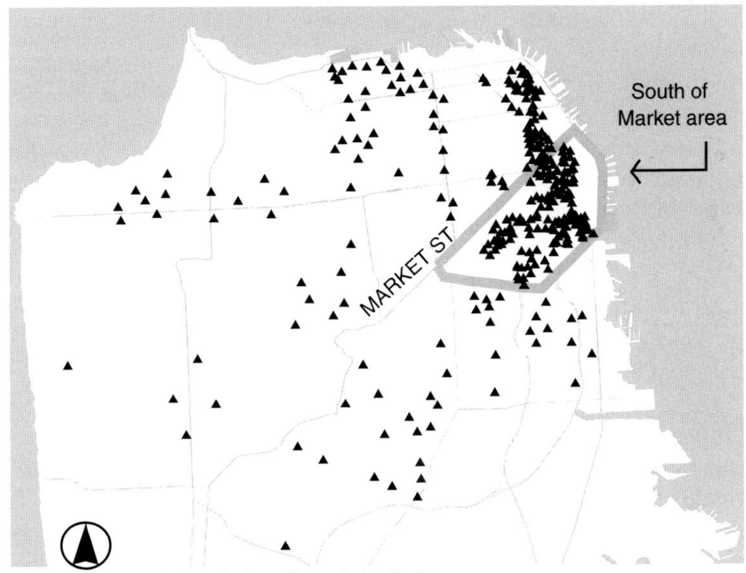

South of Market Foundation, Copyright 1997
Source: San Francisco Planning Department, Dun & Bradstreet, 1996

Figure 4.2 Location of traditional and high-tech industries.

major threat to important industries in San Francisco. In another blow to the community, the mayor proclaimed that he was not going to 'take sides' and that, as far as he was concerned, 'the displaced businesses, residents and artists could all move to another part of the city'. The reality, however, was that many firms were moving out of the city or going out of business. 'Not taking sides' was just another way of saying 'let the market go out of control and have its way'.

During the next year, realizing that the political climate had changed and that their neighbourhoods were now in jeopardy, the residents, workers, artists, non-profits and night club owners began organizing to confront the city Planning Department and Planning Commission regarding compatibility issues between small businesses and residents. They formed a coalition of artists and small businesses called SOS, for Save Our Shops.

At the same time, SOMF used its GIS model and industry research to document the impact that live–work construction was having on the local business climate. They combined, in one database, information about the location of demolition, renovation and new building permits with employment and sales information for local businesses. All businesses that were once located where demolition or new construction had taken place were telephoned to find out why they moved. If reasons given included eviction due to demolition or new construction, it was noted as 'business and job displacement due to gentrification'. Change among commercial rent prices was also mapped, as were important economic links broken due to job and business displacement (see Figure 4.3). Without GIS, this type of analysis would have been almost impossible.

In February 1998, the Planning Department proposed at a public hearing, a revised set of short-term controls for live–work by establishing Industrial Protection Zones. The proposed zones offered absolutely no protection for small businesses. Boundaries were drawn in obscure places and did not even appear in the SoMa area. In fact, some of the boundaries were drawn out into the San Francisco Bay. During that public hearing, over a hundred small business owners came out to speak of the displacement problem.

Four hours of often emotional public testimony was summed up by a city executive as: 'We did not learn anything new tonight.' The Planning Commission then voted not to apply any new controls to protect industry in the SoMa area, despite that SoMa zoning contained the only true existing Industrial Protection Zone in the city. This zone only allowed live–work development if work was the principal use of the property, and only allowed housing if it was for low-income people. The Planning Commission thus publicly agreed to ignore the Planning Code. In addition, an addendum exempted 2,300 live–work units with permit applications pending prior to the hearing, thereby excepting them from the newly adopted short-term policies.

Perhaps the most detrimental outcome of the public hearing, however, was the announcement that over the next six months, the Planning Department

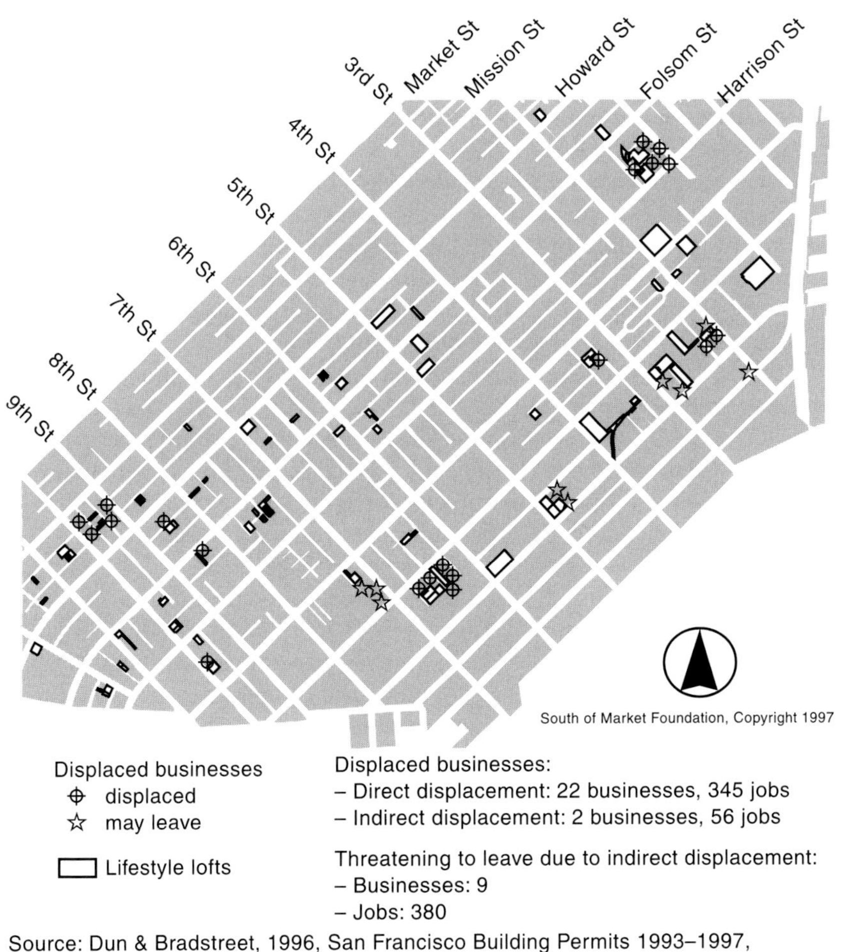

Figure 4.3 Companies displaced or threatening to leave due to lifestyle loft displacement.

would undertake a land-use study which would likely conclude that there is an abundance of industrial land in San Francisco. If this is the case, then these industrial lands will be re-zoned. Furthermore, given the short time period involved, there would be 'no community participation'.

4.4 COMMUNITY RESPONSE

The community's defeat at the hearing illustrated the political weakness of many disparate, emotional voices unable to present a unified, intelligent, fact-based argument or vision. In response to the proposed city-wide land-use and re-zoning study, which invited no community participation, the SOS group organized itself into a larger and more powerful organization called the Coalition for Jobs, Artists and Housing (CJAH). This group dedicated itself to developing a unified, sophisticated, and intelligent community voice on the issue of development.

CJAH was divided into three smaller neighbourhood groups, each responsible for reaching out to the various constituencies affected by rezoning. A network of non-profits and community activists provided technical support for the group. This included a land-use lawyer who offered legal services pro-bono, filing lawsuits against developers and assisting businesses with displacement battles and several non-profit housing developers and community activists who were very experienced in leading community-initiated referenda in San Francisco. This group had at its disposal a very powerful GIS database, which helped disprove false claims made by developers regarding job and industry displacement and assisted in identifying potential new zoning boundaries. This technical support group also served as a steering committee that met every Saturday morning in strategy sessions.

CJAH meetings were informal affairs held at various nightclubs throughout the SoMa area. Initial meetings were educational and informative in nature. The steering committee gave lessons in planning codes, economics and the history of SoMa. GIS data were key to this educational process. Computer-generated maps presented otherwise complicated statistical information in a very easy-to-understand manner.

Given the many different community interests represented at CJAH meetings, it was frequently difficult to develop consensus. Some people did not understand the complexities of a local economy. They just understood that they did not want to be displaced. In addition, everyone had his or her own preconceptions about what was happening. The living neighbourhood map became a tool that helped people move beyond their own opinions, judgements and naiveté.

Maps allowed people to see complex information more easily. Economic jargon and statistics became clearer when re-drawn as pictures. In addition, voluntary data-gathering efforts, composed of teams of people from 'opposite sides of the fence', were paramount in reconciling conflicting opinions and positions and helping everyone see the situation for what it really was. These consequences of mapping, in turn, made people much more knowledgeable at public hearings. Rather than reacting emotionally, people could present intelligent and well-informed fact-based economic arguments. Mapping also fostered a sense of connection and commitment to place. People really started to 'know' their neighbourhood and its streets.

Due to the enormous pressures and time commitments associated with countless battles before the Planning and Building Commissions, CJAH self-destructed in August 1998. However, the small technical support group stayed together and continued meeting on Saturday mornings. This group included weathered San Francisco activists who were prepared for a long fight, and this core group capitalized on the momentum established by CJAH in order to maintain the interest of the media and politicians.

By this time, public protests had forced the Planning Department to engage in its own rigorous study of the controversy, employing the same GIS methodology used by the community. In January 1999, the Planning Department finished its land-use study and came to the opposite conclusion that it had reached a year earlier. The study concluded that all industrial land in the city was being used to capacity and was needed in order to support the predicted growth in business services over the next 20 years. The study also concluded that the city was experiencing a housing supply crisis, but that there was sufficient land outside industrial areas to accommodate new residential units. The Planning Department recommended that interim Industrial Protection Zones should be established in the SoMa area, the Potrero Hill District and the Mission District. Ironically, given that this would be a hard sell to developers and some of the public at large, the Planning Department now needed the backing of the community to support them. The remaining core CJAH group used the media and mailing lists compiled by the SoMa Coalition to re-assemble a larger group of community members to come to another public hearing to advocate on behalf of the plan.

The public hearing at which the Planning Department's findings were presented to the public lasted seven hours and was attended by over 400 people from the community. Attendees who testified during the hearing were articulate and well-informed, demonstrating the benefits of community education gained through the work of the previous coalition. Community arguments were now fact-based rather than grounded in emotion. At the end of the hearing, the Planning Commission voted unanimously to approve an Interim Industrial Protection Zone. Both their decision and the Planning Department proposal reflected the complex dialogue among the neighbourhoods, developers, planners, and politicians that had occurred throughout the year, and that was presented in its entirety that evening.

This Industrial Protection Zone was a best-compromise solution. It did not cover all of SoMa, although it stretched across a fair portion of the district. The areas outside of the zone were designated as mixed-use areas encouraging development of housing and retail. Buffer zones of affordable housing were proposed between the industrial and mixed-use areas. In essence, it was a mixed-use plan designed to control gentrification.

4.5 CONCLUSION

Although the Industrial Protection Zone was a very significant step, the story of SoMa's gentrification is far from over. The proposed zone is an interim measure and it allows conversion of industrial space to office use. In truth, rent-sensitive light industry and business services will not be protected by the plan at all; too much damage has been done because most businesses have already left.

Looking back, one can find several significant outcomes of this struggle. First, a lesson was learned about which land-uses are compatible in a highly mixed-use district: some uses are appropriate neighbours, while others are not. Second, if disparate uses are to co-exist, then enforcement of the zoning code is essential.

Lastly, perhaps the most significant outcome was that the community's voice was heard and documented. It is unlikely that the Planning Department would have engaged in such a detailed study and invited public participation had it not been for the actions of a very informed and sophisticated group of community activists. At the core of this effort was the GIS-generated living neighbourhood map, which empowered the community, educated community members, and offered a means by which people could shed their individual opinions and judgements in order to see the situation for what it truly was. By both improving the quality of information available and providing a means for people to work together, the living neighbourhood map allowed people to stop reacting based on emotion, hearsay, and opinions and develop a more credible and powerful voice with which to argue in the public arena for their rights as a community. In turn, the Planning Department employed the same GIS-based methodology and came to the same conclusions. In addition, in November 2000, nearly the entire Board of Supervisors was replaced due to the outcry of the neighbourhoods against a gentrification now sweeping across much of San Francisco. The momentum of this voice and its message began in the SoMa area. The new Board of Supervisors is far more sympathetic to the culture of existing neighbourhoods than to unplanned-for market-driven development.

The SoMa story illustrates the important role PPGIS can play in an era when cities are becoming re-populated, and existing uses of land are in jeopardy of being displaced. Mixed-use zoning can lead to a very dynamic district if properly enforced. If not, uncontrolled gentrification is a very serious threat. If such a practice is to work within the context of parcel politics and a volatile market, then channels must exist for the voice of the existing community to be included in the dialogue. Clearly, GIS technologies can educate a community and help it to develop a voice that can challenge powerful market-driven interests. Thanks in part to the benefits of PPGIS, that voice – a voice that is usually left out – can no longer be ignored.

Chapter 5

Mapping Philadelphia's neighbourhoods

Liza Casey and Tom Pederson

5.1 INTRODUCTION

Since 1994, the City of Philadelphia has been working to bring GIS technology to the level of neighbourhood planners, hoping to initiate with them a PPGIS. Although it has been successful in generating enthusiasm for the applicability of GIS for this purpose, use of the technology in the neighbourhoods is still minimal. A 1995 paper by these authors documented the project with particular focus on the limits of existing mapping techniques and symbology for mapping urban neighbourhood environments. This paper documents the progress of the City's continued efforts to give neighbourhood planners access to its GIS resources and the impact of new technologies on that effort. Our finding is that although the City may now be in a much better position to distribute its GIS data through less expensive, easier to use interfaces that can effectively distribute public records, the difficulties of building effective PPGIS in urban neighbourhoods still exist.

5.2 BACKGROUND

The City of Philadelphia, which was literally on the verge of bankruptcy in the early 1990s, was rejuvenated under the leadership of Mayor Ed Rendell. The downtown has been revitalized. The new convention center is booked for years. Ben Franklin, Betsy Ross and the Liberty Bell are being more effectively promoted and Philadelphia is becoming a true destination city – even attracting the Republican National Convention in 2000.

But there is another side to the City. Philadelphia is a victim of the move away from an economy based on manufacturing. It has steadily lost jobs and people over the last 50 years. Between 1965 and 2000, the City lost over 25 per cent of its population. Just since 1988, 100,000 jobs were lost and almost 30 per cent of its residents live in poverty. As a result, many of Philadelphia's neighbourhoods are filled with vacant buildings and trash strewn lots (Figure 5.1). They are tormented by crime, drugs and unemployment.

Figure 5.1 A West Philadelphia streetscape.

The people to whom these neighbourhoods are home, are clearly marginalized communities. Working with Philadelphia's Office of Housing and Community Development (OHCD) our hope was to introduce GIS to community organizations in these neighbourhoods as a tool for strategic planning. If used correctly, GIS could help allow politicians and decision-makers see both the problems and the potential in proposed neighbourhood-based planning efforts, and to see how these neighbourhoods might be affected by their funding.

In 1994, recognizing both the appropriateness of moving neighbourhood planning back to the neighbourhoods and the applicability of GIS for this purpose, OHCD funded a pilot project to bring GIS to the neighbourhoods. The project provided equipment, software, data and training to 6 of the City's 25 Community Development Corporations (CDCs). Both authors were drawn into the activities surrounding the GIS pilot; Casey as the head of GIS for the City, and Pederson as the consultant under contract with OHCD to provide training, support and data to the CDCs.

Community Development Corporations are inner city neighbourhood organizations with a goal of neighbourhood revitalization. They emerged in the 1970s as participants of the funding and support generated by the 'War on Poverty'.

After working with the participant CDCs on almost a daily basis, we came to understand well the issues emerging from the GIS pilot. There were a number of practical and logistical problems ranging from bad addresses to problems involving the transfer of data between incompatible operating systems (DEC VMS and Windows). However, while addresses can be corrected and data transfer paths can be forged, during this pilot, we began to see problems more critical to the project – problems with the maps. A second critical issue became the lack of skills necessary to run a GIS.

5.3 PROBLEMS WITH THE MAPS

The maps we started to see as a product of the pilot could not be compared side-by-side or collectively. Each attached significance to colour differently and used its own classification schemes and categories, and symbology. We realized there were no standards and no 'symbology vocabulary' for mapping the urban environment in the way that exists in cartographic tradition for road maps or maps of natural features, such as hills or grasslands.

In addition, the maps seemed very limited in their ability to portray the qualitative aspects of a neighbourhood environment. As we were witness to maps from neighbourhoods scattered over the City and had developed site context based on our repeated visits, we were in a position to notice that the maps did not meaningfully convey the very distinct physical and social disparities we observed in the neighbourhoods. The following paragraphs are from a paper these authors wrote in 1995 that focused on the limitations of traditional mapping standards, techniques, and symbology as applied to mapping neighbourhood environments.

> If one follows the premise, that 'maps are models of the world – icons if you wish – for what our senses see through the filters of environment, culture, and experience,' then the CDCs do not seem to have sufficient tools to make appropriate models of their neighborhoods (Aberley 1993). With the parcel base maps, tax assessors data, tax delinquency and vacancy data, there does not seem to be any way, for example, to convey the beautiful old stone buildings which are such a part of Philadelphia's Germantown neighborhood. Those that were turned into multi-family dwellings are simply so coded. Those that were vacant and boarded are coded as vacant tax delinquents. There does not seem to be a means to convey the value of this wonderful architecture to the neighborhood or what it is worth as a resource. The same applies to mapping the locations of local cultural or community value, such as a famous family-owned barbecued chicken place on the corner which is a social gathering place for the neighborhood. Nor is it apparent how to map

> other elements that make the environment unique such as wall size murals or statuary created by local artists, stores selling ethnic foods and other imported goods, local restaurants, blocks of particularly well kept houses, blocks of houses with details that reflect a certain building style, or lively commercial corridors.
>
> Similarly, there is no ability to communicate the shocking degree of abandonment and dissipation in some of the neighborhoods. Crumbled buildings, burned out abandoned cars, trash strewn lots and streets, broken glass and graffiti are in evidence everywhere but not on the maps. For example, in the map of one part of West Philadelphia the neat little parcel lines, which correspond to its original development, seem to suggest some kind of active ownership interest. Whereas, in fact, whole blocks have been completely abandoned or demolished (Figure 5.2) and former owners are long gone, owing the city as much as 27 years worth of back taxes.
>
> Casey and Pederson 1995: 1

In our research for this 1995 paper, we discovered that while the problem of mapping the elements needed to portray neighbourhood environments had been recognized, there were very few suggestions of means to resolve it. Our paper proposed a three-tiered approach that included standardization, structured classification, and the development of appropriate symbology.

Figure 5.2 Entire blocks have been demolished in some Philadelphia neighbourhoods.

However, as we acknowledged, 'the answer for the CDCs is, obviously, not a simple solution that we can profile in this paper and implement through our roles as promoters and supporters of the GIS project' (Casey and Pederson 1995).

5.4 THE LACK OF NECESSARY SKILLS

Seven years have passed since the inception of the GIS pilot. The vision was that by now, scores of neighbourhood planners and interested citizens would be sitting at PCs in the CDC offices using GIS to both query the information regarding the particulars of their environments and to perform 'what if' scenarios to assist with strategic planning. This has not come to pass. If bringing that vision to reality were the only measure of the program's success, it failed. For all the distribution of PCs and software, the cleaning and organizing of the data, and the hours of training and handholding, there is still an insignificant use of GIS at the level of the CDCs.

Everyone concerned, OHCD, other city agencies watching the process, the CDCs themselves and the authors realized that one obstacle in reaching this vision far overshadowed all of the others – lack of skills necessary to use a GIS accompanied by the rapid turnover of any staff with the aptitude to learn those skills. CDCs have extremely limited budgets and their staffs do not come with training in technology. People with GIS skills, especially good conceptual and analytical skills, can easily find higher paying jobs. Our problem was that the gap between the skill level needed to navigate a Windows based GIS interface (ArcView in this case) and the skill level we would find in the CDCs was underestimated. Too much hinged on the ability of the group's designated technology enabler.

While the specific vision of PPGIS described above was not realized, the project was not a failure. On the contrary, the work that went into that pilot, the personal contacts and the 'bell ringing' about the applicability of GIS to neighbourhood planning brought, across the board, increased awareness of the potential of this technology. The best witness to this is the continuing GIS-centered activities.

The City, OHCD, and others involved responded to the problems of the GIS pilot with new strategies. Instead of continuing to fund individual CDCs, OHCD funded the Philadelphia Association of CDCs (PACDC) at a rate of about 60 thousand dollars a year to provide a 'centre' for GIS activity where CDCs could find continuing technical support for neighbourhood mapping without having to employ skilled operators. They could walk into PACDC's office and emerge with a map made to their specifications. PACDC over the last two years has created over 300 GIS-generated maps responsive to the requests of CDCs. The City made GIS data available to numerous non-profit consultants to provide maps in support of funding requests. In addition, OHCD, with the University of Pennsylvania, built a GIS-based Neighbourhood Information System.

Accompanying the realization that wholesale access to GIS was not sufficient to bring GIS to the neighbourhoods in a useful manner, have been continuous changes in the underlying technologies. The most important of which are: (1) GIS software manufacturers began to provide the capability of linking GIS systems to Internet technology; and (2) GIS software manufacturers now provide open development environments between GIS and standard database interface tools such as PowerBuilder and Visual Basic. As a result, the tools available for dissemination of GIS technology have drastically altered. The new technologies also polarized, and helped us recognize two distinct types of systems which we call *Public Records GIS* and *Neighbourhood Planning GIS*.

5.5 PUBLIC RECORDS GIS

Public Records GIS is the Internet distribution of data through GIS that a city or other government body collects as part of their administration of policy and laws, and distribution of services. It is information traditionally recorded, e.g. property owners, tax assessment, code violations and so on. It is information collected with new technologies like orthophotography. It includes data from Federal sources such as the Census Bureau or the EPA. It also includes data from utilities and from businesses that want to make their data public.

Public Records GIS covers a wide range of functionality. It includes interfaces such as a zoning application we have in Philadelphia which allows users to type in an address (or zoom in on an area) and be presented with a parcel map showing the zoning and special use codes in effect (Figure 5.3).

Philadelphia will soon be deploying another Public Records GIS from the Streets Department which will show where planned under-the-street utility maintenance is going to occur.

In 1994, the goals were to move portions of the GIS data and related city records to stand alone PCs located in neighbourhood offices for the CDCs and interested citizens to 'have at it'. Now, Public Records GISs allow the distribution of central databases using the Internet with an easy-to-use interface. This eliminates the problem of finding and keeping GIS trained personnel.

OHCD's Neighbourhood Information System is an example of a Public Record's GIS with neighbourhood information. The Cartographic Modeling Lab of the University of Pennsylvania developed it with grants from the Pew and William Penn Foundations to help with the analysis of potential

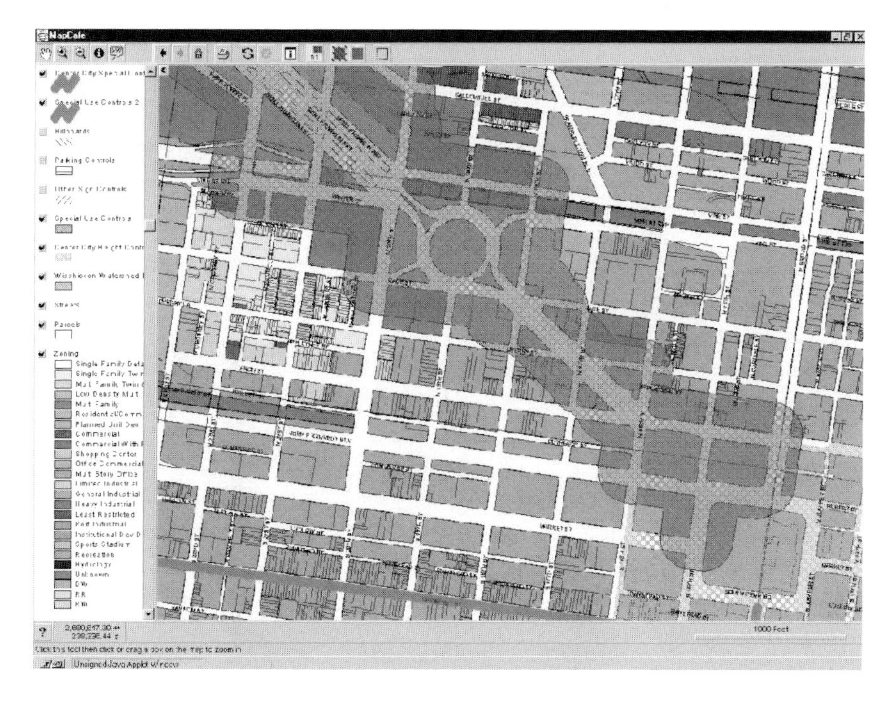

Figure 5.3 License and inspections zoning application.

redevelopment of vacant land and buildings in Philadelphia (Figure 5.4). It is based on the parcel map. The user selects a parcel and is provided with information from the tax assessor, the department of Licenses and Inspections, the utilities, and data from the US Census. It includes the ability to aggregate certain of these data to various geo-political boundaries. As we go to press, this system has been operational at the university for over a year with access provided to city employees and people working in a community development capacity.

5.5.1 Strengths of Public Records GIS

Public Records GIS systems have the potential to be very useful for marginalized urban communities. For example, in connection with work we recently completed for The United Way for their Youth at Risk programme, we created a very powerful map using City data and Census information. We were helping to determine where the programmes should be focused. We put together a few of the pieces of data that had been made available and soon we were looking at a map showing a Census Tract in Philadelphia where last year, 4 kids under 16 years were murdered, 5 girls from 10 to 14 years old

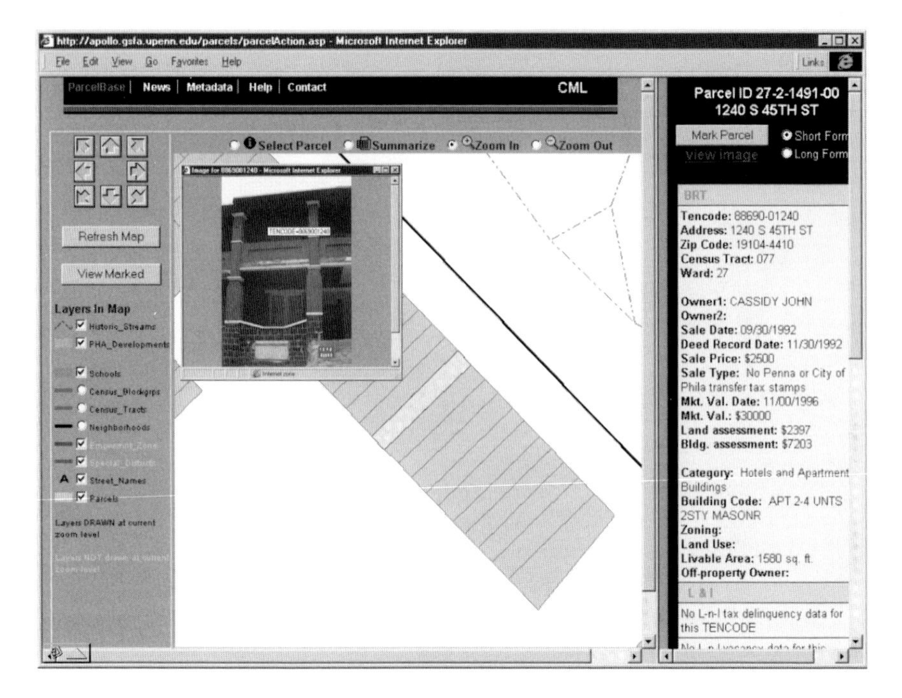

Figure 5.4 The neighbourhood information system.

had babies, and 251 single woman with children 5–17 were living below poverty. Somehow, without further research, it seems clear that if the people in this neighbourhood were organized and had access to this information, they could use it to cry out that Youth at Risk money would be well spent for their community.

It is easy to envision a series of web-based, neighbourhood-oriented, user-centred, task-driven interfaces like the Neighbourhood Information System geared to a variety of intents. These would allow unskilled users to select parameters and, with a few tools to control scale and extent, to create maps to help bring attention to their needs. We feel certain that Philadelphia will continue to expand the use of systems like this.

Besides ease of use, there are other advantages to the new technologies. The cost to the user is reduced to the cost of a PC and an Internet web browser that can usually be acquired at no cost. With the data sets stored centrally and the processing being done on the server, the need for hefty PCs to accommodate the large files and complex processing is eliminated. A very basic PC is sufficient.

Another advantage is *de facto* standards in neighbourhood mapping, which we have been championing for some time now, will be enforced.

The themes, symbology and classifications will all be pre-set and unalterable. The inclusion of these standards as part of an interface allowing access to data needed for neighbourhood planning will make all users interested in that same data familiar with the symbology vocabulary and encourage its use.

5.5.2 The limitations of Public Records GIS

There are still problems with a reliance on web interfaces and Public Records GIS for neighbourhood planning. The obstacles to publishing this data caused by the limitations of the technology may have faded, but in Philadelphia, as in many other places, the issue of distributing the data evokes numerous strong opinions regarding legal implications and political ramifications. Giving the data to CDCs for their own use is not the same as publishing it on the web. City Hall may not want to make it so easy for the public in general to be able to find that 25 per cent of the properties in the City are tax delinquent. The City will have to establish distribution policies that will address specific databases and even data elements within those databases. The City is in the process of establishing an internal review board to sort through these issues which will be resolved, but not without time and effort.

Much more complex than distribution policies will be the difficulties we found hard to resolve in mapping neighbourhood environments in general. The problem at issue in the 1995 paper, namely that existing mapping techniques and symbology are inadequate to map the qualitative aspects of neighbourhoods, has not gone away. In addition, the Public Records GIS, which eliminates dependency on the high-level skills required to operate a stand-alone GIS, does so with sacrifices. It limits the available data, the tools for presentation and analysis, and the features that can be manipulated. The components that are included in the interface between the user and the City's data become the only components available. A great deal of research and architechting will have to go into the development of the Public Records GISs for them to be truly effective.

In addition to the component set, the manner of presentation needs to be carefully considered. If map themes are limited to 'hard wired' depictions, it will have a direct impact on the portrayal of various factors. Information coded as cross hatches could be used in conjunction with information coded with solid colours, but the possibilities become more limited when the need is to depict solid colour-coded themes with other solid colour-coded themes. The order of the themes, what is displayed on top of what, would also be preset and limit the ways in which the data could be viewed. A particular interface could present the data in a biased fashion and unintentionally misrepresent a neighbourhood's assets and resources.

5.6 NEIGHBOURHOOD PLANNING GIS

Neighbourhood Planning GIS can take advantage of all the same data in a Public Records GIS, but it adds two other things. It adds community-based data and it adds the facility for manipulating and analysing the data.

5.6.1 Place-based knowledge

Features that make a neighbourhood unique, such as cultural characteristics and architecture, as well as places that have community value are not collected in the normal course of a city's record keeping. The things of value in a neighbourhood such as architecture or home grown community gardens, as well as the things of negative value such as garbage-strewn playgrounds and crack houses, are not line items in any city database. In fact, knowledge of the value of neighbourhood features is something that can only come from the neighbourhood itself. This was precisely the thought behind OHCD's strategy to put this tool in the hands of the CDCs. It follows along the lines of what Doug Aberley expresses in *Boundaries of Home: Mapping for Local Empowerment.*

> If images of our neighbourhoods, our communities, and our regions are made by others, then it is their future that will be imposed. But if maps are made by resident groups, individuals who have quality of life as a goal, then images of a very different nature predominate.
>
> Aberley 1993: 130–131

5.6.2 Data manipulation facility

The stringent limits to the manoeuverability of features and themes that a Public Record's GIS imposes means that the experiments or 'what-ifs' that are required for neighbourhood planning are severely curtailed. In Philadelphia, where we are losing population, a planner could want to show the effect of moving a few active residents into an otherwise vacant area, creating a new neighbourhood and freeing a large unutilized area for re-development. Or, a neighbourhood might want to show the impact of a new business on support service businesses. Or, a neighbourhood might want to propose to be the site for a new City-sponsored mural (a growing phenomenon in Philadelphia) and demonstrate to the selection committee the path a tourist bus might take.

An effective Neighbourhood Planning GIS would need to go beyond a fixed package of data and be able to pull the data together in new ways. An interface that allowed a user access to the facilities that would permit these types of analyses would be, realistically, as complex as the ArcView software with which we started out. This would result in exactly the same problem we witnessed in the OHCD pilot: The skills to operate a system with that level of sophistication just are not available in Philadelphia's marginal neighbourhoods.

This doesn't mean that in Philadelphia we are just going to stop trying to use GIS for neighbourhood planning. We haven't stopped – the effort continues. What we are doing is, reassessing where we put our technology resources and trying to work through some of the issues that were uncovered.

For Neighbourhood Planning GIS it seems that the City, for the time being, needs to focus on skill centers, similar to the example of PACDC, where community groups can voice their ideas and hopes for their neighbourhoods and find the skills necessary to *have it mapped for them.* We might, also, better employ resources by establishing town meeting types of events where the what-ifs can be depicted with groups of residents providing their input, and facilitators and GIS technicians mapping the community's feedback.

5.7 CONCLUSION

In Philadelphia's marginal neighbourhoods, the needs for creative planning are in unquestionable demand. Mapping techniques and symbology, the symbology vocabulary, still need major attention before GIS can effectively address the condition of Philadelphia's urban environments. The City needs to ascertain the most efficient ways in which GIS can be used to capture community opinions and place-based knowledge. At the same time, the component sets in our Public Records GISs need considerable attention and examination before they can be used effectively by neighbourhoods.

In both Public Records GIS and Neighbourhood Planning GIS, the expectation is that success will be based on a highly iterative process taking place between those with design and programming skills and those with knowledge of the neighbourhoods, with that process being informed by academic research. It is likely that the limits of even today's technologies will frustrate the process and we will have to wait for still other technical capabilities to meet the needs of Philadelphia's marginal communities.

Of the activities we have discussed, Neighbourhood Planning GIS may be the activity that falls within the definition of PPGIS, or maybe, to a certain extent, they both will. In the meantime, it is important not to confuse Public Records GIS and its limitations, with Neighbourhood Planning GIS and its ability to map place-based knowledge and the ability to analyse a wide variety of scenarios. Above all, it is important to continue to recognize GIS as an effective tool for neighbourhoods in Philadelphia and other cities across the country and around the world.

REFERENCES

Aberley, D. (ed.) (1993) *Boundaries of Home: Mapping for Local Empowerment*, Philadelphia, PA: New Society Publishers.

Casey, L. and Pederson, T. W. (1995) 'Urbanizing GIS: Philadelphia's strategy to bring GIS to Neighbourhood Planning', *Proceedings of the Environmental Systems Research Institute User Conference*, http://www.esri.com/library/userconf/proc95/to150/p107.html

Chapter 6

The impacts of GIS use for neighbourhood revitalization in Minneapolis

Sarah Elwood

6.1 INTRODUCTION[1]

The use of GIS by community organizations working to improve conditions in US urban areas, particularly distressed inner-city neighbourhoods, has been a central focus in debates about the potential of this technology for marginalization and empowerment (Sawicki and Craig 1996; Barndt 1998; Elwood and Leitner 1998; Ramasubramanian 1998; Ghose 1999; Elwood 2000). The literature provides rich description of the ways in which these community groups experience unique needs and constraints with respect to gaining access to and using this technology. Nonetheless, the use of GIS by these organizations is proliferating rapidly. The primary purpose of this chapter is to describe the use and impacts of GIS in neighbourhood improvement efforts. This information is drawn from participant observation, archival research, and intensive interviews I conducted with a Minneapolis inner-city neighbourhood group. After a brief description of neighbourhood conditions and concerns, I describe the organization's strategies for gaining access to GIS, outlining factors that have facilitated and limited their access to and use of GIS and digital data. Then I examine their application of these tools to critical issues, and discuss some of the impacts of these efforts on participation and power within the neighbourhood.

6.2 GIS USE IN COMMUNITY-BASED HOUSING IMPROVEMENT

Powderhorn Park is a neighbourhood in south central Minneapolis, centered around a large park with a lake (Figure 6.1). With the exception of its northern border, which is a major commercial corridor for the city, the neighbourhood is largely residential. Powderhorn Park's approximately 8,000 residents are multi-racial and multi-ethnic, and the neighbourhood is home to a growing number of recent immigrants from around the world. The neighbourhood faces a familiar set of inner-city concerns, including loss

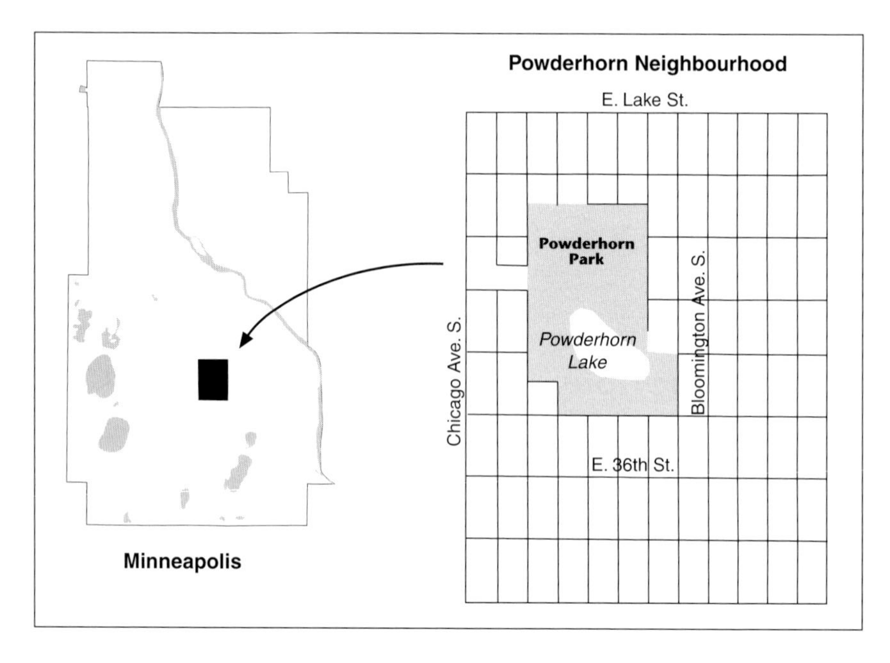

Figure 6.1 The Powderhorn Park neighbourhood is south of downtown Minneapolis.

of business and employment opportunities, a relatively high crime rate, and declining conditions in its ageing housing stock. In spite of these problems, the neighbourhood is a vibrant and active place in which many residents are active in trying to improve conditions. Residents have created flower and vegetable gardens and art parks on many of the neighbourhood's vacant lots. The neighbourhood hosts numerous community events intended to build ties among residents – an annual May Day parade, summer campouts in the park, and several cultural festivals throughout the year. The Powderhorn Park Neighbourhood Association (PPNA) is the primary institution through which these and other neighbourhood organizing and improvement efforts are coordinated.

PPNA is directed by an elected board of neighbourhood residents, and residents serve on various committees that work on multiple issues in the neighbourhood, including housing improvement, community and economic development, environmental improvement, transportation planning, and family and youth support. The organization employs six staff members, some of whom are neighbourhood residents, and others who are not. PPNA's efforts are funded through a combination of foundation grant support, community development block grant funds, and funds from Minneapolis' Neighbourhood Revitalization Program (NRP).[2] As part of NRP, the neighbourhood created a comprehensive neighbourhood revitalization plan addressing the issues mentioned above, and PPNA has been coordinating the implementation of this multi-faceted plan.

In the past five years, PPNA has begun to use a range of different information technologies, including GIS, digital databases, the Internet, and e-mail to support a wide range of activities in the neighbourhood. In this chapter, I will focus on their use of GIS and digital databases in housing improvement efforts, a major emphasis of PPNA's activities in the neighbourhood. As part of its NRP revitalization plan, PPNA obtained MapInfo GIS software, and developed a digital database that is referred to as the housing database. PPNA's housing database includes administrative information gathered from local government sources that describes land and structures in the neighbourhood, names and contact information for individuals connected to these properties, and textual notes concerning PPNA's involvement with these properties or individuals (see Table 6.1 for a more detailed listing of these data attributes). The textual notes may contain, e.g. notes that a resident has volunteered to translate for non-English speakers at PPNA meetings, or that a house was recently re-roofed using a PPNA housing repair grant.

It is important to note that the housing database includes information obtained from the City of Minneapolis as well as information gathered from residents of Powderhorn Park. A major concern raised in critical studies of information technologies has been the exclusion of such 'local knowledge' from digital databases and GIS and by extension, exclusion of local residents from planning processes that make use of these technologies (cf. Rundstrom 1995). A major element of Harris and Weiner's (1998) conceptualization of a community-integrated GIS is the notion that such a GIS will incorporate local expertise, rather than privileging the 'expert' knowledge of government actors or professional planners. PPNA's efforts to include both government-generated data and local knowledge in their digital databases represent one way in which their GIS efforts might be considered to be community-integrated.

While many small community organizations gain access to GIS and digital data through secondary agencies that assist or collaborate with them, PPNA's GIS use has been an independent, neighbourhood-guided initiative. The organization has funded acquisition of its own hardware and software, and residents and staff members have gathered primary and secondary source data for PPNA's databases. PPNA's success in gaining access to GIS and digital data speaks to the comparative advantages of the local context in which they are situated. The City of Minneapolis strongly supports neighbourhood planning and organizing, and as a consequence Minneapolis neighbourhood groups are relatively well-funded and are included as valued contributors in urban planning efforts. Both the City of Minneapolis and Hennepin County have been willing to share their digital data with neighbourhood organizations. These aspects of the local context

Table 6.1 Data attributes of PPNA housing database

Property	*Involved individuals*	*Activities/Problems*
Lot size*	*Owner/Taxpayer* • Name* • Address* • Phone number* • PPNA involvement • Volunteer skills	Past problems
Zoning*	*Rental License Holder** • Name* • Address* • Phone number* • PPNA involvement • Volunteer skills	PPNA actions
Property ID number*	*Caretaker/Manager* • Name • Address • Phone number • PPNA involvement • Volunteer skills	Staff/Resident observations
Age of structure*	*Block Leader* • Name • Address • Phone number • PPNA involvement • Volunteer skills	
Condition code*	*Tenants* • Name • Address • Phone number • PPNA involvement • Volunteer skills	
Legal description*		
Tenure status*		
Tax delinquent status*		
Sales history		

* Indicates attributes for which data are obtained from local government sources and are maintained for all neighbourhood properties. All other information is locally collected and is known for some, but not all, properties in the neighbourhood.

have fostered PPNA's ability to obtain its own hardware, software and data, as well as their success in using these IT resources to address critical housing concerns in the neighbourhood.

Regardless of these comparative advantages, PPNA still experiences some difficulties in its use of GIS. Technical problems in the design of the housing database have meant that the process of importing these data for use in MapInfo is quite complicated and time-consuming. Only one PPNA staff member received extensive training in using MapInfo, and this staff member has had little opportunity to teach staff members or resident volunteers about GIS. Staff members did not receive training from the housing database programmer as to how to import updated government information into the database. These problems emerge from a complicated nexus of financial, time, and training barriers experienced by many community and non-profit organizations that seek to use ITs. PPNA staff members who have been involved in design and use of the organization's databases and GIS do not have extensive experience in these areas, and they have limited time to devote to learning new procedures and techniques for information management and analysis. These barriers have constrained PPNA's use of its GIS software, and much of their 'analysis' of data held in the housing database occurs through simple direct querying of the housing database and examination of the resulting selections of data.

In spite of these limitations, PPNA has used the housing database and GIS successfully to address several critical housing issues in Powderhorn Park. They have been used most extensively in making plans for the housing improvement in the neighbourhood. For example, in 1998, PPNA's Housing and Land-Use committee relied extensively on analysis of these data to inform their efforts to design different kinds of grant and loan programmes to assist residents in making housing improvements in the neighbourhood. Since PPNA's digital databases have been used to maintain data on housing conditions for all properties in the neighbourhood, the organization was able to, for the first time, conduct comprehensive geographic analysis of housing issues in the neighbourhood. This comprehensive geographic analysis strongly informed the programmes designed by the Housing and Land-Use committee. For instance, a staff member's analysis showing the concentration of dilapidated rental properties along the neighbourhood's major transportation corridors inspired the committee to design assistance programs specifically targeted to the improvement of rental properties – not only through provision of repair funds but also by creating a rental property owners' forum in which landlords could meet to discuss strategies for resolving common problems.

In another instance, the Housing and Land-Use committee relied on data from the housing database to design housing improvement strategies that they felt targeted areas of the neighbourhood in greatest need. The housing organizer explained,

> ...the committee looked at maps that showed housing condition decline and housing stability and they looked at the numbers of boarded and vacant [houses], and the number of absentee rental properties. And what happened from that was a direct campaign to improve the 3000 and 3100 blocks of the neighbourhood. So we created a special housing repair program for that area, and focused [a housing program] there. So housing development, housing information, and housing grants and loans were focused in that area.
>
> Meghan,[3] personal interview 1999

In addition to analysing data to support housing improvement planning, PPNA has used the housing database to monitor conditions in the neighbourhood. A year after completion of the first housing repair grant that targeted particular areas and owners in the neighbourhood, the Housing and Land-Use committee returned to the updated housing database to examine whether property values and conditions had changed. Quantitative information on changes in property values and conditions, as well as qualitative information capturing the perceptions of residents regarding changes in conditions were both used in this evaluation of the impacts of their repair grant programme.

PPNA has also utilized its housing database and GIS in addressing long-standing concerns of residents about the provision of adequate amounts of housing in the neighbourhood. Residents have long been concerned about what they perceive as a lack of housing, and have debated a range of factors purported to cause loss of housing. In particular, PPNA staff and residents have long argued that many of the neighbourhood's lots are narrower than the minimum lot width now required by Minneapolis housing code. Houses were built on these lots at a time when the minimum width was smaller, and they represent potential for significant housing loss, because when a dilapidated house is torn down on such a lot, it is difficult to obtain permission to rebuild. While residents active in PPNA have long perceived this situation as a source of housing loss, they have had difficulty convincing the city officials of the severity of the problem or gaining permission to build on substandard size lots. PPNA used its database and GIS to conduct analysis challenging the arguments made by city officials, as one staff member described,

> The City said, 'Why should we lower the lot size, because there's only 5000 lots in the city that are [substandard size]?' So we could say, 'Actually, according to our database, there are 2300 in our neighborhood alone. That would mean that half of the undersize lots in the city are in Powderhorn Park alone. We think that's wrong.' It forced the city to go back and re-evaluate [their own data].
>
> Jeremy, personal interview 1999

PPNA's use of their GIS and database to challenge the City's interpretation of this particular housing problem was instrumental in facilitating a significant change in the City's application of housing policy in Powderhorn Park. While using their data to illustrate to city officials the severity of housing loss due to substandard size lots, PPNA also proposed a novel way of beginning to address the problem. They proposed conducting an architectural design competition to develop designs for houses that might be built on these small lots, and sought an agreement that city officials would grant them permission to build the winning designs on small lots in Powderhorn Park. In the fall of 1998, PPNA conducted this design competition and successfully gained city support for the competition and eventual construction of houses on small lots in Powderhorn Park.[4] In the spring of 1999, the winning design was constructed on a small lot in the neighbourhood, and both the PPNA and city officials are planning that similar houses will be constructed.

While the previous examples illustrate housing improvement efforts spearheaded by PPNA, the organization has also used its GIS and digital data to support the efforts of individual residents to resolve housing concerns and improve conditions in their immediate surroundings (Figure 6.2). When residents made plans to develop the housing database, one of their goals was to be able to provide housing information directly to residents. Individual residents and block clubs frequently contact PPNA for information from the housing database that they use in taking action to resolve their own housing concerns. For instance, block clubs working to resolve conflicts among neighbours obtained names and contact information for owners and tenants, enabling them to invite these individuals to a block club meeting to try to resolve the conflict. In another example, two residents who wished to divide ownership of a tax forfeited and vacant lot between their homes sought information from the database regarding the ownership history, zoning, and legal description of the lot. They used this information in negotiating with city and county officials regarding the property.

Figure 6.2 One of the primary benefits of the PPNA's housing database has been the ability to make information more readily available to neighbourhood residents.

Another primary motivation behind the formation of PPNA's housing database was the desire of staff and residents to centralize data obtained from numerous city or county agencies in a single location familiar to residents – the PPNA office. They felt that this would lower some of the barriers experienced by residents in gaining access to housing information. Some residents did not have transportation to the government offices where information could be obtained, and others did not know which office to contact or felt intimidated to do so. The original intent was that residents could visit the PPNA office and do their own querying and mapping of the data at a publicly accessible computer terminal. For a number of reasons, including the previously described technical difficulties in the database itself, this particular aspect of their vision has not been realized. Instead, staff members familiar with the database and MapInfo generate information for residents. Nonetheless, the capacity to provide comprehensive housing information directly to residents has changed information access in Powderhorn Park, and given some potentially marginalized residents a new sense of efficacy in improving their surroundings. In the words of one elderly resident who uses information from the housing database:

> I think a lot of seniors don't know who to call if they're having problems.... If PPNA gives them information that helps them figure out what to do, it makes such a difference. As you gain expertise in knowing what you can do, it makes you so much more powerful. And the reward is knowing you're getting something done, that you're making a difference the place where you live!
>
> Chenda, personal interview 1999

6.3 NEGATIVE IMPACTS OF GIS USE IN POWDERHORN PARK

Based on the examples given so far, we might conclude that PPNA's use of GIS and digital databases has had largely positive implications for the neighbourhood and its residents – the neighbourhood has used these tools to address critical housing issues in successful and creative ways, and barriers to information access for many residents have been lessened. Local knowledge and priorities are included as a valued part of the neighbourhood's IT efforts. Simultaneously, however, the use of these technologies is fostering changes that have simultaneously negative implications with respect to the participation and power of neighbourhood residents.

PPNA's housing database makes a vast array of local government housing data easily available to staff and residents for use in their planning and decision-making efforts. In spite of their success in including local knowledge, the use of 'official' government data has become increasingly prominent in

PPNA's planning efforts. This shift in the information used for planning has been accompanied by a change in neighbourhood discourse around housing, which increasingly reflects the priorities and perceptions captured in the local government's own data. Neighbourhood deliberations about housing improvements increasingly utilize language that requires detailed knowledge of City or County housing policies. For instance, the City of Minneapolis uses a complex coding system to record information about property conditions. Many participants in PPNA's discussions of housing conditions have come to use these codes. A vacant and boarded house is described as being 'on the 249 list'. In the past, residents relied on visual attributes in their descriptions of housing conditions, describing characteristics of a roof or foundation, for instance. Another resident, describing the shift in neighbourhood discourse around housing, explained,

> [At meetings now], there is an awful lot of recitation of what the City allows or requires or whatever. There is no question in my mind that things are framed with a more bureaucratic tone to them now. It's good for the people who can say things in ways that are acceptable to the bureaucracy!
>
> Melisssa, personal interview 1999

As this resident alludes to, the increasing presence of such language in PPNA's deliberations about housing conditions raises the level of expertise and knowledge that residents need to understand and participate in these discussions. For residents who have experience in planning or housing issues, either through their community activities or their professional employment, these changes are not problematic. For residents without such expertise, these changes constitute a significant barrier to their participation in PPNA's planning efforts. In Powderhorn Park, the residents who are most affected by this shift in language and expertise are those who have traditionally been marginalized from neighbourhood organizations – people of colour, renters, senior citizens, and non-native English speakers.

This reinforcement of existing barriers to participation is not occurring unremarked at PPNA, but rather, is a source of significant discussion and conflict in the organization. Calling attention to changes occurring in PPNA's planning efforts, one resident said,

> If I, a person raised and educated in mainstream society, feel alienated and intimidated by 'The Process', how can you ever hope to recruit, or better yet, keep involved, those very people you wish to represent?
>
> Gwen, personal interview 1999

In contrast, other residents have argued that the benefits of PPNA's access to and use of local government housing data outweigh these potential drawbacks. One proponent disagreed with claims about increasing exclusion and argued,

> What we've done is to develop some tools to help us make consistent and fair decisions that are based on good solid information.
>
> Michael, personal interview 1999

These disagreements have spawned a critical awareness within PPNA of the potential impacts of their use of GIS and digital databases. When asked about the impacts of these technologies on planning at PPNA, one resident explained that there was an increasing need for the organization to be attentive to helping all residents participate, regardless of their expertise or experience in planning and housing. Committee chairs and other residents, she argued, should avoid the use of 'expert' language and explain any local government policies being discussed. Clearly, PPNA's use of the housing database and GIS are altering their housing improvement efforts in ways that have the potential to marginalize some residents. However, the organization is actively engaged in trying to ameliorate some of the negative implications with respect to participation and power of neighbourhood residents.

6.4 CONCLUSION

The case of PPNA highlights several issues with respect to the use and impacts of GIS for CBOs. First, it illustrates the important role that community dialogue plays in shaping the impacts of this technology on participation and power. GIS and digital databases have the potential to be used in ways that enhance as well as limit the participation and power of some residents. Organizations like PPNA that foster rich community dialogue have a forum in which to debate these impacts and consider ways of minimizing negative impacts. For instance, as a result of the conflicts described above, concerning GIS use and barriers to participation, organization members developed several strategies for ameliorating some of the emerging exclusions they identified. The Housing and Land-Use committee, e.g. agreed to expand its efforts to gather housing information from residents as well as from city and county sources by coordinating volunteers to walk the neighbourhood and make observations, and visit residents in their homes to gather information about their observations and housing concerns.

Further, the case of PPNA illustrates the continued need to consider a range of strategies and institutional arrangements through which small, relatively resource-poor community organizations might gain access to and use GIS. While PPNA has been able to use GIS and develop its own digital databases, this study has revealed ongoing limitations to their GIS efforts because of the limited time, training, and financial resources that can be devoted to these efforts. The continued presence of such constraints on independent community-driven GIS efforts suggest that community organizations seeking to use GIS should consider collaborative arrangements that enable them to complement their own resources with the resources and expertise of collaborative partners. Such collaborations bear investigation in future research, because the involvement of collaborative partners, whether they are academic researchers, government institutions, or other community organizations, initiate complicated networks of relationships and interactions that will inevitably shape the use and impacts of GIS by community organizations.

NOTES

1. I am grateful for funding support from the University of Minnesota's Center for Urban and Regional Affairs and Department of Geography; and also for the generous involvement of the Powderhorn Park Neighbourhood Association and neighbourhood residents in this project.
2. Minneapolis' Neighbourhood Revitalization Program (NRP) is a 20-year programme begun in 1990 that redirects tax-increment funds from downtown development into neighbourhood improvement efforts. See Nickel (1995) and Fainstein and Hirst (1996) for further description and discussion.
3. In quotations from interviews, staff and residents are identified by pseudonyms, given the preference of some participants not to be identified.
4. Thinking beyond PPNA's use of digital databases and GIS to their use of other information technologies, e-mail and the Internet were critical in implementation of the design competition. Architects from across the US and Canada submitted designs, and PPNA relied extensively on e-mail and its website to communicate with these designers. Staff members posted maps of the neighbourhood and information gathered from residents on PPNA's website to help distant designers learn about neighbourhood conditions and priorities.

REFERENCES

Barndt, M. (1998) 'Public participation GIS: barriers to implementation', *Cartography and Geographic Information Systems* 25(2): 105–112.

Elwood, S. and Leitner, H. (1998) 'GIS and community-based planning: exploring the diversity of neighborhood perspectives and needs', *Cartography and Geographic Information Systems* 25(2): 77–88.

Elwood, S. (2000) *Information for Change: The Social and Political Impacts of Geographic Information Technologies*, Ph.D. Dissertation, Department of Geography, University of Minnesota.

Fainstein, S. and Hirst, C. (1996) 'Neighborhood organizations and community planning: The Minneapolis Neighborhood Revitalization Program'. In Keating, Krumholz, and Star, (eds) *Revitalizing Urban Neighborhoods* (Lawrence: University Press of Kansas), pp. 96–111.

Ghose, R. (1999) 'Use of information technology for community empowerment: transforming geographic information systems in community information systems',

presented at the First International Conference on Geographic Information and Society, 20–22 June, Minneapolis, MN.
Harris, T. and Weiner, D. (1998) 'Empowerment, marginalization, and "community-integrated" GIS', *Cartography and Geographic Information Systems* 25(2): 67–76.
Nickel, D. (1995) 'The progressive city? Urban redevelopment in Minneapolis', *Urban Affairs Review* 30(3): 355–377.
Ramasubramanian, L. (1998) *Knowledge production and use in community-based organizations: examining the impacts and influence of information technologies*, Ph.D. Thesis, University of Wisconsin-Milwaukee.
Rundstrom, R. (1995) 'GIS, indigenous peoples, and epistemological diversity', *Cartography and Geographic Information Systems* 22(1): 45–57.
Sawicki, D. and Craig, W. (1996) 'The democratization of data: bridging the gap for community groups', *Journal of the American Planning Association* 62(4): 512–523.

Chapter 7

The Atlanta Project: reflections on PPGIS practice

David S. Sawicki and Patrick Burke

7.1 INTRODUCTION

In October of 1991, former President Jimmy Carter announced his support of a comprehensive initiative to assist Atlanta's inner-city poor, The Atlanta Project (TAP). TAP's approach was based upon principles of greater democracy and community empowerment. Throughout its history, the programme has sought to give residents a voice and a sense of power in determining the future of their communities. To achieve these objectives, the organization adopted a structure to support a neighbourhood-based model of community change. TAP also made available centralized technical resources to support the community-building process.

The Data and Policy Analysis group (DAPA) plays an important role in TAP's strategy to share expertise and resources with community organizations. DAPA is a partnership between The Atlanta Project and the Georgia Institute of Technology's graduate City Planning programme. It was one of the first neighbourhood data intermediaries established during the new era of microcomputer-based GIS technology. Our organization has worked for nearly a decade to increase access to GIS technology so that underserved residents and community organizations might utilize it for their own empowerment.[1]

In this chapter, we discuss two fundamental issues regarding PPGIS, community accessibility to GIS technology and the role of GIS within community decision-making. The method of access and level of citizen engagement directly impact measures of empowerment. Second, we examine the role of GIS technology within the context of community decision-making and empowerment. We describe our work with two sample case studies.

Clear tensions exist between community decision-making models that utilize and rely upon technical expertise and ones that advocate a participatory approach. In the long debate in urban planning regarding models of decision-making, some suggest that these approaches are difficult to reconcile (Peattie 1968; Fox Piven 1970; Peattie 1994). We believe that these models are not mutually exclusive but that they can coexist. Empowerment through technology implies that citizens must be more than simple consumers of information produced for them. They should be partners in the use of the technology for the production and communication of information and the knowledge that results (Beamish 1999).

7.2 THE CONTEXT OF THE ATLANTA PROJECT

The Atlanta Metropolitan Statistical Area (MSA) had the twelfth largest MSA population in the nation in 1990. The 1990 Census indicated that it had a 1989 median household income of $36,051, almost 20 per cent above that of the nation. However, as with many metropolitan regions, Atlanta is a tale of two cities. Twenty-seven per cent of the households living in the City of Atlanta had incomes below the poverty level in 1989. The City consistently ranks among the ten poorest in the nation, with extremely high rates of infant mortality, teenage pregnancy, homelessness, and crime. The physical conditions of neighbourhoods in and near the downtown serve as a testament to the results of this poverty with housing conditions as bad as any in the nation.

The history of TAP begins with an epiphany experienced by President Carter when he visited the neonatal clinic at Grady Memorial Hospital, Atlanta's large inner-city hospital. During his tour, the former President met baby 'Pumpkin' born to a mother on cocaine and alcohol, four months premature, weighing barely a pound. Carter claimed to have been unaware until that moment of the depth of poverty in his own home state's inner-city (Thompson 1993). The experience became the seed of TAP. The former President consulted a number of persons about how to attack poverty problems in Atlanta, and within a year, the basic structure was in place.

The area of TAP includes 500,000 people experiencing the highest teenage birth rate, highest poverty rate, and the densest concentration of female-headed households within the Atlanta region. The area is mostly concentrated in the southern portion of the City of Atlanta, but it includes also urbanized portions of DeKalb and Clayton Counties. This TAP region was subdivided into 20 distinct 'clusters' of roughly 25,000 people each centred on a high school. These clusters were the organizing component of TAP's community-based decision-making model. Each cluster had a coordinator and its own committee structure. Cluster coordinators were charged with the responsibility of organizing and mobilizing residents within their communities. Each cluster was partnered with one or more major corporations and an institution of higher learning.

To provide central support for the effort, the entire TAP was led by an executive director and secretariats in seven key policy areas, namely, housing, economic development, public safety, health, education, children and families, and arts and culture. These secretariats provided resources and informed decision-making in collaboration with the clusters. TAP attracted millions in gifts and in-kind donations from dozens of foundations and Atlanta corporations, including IBM, Cox Enterprises, Equifax, Trust Company Bank, John Harland Company, Marriott Corporation, Arthur Anderson Consulting, Coca-Cola, Atlanta Gas Light, NationsBank, United Parcel Service, Delta Airlines, Georgia Power, AT&T, First Union, Home Depot, Wachovia, Turner Broadcasting, BankSouth, SouthTrust Bank, Northern Telecom, Equitable Real Estate, Prudential, Kroger, and Bell South.

7.3 TECHNOLOGY AND COMMUNITY DEVELOPMENT: THE OFFICE OF DAPA

From its beginning, technology played a role in how TAP addressed its mission. IBM contributed over one million dollars in hardware and software, including the furbishing of a state-of-the-art 'collaboration room' that permitted the use of facilitation software for community meetings, discussions, and planning sessions. While IBM was building the room, it was also searching for other ways to assist this massive local 'war on poverty'. It was at this point that they found David Sawicki, a Professor of City Planning and Public Policy at Georgia Tech and among a handful of faculty who were using microcomputer technology and desktop GIS in the classroom and in his research.

Sawicki was joined by graduate students who took courses for credit under his direction. Within a year, Sawicki had articulated a structure that was to serve this matrix organization of 20 clusters by seven secretariats. DAPA was a quick-response team able to provide leaders with data, maps, and information and soon clients using DAPA spread well beyond TAP organization proper. Clients included governmental departments, non-profits, and even private corporations working with TAP. TAP leaders began to provide an annual budget of roughly $100,000 and gave Sawicki the go-ahead to serve anyone with a reasonable request whose objective was to serve the residents within the TAP geographic boundary. Sawicki soon hired full-time staff but continued to employ Georgia Tech graduate students.

DAPA is built on a data intermediary model of providing access to GIS technology. The application of technology is use-driven with work defined by our clients' demands. The model is consistent with those who suggest that GIS for community empowerment must be demand-driven and not technology-driven, requiring that technology be taken out of a conventional top-down context (Hutchinson and Toledano 1993).

Our approach is to centralize the operations of a data centre and share that capacity among all non-profits working within a geographic area. DAPA's centralized data intermediary model permits the development of scale economies in the areas of data warehousing, personnel, and access to advanced information technology (Sawicki 1993; Sawicki and Craig 1996; Leitner *et al.* 1998; Burke 1999). A primary role of DAPA is the collection and warehousing of data that may be used in a neighbourhood-planning context. Information might include data on land-use, tax delinquency of property, crime, CDBG investment, housing code enforcement, and institutional assets to name a few. Many of these data are collected and maintained by private, public, and non-profit agencies. As an intermediary, we serve as a central conduit for these agencies to end-users. We negotiate agreements for release of data among these agencies, process and clean data, and integrate information across all databases. Types of data collected are driven by the work, and the uses desired by the clients.

A second benefit of our approach is that it permits the development of a sustainable core of competent GIS professionals. Non-profits and CBOs find it difficult to maintain competent technical staff for any IT needs. It seems appropriate that non-profits share resources to develop a centralized shared-technology capacity.

Finally, we experience scale economies created by centralizing software and hardware purchases. We utilize a variety of GIS, database, programming, and statistical software packages. DAPA has access to a variety of media and output devices, including an E-size plotter, various colour printers, a nine-track tape reader, a large scale digitizing tablet, CD-writers, Jaz drives, Zip drives, and a variety of other devices.

DAPA has become a resource of community-based planning and decision-making within inner-city Atlanta. During FY 1999–2000, DAPA provided research, data analysis, and planning services to over 75 public and non-profit agencies.[2] We believe that DAPA's success has stemmed from a unique set of circumstances that existed during its first five years of operation. First and foremost, we enjoyed the cachet of former President Carter's enormous popularity in corporate- and governmental-Atlanta. Not only did it bring us clients, but it also provided access to resources and data. Second, Professor Sawicki was given a substantial budget and a free hand at deciding what constituted worthy projects. Third, DAPA's products were given away. Once DAPA staff agreed to support a project for a client, the work was completed very quickly. A small academic research group with microcomputers compared favourably to people's previous experience with large public bureaucracies and mainframes, most requiring reimbursement.

To explore further the potential impact of GIS technology on the goals and effectiveness of community-based planning and neighbourhood change in Atlanta, we provide two case studies illustrating specific DAPA efforts. The first case depicts the role of GIS within the context of housing code enforcement and is an example of the rare occasion when citizen participation results in real system change. We believe the role of technology access was invaluable in communicating the community's concerns. The second case is more typical of the work we perform, working with an organization representing the interests of a community to explore inequities and opportunities for policy intervention.

7.3.1 GIS and housing code enforcement

The condition of a neighbourhood's housing stock is a measure of the community's health, vitality and quality of life. An inadequate judicial system, shortage of code inspectors, and declining city budgets allow negligent house owners to go unmolested. In 1994, Atlanta residents launched a community-based effort to improve local enforcement of housing maintenance codes. Residents approached TAP with the hope that the organization could work with them to reverse the decline of housing conditions.

The Housing Resource Team (HRT) of TAP responded to the residents by organizing an initiative to address their concerns. The HRT consisted of advisors from corporate, academic and community organizations (The Atlanta Project 1994). The DAPA group supported the initiative. The strategy adopted was consistent with TAP's philosophy of self-determination. We worked with residents to build their information capacity.

The capacity-building process was twofold: increase residents' education about the issue and apply technology to information collection and dissemination. The first step in the process was a community education effort aimed at training neighbourhood leaders and residents to identify code violations in their neighbourhoods. Staff prepared a pamphlet explaining the various code violations as defined in the Atlanta ordinances. TAP personnel distributed these materials through the cluster communities.

With training, residents from two neighbourhood planning units[3] (NPUs) surveyed their neighbourhoods for code violations using a standard data sheet. Staff from DAPA cleaned and organized the data into clear reports. We used GIS technology to map the code violations recorded by programme participants. As information was entered from resident surveys into the GIS programme, records were automatically given *x–y* coordinates based upon the address, facilitating production of a map of code violations (Figure 7.1). We were able to correlate the address-based information with a database that included ownership and other relevant tax information.

Working with residents and the HRT, DAPA prepared a report presenting information and an analysis of the data for the City of Atlanta Department of Housing and Community Development (HCD). City officials did not react to the effort. When the HCD failed to respond to the violations identified, City Councilwoman Gloria Tinubu, who represented the area, became involved.

In the summer of 1995, Councilwoman Tinubu, TAP's HRT coordinator and NPU citizen representatives joined together with other involved parties

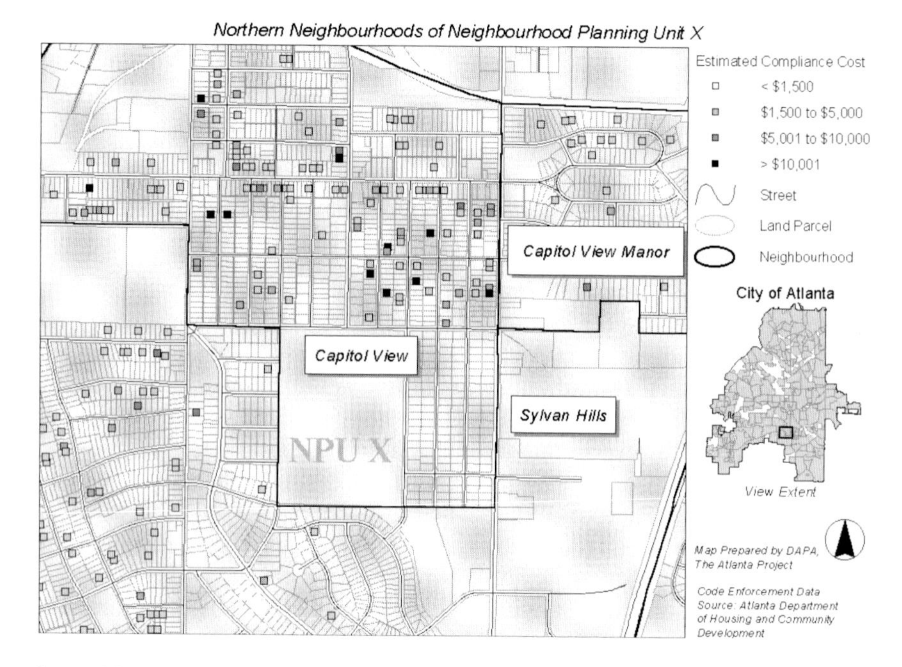

Figure 7.1 Residential code enforcement violations and estimated compliance cost.

to form the Code Enforcement Task Force (CETF). DAPA, on behalf of the Task Force, conducted a study of the process of code enforcement employed by the City and of their capacity to handle the code violation concerns. The study concluded that the HCD, though responsible for code enforcement, did not have sufficient technology or human resources to address the enormity of the code enforcement problem. Based on these findings, the CETF made recommendations to the City for improving the functioning of its code enforcement process.

In response to the study and continued interest, the CETF and HCD developed a number of proposals for improving code enforcement. The two groups listed nine action items targeted at improving code enforcement in the City of Atlanta. Key action items included an upgrade to city computer capacity; an increase in efforts to educate the public about code enforcement; and a proposal to develop a programme of neighbourhood-based code enforcement representatives or 'neighbourhood deputies'.

Between January and March 1997, the Task Force, along with input from various city departments, defined a neighbourhood deputies programme that would operate throughout the City. On May 20, 1997, the Task Force met the City of Atlanta's Mayor, Bill Campbell, to get his commitment to fund the programme. The Mayor agreed and provided funding for the first year. The plan provided a pilot in two neighbourhood planning units during the first year and an additional three to six NPUs during the second year.

The city Department of Planning and Community Development provided the first year funding. Funding support for the neighbourhood deputies programme in the second year came from Community Development Block Grant (CDBG) funds. By August 1998, the funded initiative operated in 17 of the City's 24 neighbourhood planning units, with 78 trained volunteer deputies (Reid 1998).

This case illustrates how TAP and DAPA operate: providing resources to community-driven initiatives, linking individuals to information to affect community change. TAP was a capacity builder, strengthening the voices of concerned citizens through the provision of education and information on code enforcement. We conclude as others have that the existence of a strong, independent citizen action group may be a precondition to a successful citizen participation effort (Suskind and Elliot 1983). The combination of DAPA's technical capacity, organization, and the community's political weight allowed TAP to speak in such a way that the message could be heard by those responsible for taking action.

The information and analysis produced through the application of GIS technology supported the position of the citizenry. In the end, the analysis itself was not the key to the success of the effort. Rather it was the community building and organizing that resulted in a consolidated and sustained effort to change the system. The application of GIS technology and the involvement of residents contributed to the validation of citizens' perceptions of the City's troubled state of code enforcement. The informed analysis presented to the City provided citizens with the ammunition necessary for achieving accountability.

7.3.2 GIS as a tool for strategically planning early childhood development initiatives

Access to early childhood development programmes is a salient issue among inner-city residents. Both education professionals and parents agree that programmes such as Head Start and pre-kindergarten have positive impacts on the long-term learning outcomes of children. Many also believe that children living in low-income households are less likely to have a quality early education experience because they lack access to affordable programmes. Between 1996 and 2000, TAP made the issue of access to quality early learning programmes a policy priority. Staff from DAPA provided support to initiatives during this period. We helped strengthen the cases made to state and federal agencies for more classrooms in TAP's poorest neighbourhoods.

Georgia's lottery began funding a statewide pre-kindergarten (pre-k) programme during the 1992–93 school year. In the early years of implementation, the state targeted pre-k funding to the most needy low-income children. During 1992–93, the pre-k programme served 750 low-income children statewide. By 1995, the Georgia Office of School Readiness (OSR), which administers the programme, had begun to grant funding to programmes serving all four-year-olds, regardless of family income. During the 1996–97 school year, pre-k funding had reached $205 million and supported almost 57,000 four-year-olds statewide. By 2000, the programme served approximately 60 per cent of the state's four-year-old population.

Despite the rhetoric of elected officials claiming that every parent who wanted a pre-k slot for his/her child would have a seat in a classroom, DAPA staff estimated there were approximately 5,000 four-year-olds within TAP's catchment area who did not have access to pre-k, Head Start, or alternative programmes.

During the fall of 1996, TAP created a coalition around the issue of accessibility to early learning programmes. The coalition began by targeting expansion of pre-k programmes for the 1997–98 school year. Members of the coalition included over 60 individuals representing 43 different organizations including neighbourhoods, government agencies, and service providers. TAP was also able to involve former President Carter and Mrs Carter in the dialogue.

DAPA prepared an analysis for the coalition that examined pre-k capacity within TAP's borders, the availability of alternative programmes such as Head Start, and concentrations of four-year-old children. Using this supply–demand analysis, we were able to identify and rank under-served neighbourhoods within TAP (Figure 7.2). Potential providers were recruited and DAPA used GIS analysis to prepare detailed supply–demand studies to support each of their applications to OSR for pre-k funding. We also prepared summary statistics on the socio-economic status of these areas to illustrate that it was not feasible for many of these families to afford or travel to private alternatives.

In March of 1997, TAP staff were stunned when they discovered that the State did not plan to fund additional classes in either DeKalb or Fulton Counties, thus eliminating virtually all TAP proposals. The OSR had given priority for new pre-k classes to counties whose current classes had capacity to serve less than 50 per cent of their county's four-year olds. The problem was classically suited to GIS analysis. Georgia has 159 counties but they range in size from Fulton at almost 700,000 people to many around 2000; DeKalb County is also large. There are often significant problem pockets of any phenomenon in the larger counties even though the county as a whole is above the threshold. Legislative formulas that employ average statistics usually favour small, rural counties. This case was no exception. Though greater than 50 per cent of four-year-olds in Fulton and DeKalb had access to pre-k, we believed that there were relatively few sites in inner-city neighbourhoods. Furthermore, the majority of four-year-olds in TAP

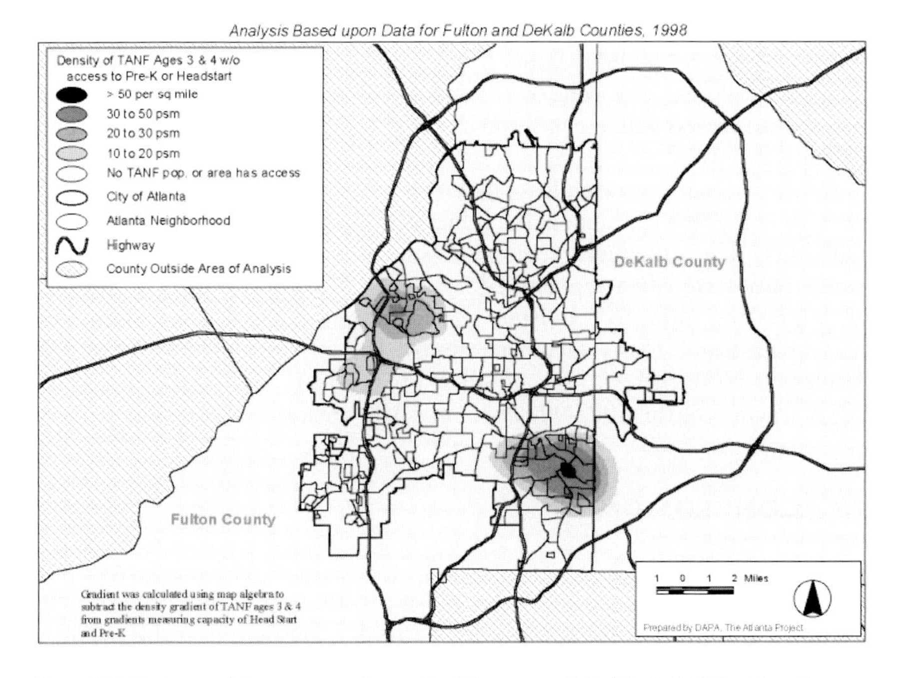

Figure 7.2 Regions with concentrations of children ages 3 & 4 in TANF (welfare) households without access to Head Start and/or Pre-Kindergarten.

neighbourhoods lived in low-income families who arguably were in the greatest need of such programmes.

In May of 1997, TAP's coalition met to revise the work plan and develop a new set of strategies. The new work plan included an expansion of DAPA's analysis, collection of research to document the value of pre-k for children in poor urban neighbourhoods, and development of citizen education programmes to inform parents and public officials of the need for pre-k within the city. The new analysis was provided again to OSR. But more importantly, the plan included a personal call for action from former President Carter to the Georgia Governor and his OSR.

Within a month, eight of the applications that TAP supported and one additional programme located in TAP's region, were approved for funding for the 1997–98 school year. By the end of the 1998–99 funding cycle, staff from TAP and DAPA assisted in the expansion of 16 total programmes (including Head Start services at two sites). By 1998, 29 new classes were funded with a capacity of 20 seats in each class resulting in 580 seats in some of the most under-served areas of TAP.

During the pre-k initiative, GIS technology helped guide decision making for the coalition. We developed systems and information to provide knowledge to the coalition regarding under-served areas of TAP and also to validate their position with the state administrative agency. We also worked closely with potential program administrators to provide quantitative analysis to support their applications. The role of technology in the decision making process was to inform and validate positions held by residents and community organizations.

PPGIS has within its philosophy concepts of capacity building and empowerment. These are democratic principles of self-determination, improving individual lives through greater power and understanding. However, communities are often represented by a few. The pre-k coalition recruited an impressive, informed, and fair representation of the TAP population as part of its team. Thirty members, almost half of the coalition, were TAP residents. Nevertheless, the number of residents involved represented a tiny fraction of the 500,000 people in TAP's area.

7.4 CONCLUSION

Since the publication of Davidoff's work on advocacy planning (1965), providing a link between technical expertise and communities historically excluded from the planning process has been a core ethic of the planning profession. At the time of his work, the synoptic rationalist paradigm dominated the practice of planning. Under this paradigm, using technology and technical expertise to develop public interest and planning alternatives is often too centralized to permit broad citizen participation and tends to overlook the needs of those on the periphery of society (Friedmann 1987). DAPA's approach is to overcome that barrier by providing broad access to technical knowledge through a data intermediary.

Some critics believe that using technology and technical expertise in community-based planning not only does little for truly representing the interests of the poor, it potentially co-opts them. Fox Piven believes that technical skill is only one small aspect of the power discrepancy and that the use of technical approaches to planning diverts attention 'from the types of political action by which the poor are most likely to be effective' (Fox Piven 1970).

Our case studies support the importance of political mobilization and use of existing political power for community change. We illustrate that there is no fundamental incompatibility between the use of technology and community empowerment. In the code enforcement case, citizen mobilization was the determining factor in the successful change in the City's approach to enforcement. The function of GIS was to communicate citizen concerns to the city bureaucracy and to create greater accountability through better communication. In the pre-k case, GIS functioned both to communicate concern as well as inform decision-making regarding the allocation of resources.

Peattie questions the role of the intermediary as an instrument of advocacy. She believes that advocate planners do not represent all interests, just the ones with an interest compatible with their own (Peattie 1994). This calls into question the values inherent in the use of GIS systems for public decision-making. Whose interest is ultimately represented in GIS applications, those of the community for which we advocate or our own? Is the application of GIS technology the 'hammer looking for nails', or is it a responsive method being employed in an appropriate context?

As Howell Baum warns, when planners make decisions in a technical world, they 'lose sight of two things that are very important to nearly everyone else: social interest and personal feelings' (Baum 1996). We implore users of GIS technology for community decision-making to focus on the values and concerns of the citizens they seek to serve, not on the gee-whiz technology. Our model and two cases illustrate that successful collaborations between intermediaries and communities begin and end with a community-driven process. GIS tools can be powerful if used within a process where participants increase their mutual understanding, knowledge, awareness, empathy, and compassion regarding an issue.

NOTES

1. Dr Michael Elliot of the Georgia Tech deserves recognition for providing the context of this reflection on DAPA's practice and issues of democracy. One of Patrick Burke's questions from his Ph.D. comprehensive examination required a theoretical reflection on the dichotomy between technical planning and egalitarian approaches to planning.
2. Once the first phase (1991–1996) of The Atlanta Project ended, DAPA broadened its base of support to include other agencies and foundations. TAP confined DAPA to specific projects, curtailing its freewheeling style of picking clients. The Annie E. Casey Foundation, however, provided some funding for keeping the open agenda alive, though on a reduced basis.
3. The City of Atlanta utilizes a system of neighbourhood planning units to plan and implement its comprehensive development plan. Neighbourhood planning units or NPUs are collections of neighbourhoods. In 1990, each NPU represented between 6,348 and 31,064 persons with an average size of approximately 16,417.

REFERENCES

Baum, H. D. (1996) 'Practicing planning theory in a political world', in Mandelbaum, Mazza and Burchell (eds) *Explorations in Planning Theory*, New Brunswick, New Jersey: Center for Urban Policy Research Press, pp. 365–382.

Beamish, A. (1999) 'Approaches to community computing: bringing technology to low income groups', in Schon, Sanyal and Mitchell (eds) *High Technology and Low Income Communities: Prospects for the Positive Use of Advanced Information Technology*, Cambridge: MIT Press, pp. 349–369.

Burke, P. (1999) 'Democratizing information: community capacity building through neighborhood-based planning information systems', presented at the International Community Development Society Conference, July 1999, Spokane, Washington.

Davidoff, P. (1965) 'Advocacy and pluralism in planning', *Journal of the American Institute of Planners* 31(4): 331–338.

Fox Piven, F. (1970) 'Whom does the advocate serve?' *Social Policy* (May/June): 32–37.

Friedmann, J. (1987) *Planning in the public domain: from knowledge to action*, Princeton, New Jersey: Princeton University Press.

Hutchinson, C. F. and Toledano, J. (1993) 'Guidelines for demonstrating geographical information systems based on participatory development', *International Journal of Geographic Information Systems* 7(5): 435–461.

Leitner, H., McMaster, R., Elwood, S., McMaster, S. and Sheppard, E. (1998) 'Models for making GIS available to community organizations: dimensions of difference and appropriateness', presented to the NCGIA specialist meeting on Empowerment, Marginalization and GIS, October 1998, Santa Barbara, CA.

Peattie, L. R. (1968) 'Reflections of an advocate planner', *Journal of the American Institute of Planners* 31(1): 13–22.

Peattie, L. R. (1994) 'Communities and interest in advocacy planning', *Journal of the American Planning Association* 60(2): 151–153.

Reid, S. A. (1998) 'Community: city faces: An eye on neighborhood pride: New position: her appointment gives a West End woman a chance to act on her desire to clean up her community', 10 September 1998 *The Atlanta Constitution*. p. 2JD.

Sawicki, D. (1993) 'Developing The Atlanta Project's information system', *Proceedings of the Urban and Regional Information Systems Association II*, 129–145.

Susskind, L. and Elliot, M. (1983) *Paternalism, Conflict and Coproduction*, New York: Plenum Press.

Thompson, P. (1993) 'Technology & The Atlanta Project', *Georgia Tech Alumni Magazine* (Spring 1993): 26–35.

The Atlanta Project (1994) Housing Resource Group Fact Sheet, The Carter Presidential Center, Atlanta, Georgia.

Chapter 8

Web-based PPGIS in the United Kingdom

Richard Kingston

8.1 INTRODUCTION

The Internet and the World Wide Web (WWW) have created many opportunities for those involved in GIS and decision support research. Recently many GIS products and applications have appeared on the web (Carver and Peckham 1999), and GIS applications are becoming more frequent in many fields (Doyle *et al.* 1998). These systems tend to vary in nature from simple demonstrations to more complex on-line GIS and spatial decision support systems. With the increased availability and use of GIS applications, previous criticism of GIS as an elitist technology (Pickles 1995) may no longer be valid. GIS and the WWW are ever evolving technologies with the potential for increasing public involvement in environmental decision-making. To gain an understanding of the potential benefit of web based PPGIS, a real decision-making problem was used to develop, live test, and monitor public participation in local environmental decision-making. Traditional methods of public participation were examined by working closely with several organizations in the United Kingdom (UK). The specific aims of this research have been to:

- develop an example web based-PPGIS using a real decision problem,
- analyse user responses to web-based PPGIS in order to evaluate the potential of these systems to democratize the decision-making process, and
- theorize the future role of web-based PPGIS in improving public involvement and policy maker accountability in environmental decision-making.

Opportunities for direct public involvement in environmental decision-making is currently limited in the UK. This is despite the fact that public participation in environmental decision-making in the UK has a relatively lengthy history. Ever since the first Town and Country Planning Act in 1947, varying degrees of public participation have existed although it was not until 1969 (Skeffington 1969) that widespread public participation became embedded in the process. Given the appropriate political will and sufficient public interest, the theory, methods, and practical applications developed here can contribute to radical improvements in future decision-making processes and policy formulation.

8.2 TRADITIONAL VERSUS ON-LINE PUBLIC PARTICIPATION

With traditional methods of public participation, those who are interested attend public planning meetings which often take place in an atmosphere of confrontation. This can discourage participation by a less vocal majority resulting in public meetings that are dominated by vocal individuals who may have extreme views. These views may not represent the opinions of local people who may resist expressing their concerns, opinions and viewpoints, and who therefore 'rarely if ever emerge as definable actors in the development process' (Healey *et al.* 1988). Planning meetings often take place in the evening, limiting the number of people who are able to attend. The actual location of and physical access to public meetings can further restrict the possibility of widespread attendance, particularly for those who are disabled or without access to transport.

In contrast to traditional methods, new web-based forms of participation are beginning to evolve in the UK. Although these are in the early stages of development in the UK, experience from North America (Howard 1998) suggests that there are many advantages to web-based approaches to participation, including:

- the meetings are not restricted by geographical location,
- access to the information is available from any location that has web access,
- the information is available at any time of the day, thus avoiding the problems associated with holding meetings in the evenings, and
- the concept of '24/7' access opens up opportunities for more people to participate.

The use of the WWW has the potential to break down some of the barriers to participation by taking away certain psychological elements which the public face when expressing their points of view at public meetings. For example, with a web-based system the public can make comments and express their views in a relatively anonymous and non-confrontational manner compared with the traditional method of making a point verbally in front of a group of relative strangers. As Graham (1996: 2) argues, the Internet generates 'a new public sphere supporting interaction, debate, new forms of democracy and "cyber cultures" which feed back to support a renaissance in the social and cultural life of cities'.

To achieve greater involvement in environmental decision-making, the public need to be provided with systems that allow them to create virtual spaces. Such systems should allow participants to proceed through the following four stage model:

1 explore the decision problem,
2 experiment with choice alternatives,
3 formulate one or more decision choices, and
4 provide feedback and evaluation of the system.

Exploration of the decision problem is an essential part of the user's learning process. Having direct and easy access to the information relating to a decision problem is a key element in learning about its various facets. In this context, information should be available on the spatial and aspatial aspects of the decision problem, and should convey the historical and policy context of the decision problem as well as its physical, social, cultural and economic setting. Existing community or individual ideas and perspectives on the decision problem should be presented where known. Through learning about all aspects of a decision problem, the user can begin to modify existing ideas and generate new ones that can be fed back into the system.

Experimentation with choice alternatives is also an essential part of the learning process. These 'What if?' scenarios are fundamental to many analyses undertaken by a GIS. With this in mind, web-based GIS should also be capable of allowing the user to:

- test basic theories or hypotheses regarding their decision alternatives,
- develop decision models or pathways applicable to the decision problem, and
- approach consensus or compromise through comparison and trade-off with users' ideas.

Formulation of decision choices should aim to maximize consensus and minimize conflict. This is often difficult to achieve particularly if the decision problem is complex, but it may be possible to identify the best compromise solution and thereby maximize the acceptability of the final decision. The ability to formulate decision choices based on exploration and experimentation is an essential part of any web-based decision support system.

Finally, the system should allow for feedback and evaluation by the public. Feedback by the public throughout the decision process can inform the local authority how and why particular choices are made. Also in a reverse scenario, the local authority can provide feedback to the public so they know how and why certain decisions are taken. This two-way process keeps the local planning authority aware of how users are formulating their decisions. This will also allow designers to improve future systems.

8.3 VIRTUAL SLAITHWAITE

The Planning for Real® (PfR) exercise arranged in the village of Slaithwaite in June 1998 by the Colne Valley Trust (CVT), a local community action group, emerged as a good case study to investigate the potential of a web-based PPGIS. PfR was developed as a means of getting local people more involved in local planning decisions through interaction with large scale physical models of their community. The Slaithwaite PfR exercise was coordinated for the CVT by planning consultants from The Neighbourhood Initiatives Foundation[1] (NIF) and was partially funded by the local council.

A 1:1000 scale three dimensional model of a 2 km^2 area of the village was constructed by CVT and the planning consultants with the help of local school children. This was used as a focus for local discussion about planning issues within the village. Local people were invited to register their views about particular issues by placing flags containing written comments at any location on the model. The results of this exercise were then collated by the NIF consultants, and subsequently fed back into the planning process through appropriate policy documents and plan formulation mechanisms. One of the main aims of CVT was to consult with local people to find out their views, and involve them in local decision-making. The main features of the PfR method include:

- providing a large scale model of the chosen area on which the public can place ideas and comments about their community,
- offering a completely open-ended approach in which anything can be said or suggested,
- allowing the community to assume leadership of the input process,
- providing a mechanism for input open to all members of the community at a time when most can participate, and
- providing information and local opinion that can be of use to both the community and local authorities in future planning.

The Slaithwaite PfR exercise provided an ideal opportunity to develop and live test a simple PPGIS that mirrored the physical PfR model. This system was called 'Virtual Slaithwaite' and was still available on-line at the time this book was published. The virtual version of the exercise was launched on the web and displayed alongside the physical PfR model at a local village event organised by CVT called 'Shaping Slaithwaite'. Eight networked Windows NT machines with Netscape Communicator installed were made available for public use in the local sports hall. This helped to overcome problems associated with access to GIS technology.

8.3.1 System design

The Virtual Slaithwaite PPGIS was arguably among the first such system available to the general public in the UK as part of a real public participation process. The web browser window consists of four frames, each containing particular pieces of information (see Figure 8.1). The system design revolves around a Java map application called GeoTools (Macgill 2000). Using this Java map applet, users can view a map of Slaithwaite, perform zoom and pan operations to assist in visualization and navigation, perform simple spatial queries (e.g. 'What is this building?' or 'What is this road?'), and then make suggestions or comments about specific features identified from the map. All user input is stored in the web access logs for future analysis and feedback into the planning process. In this manner a community database is created, representing a range of views and feeling about planning issues in the village. User responses were handled using perl server-side scripts and html forms. The map applet displays ESRI® ArcView shape files and allows the retrieval of attribute information from the associated dbf file.

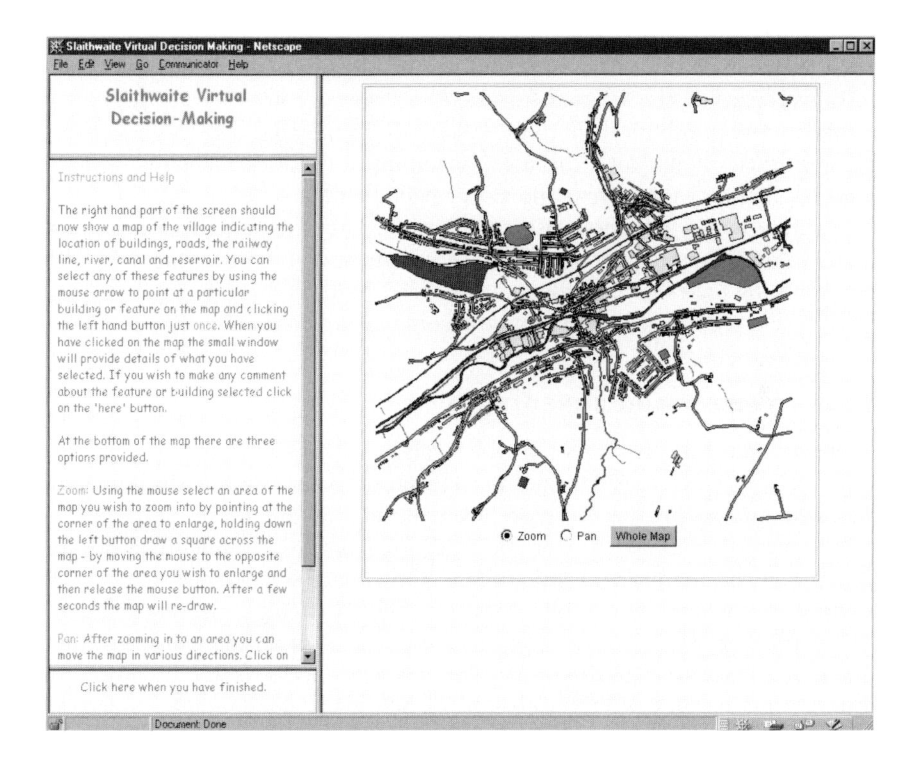

Figure 8.1 Virtual Slaithwaite website.
(Source: http://www.ccg.lecds.ac.uk/slaithwaite)

When users first enter the site, they see an initial welcome window, and then are prompted to fill in a profile. This was seen as an essential part of the system design as it could be used to build a database of users to help validate responses and analyse the type of people who were using the system. Of course, this assumes that users enter correct information about themselves, and collated evidence suggests that not everyone was truthful. However, it is possible to cross-check certain information such as age and occupation. For example, a nine-year-old professional can be assumed to be an invalid profile. Then again, on this evidence alone should the suggestions provided by this person be ignored? It may be a genuine error, or maybe the person felt such information was too personal and therefore filled in the form incorrectly. Issues surrounding privacy and intellectual property rights in the use of PPGIS require further research.

Once the profile is completed and submitted, the map of the village and the associated attribute datasets are downloaded. The frame to the left of the screen contains 'Instructions and Help' information that can be read while the map loads. Once the map is displayed, the user is free to select any feature on the map, including buildings, roads, open spaces, the river, or the canal. When a feature on the map is selected, the small frame in the top left hand corner of the screen displays what the feature is and the original 'Instructions and Help' window changes to a form that can be filled in with comments or suggestions regarding the selected feature. Once they are happy with their comments, users can submit them to the system for future analysis. This effectively registers their views with the local planning authorities. When they have finished, they exit the system and are provided with a series of questions asking them how they felt about using the system. They are also given the opportunity to make any further comments. A comment map is also generated with dots marking the exact location(s) where users made comments. Each dot can then be selected in order to display the comments recorded there.

8.3.2 Web-based advantages

There are several advantages to this web-based method compared to the traditional PfR exercise. The ability to instantaneously update the comment database and to profile users on-line was seen as one of the most useful advantages of the system over the traditional PfR technique. The on-line system can be maintained indefinitely allowing people to use the system anytime, anywhere. The public does not need to attend a meeting at a particular time or place. This is often the single most inhibiting factor in traditional methods of participation. The system allows faster collation of results from log files and the web site can be used to disseminate results and provide feedback. The traditional PfR requires facilitators to periodically remove participants' suggestion flags from the physical model and then enter this information into a database for future analysis. The on-line system avoids this problem and facilitates a quicker turnaround of results. Unlike the physical PfR model, it was decided not to allow users to view other people's comments to encourage imaginative responses. This avoids 'leading' members of the public into making particularly common suggestions in response to seeing a cluster of flags on the model where many other people have made the same or similar comments.

8.3.3 User responses and interaction

Results from the Slaithwaite study suggest that among certain sections of the population, the web-based system was found both useful and popular. At least 126 people used the system, largely during the CVT-sponsored Shaping Slaithwaite event. Although the system was easily available, actual use of the system was clearly skewed toward particular demographic groups. There was a strong (70.6/29.4%) male to female bias among users. The occupation information collected in the database suggests greater usage by those in professional/managerial or educational positions, while the age distribution of users is heavily skewed towards schoolchildren. The latter is partly a result of educational trips to local primary schools made prior to the event, and partly reflects the inability of schoolchildren to use the three-dimensional map, which was too high and wide for them to reach. Although data were not collected on the mobility of the users, it was clear at the event that the PC-based maps also attracted a number of adults who found the three-dimensional map difficult to use. Given the age distribution of users, it may be worth noting that once the age data were stripped from the comments, it was impossible in most cases to guess the age of the users from their suggestions, reflecting the genuine interest of all users in their local environment.

On the whole, it appears that the public response to the system was positive. All users seemed to prefer the ability to type any amount of information on any subject into the comment areas. This contrasts with the traditional PfR method which limits contributions to a few lines classified by category based on the types of planning problems anticipated by CVT and NIF. In terms of evaluating both systems, only 29 people completed evaluations of the PfR method. This poor rate of return was partly due to the fact that many people left the exhibit once they had made their suggestions (CVT 1999). There was a slightly better evaluation response rate to the web-based version, as Table 8.1 shows.

During the 'Shaping Slaithwaite' event, it was possible to view the public using the system. A high degree of proficiency in map usage among all the users was observed. Users who could not immediately locate the area they wished to comment on simply found a building or road they recognized and then moved along the path that they would on the ground, querying features by clicking on them until they reached the area. Far more problems were experienced with the use of the computers themselves,

Table 8.1 User evaluation of traditional Slaithwaite PfR and web-based virtual Slaithwaite models

	Traditional		*Web-based*	
	No.	*%*	*No.*	*%*
I have full control	0	0	3	9
I have some power for making changes	8	28	9	26
I have voiced my opinion, but have no power to make changes	9	31	13	37
I have been asked what I think	7	24	10	28
I have been told what changes will happen	5	17	0	0
I have no involvement in changes	0	0	0	0
I have no opinion	0	0	0	0
Total	29	100	35	100

particularly the mouse-controlled interface. When one of the research staff was not available, younger members of the community often helped those experiencing difficulties with the computer, or entered data for them.

8.4 CURRENT DIFFICULTIES

This case study provided useful feedback about how people interact with on-line systems. This is perhaps one of the most important achievements of the project, and will enable future systems to be upgraded and improved.

8.4.1 Training requirements

One of the main obstacles to developing PPGIS has been the general lack of familiarity with the technology involved. In particular, many people, especially those from older age groups or manual trades, had never used a mouse before. A much smaller number of people had difficulty understanding the map itself. This provided very useful insights into how people perceive a two-dimensional map (MacEachren 1995) and how subsequent versions of the on-line system could be improved. However, these issues might become less important as more and more people become familiar with using computers through work, leisure or education.

8.4.2 Access to the Internet

Access to the Internet is increasing. National Opinion Polls (NOP) estimated 7 million Internet users in the UK in December 1997 (NOP 1997), with market saturation likely within a decade. A survey in 1999 estimated that the Internet was attracting 10,900 new adult users in Britain every day (NOP 1999). While these figures suggest that the Internet and the WWW are becoming popular, there is the potential that an information underclass is emerging. One method of combating this problem is providing public access terminals in libraries, community centres and even bars and restaurants. In Slaithwaite, there was only one public access terminal available in the public library for local people to use. This is a completely inadequate situation if local authorities wish to increase levels of participation on-line, and further access points would be required to give improved access to all the community. This highlights an important issue for future implementation of on-line systems. It has to be recognized by local authorities that such systems should be in place to enhance and offer alternative means of participation, but they should not replace traditional methods. The advent of free local telephone calls (something not widely available in the UK) over the next few years will also help to alleviate this problem.

Another development that may also circumvent the computer-literacy problem is digital television. Over the next five years, digital television channels devoted to Internet-type access could provide a direct portal to the types of on-line PPGIS systems described here without the need for a computer with an Internet connection. Analogue broadcasting is due to be phased out by 2006 in the UK. This effectively means that the majority of households will have a digital television and, hence, should have access to Internet-type channels, some of which may provide public services such as online voting, public information and participatory democracy.

8.4.3 Copyright issues

Although many of the technical obstacles with PPGIS that were first encountered have been overcome, an important legal issue remains unresolved. This relates to the copyright for data contained in the system. The ownership of each individual piece of information or datum within an on-line system can cause major problems. Any system that is map based could potentially create complex copyright and legal issues. The major problems encountered so far relate to Ordnance Survey (OS) maps being distributed via the Internet (OS 1997; 1999). The OS is the UK's national mapping agency, which holds the copyright for most maps. The cost of paying copyright fees for on-line maps could make the whole exercise prohibitively expensive. This is particularly true for a public organization such as a local authority or trust with limited funds. Under present information copyright laws, copyright issues may prevent publicly funded organizations and projects from developing web-based PPGIS. A possible solution that could protect OS data without imposing copyright fees is the use of encryption and coding software in order to transmit the data in a form that cannot be imported into a proprietary GIS (Kingston *et al.* 1999).

8.5 CONCLUSIONS AND DIRECTIONS FOR FUTURE RESEARCH

The on-line PPGIS experiment in Slaithwaite has provided evidence that it is possible to develop systems which allow the public to interact with real world representations without necessarily being in a particular physical location and at a particular time. A number of potential advantages over the existing traditional approaches to PfR have been identified in the virtual PfR. These can be summarized as:

- ability to customize the map images or display by adding and removing layers,
- ability to interactively zoom and pan through the data,
- ability to interrogate map features to retrieve a description and/or attributes,
- ability to instantaneously add new attribute information to the map database,
- ability to profile users,
- longer residence times of the virtual PfR model (i.e. it is available 24/7),
- faster collation and turn around of results from the PfR exercise, and
- availability of the PfR website to disseminate results and feedback from the PfR exercise.

8.5.1 Public access

If planning authorities and other decision-making organizations wish to see an increase in public participation, they have to realize the need to provide public access points which the general public can easily access. The provision of public access points in council offices, libraries and community centres are likely to overcome these concerns. In particular planning problems and 'policy formulation process participatory on-line systems' will become a useful means of informing the public and to allow access to data and planning tools such as on-line GIS as an additional means of public participation in the UK planning process (Kingston *et al.* 2000). These will provide mechanisms for the exploration, experimentation and formulation of decision alternatives by the public in future planning processes and have the potential to move the public further up the participatory ladder.

8.5.2 Effects of scale

Early evidence emerging from current and on-going research is focussing on the *effects of scale*. While this case study has investigated local issues which tend to interest the majority of people living locally, as problems increase spatially less people become interested, even though the decision problems potentially become more important. A good example of this is nuclear waste disposal. While the issue is initially a national one, as the focus returns to the local when a potential storage or disposal site is identified, everyone locally is interested. In the first instance, only people already interested in the problem at the national scale may participate but once a site is identified it becomes a local problem. The point at which a problem is *perceived* to be a local one is an area for future research.

8.5.3 Fuzziness

A further aspect of the problem concerns the distinction made between discrete and fuzzy definitions of spatial objects or regions. Many aspects of peoples' everyday lives involve fuzzy entities which are not bounded by neat lines which are the mainstay of traditional maps and digital representations. One of the most important elements of a future PPGIS scenario is how to elicit this soft, fuzzy, possibly non-spatial information from the public. Methods need to be developed which allow aspects such as *kind of over there* or *up there somewhere* to be represented on maps. From a technical aspect, the crisp clean data represented on a traditional map can now be distributed on the web as more 'off the shelf' packages become available. The real challenge of future web-based PPGIS, and an area for further research, is how to elicit, represent and handle user-defined fuzzy information which is difficult to represent on a map.

ACKNOWLEDGEMENTS

This research was undertaken at the Centre for Computational Geography, School of Geography, University of Leeds and was funded by the Economic and Social Research Council's *Virtual Society?* Programme award No. L 132 25 1014. The Programme aims to examine, if there are fundamental shifts taking place in how people behave, organize and interact as a result of emerging electronic technologies.

NOTE

1. The Neighbourhood Initiatives Foundation (NIF) is a National Charity, founded in 1988, with the aim of maximizing the participation of local people in decisions that affect their neighbourhoods and quality of life. The founding director, Dr Tony Gibson, devised 'Planning For Real'® in the 1970s as a technique employed by the NIF fieldwork team. NIF has continued to develop and adapt this primary tool to meet both local and strategic consultation needs and it has become an essential tool in community development programmes. NIF fieldworkers usually facilitate the process using large 3D scale models of the local area.

REFERENCES

Carver, S. and Peckham, R. (1999) 'Internet-based applications of GIS in planning', in Geertman, Openshaw and Stillwell (eds) *Geographical information and planning: European perspectives*, Springer-Verlag, pp. 361–380.

CVT (1999) 'Shaping Slaithwaite, part one – the process, Huddersfield: Colne Valley Trust'.

Doyle, S., Dodge, M. and Smith, A. (1998) 'The potential of web-based mapping and virtual reality technologies for modelling urban environments', *Computers, Environment and Urban Systems* **22**, No. 2, 137–155.

Graham, S. D. N. (1996) 'Flight to the cyber suburbs', *The Guardian* 18 April, 2–3.

Healey, P., Mcnamara, P., Elson, M. and Doak, A. (1988) *Land use planning and the mediation of urban change*, Cambridge University Press.

Howard, D. (1998) 'Geographic information technologies and community planning: spatial empowerment and public participation', *Empowerment, Marginalisation and Public Participation GIS meeting*, 15–17 October, Santa Barbara, California.

Kingston, R., Carver, S., Evans, A. and Turton, I. (1999) 'A GIS for the public: enhancing participation in local decision-making', GISRUK Conference Series, 14–16 April, University of Southampton.

Kingston, R., Carver, S., Evans, A. and Turton, I. (2000) 'Web-based public participation geographical information systems: an aid to local environmental decision-making', *Computers, Environment and Urban Systems* 24(2): 109–125, Elsevier Science Ltd.

MacEachren (1995) *How maps work: representation, visualization, and design*, London: Guilford Press.

Macgill, J. (2000) 'The geotools mapping toolkit', http://geotools.sourceforge.net.

NOP (1997) 'One in twenty-five British households now linked to the Internet', http://www.nop.co.uk/internet/surveys/in07.htm NOP Research Group.

NOP (1999) 'Internet research', http://www.nop.co.uk/internet.asp/ NOP Research Group.

Ordnance Survey (1997) *Developments for the Web*, Information paper 13/1997, Southampton: Ordnance Survey.

Ordnance Survey (1999) *A new pricing policy for mapping on the Internet*, Information Paper 1/1999 Version 2, Southampton: Ordnance Survey.

Pickles, J. (1995) *Ground truth: the social implications of geographical information systems*, New York: Guildford Press.

Skeffington, A. (1969) *Report of the committee on public participation in planning: people and planning*, London: HMSO.

Chapter 9

GIS-enhanced land-use planning

Stephen J. Ventura, Bernard J. Niemann, Jr., Todd L. Sutphin and Richard E. Chenoweth

9.1 INTRODUCTION – INFORMATION IN LAND-USE PLANNING

The role of spatial information technologies in decision-making has been debated almost since the inception of their use in local government land information systems. An important question that often arises is whether land information is used to help make decisions, or is it used to justify decisions made for many other reasons? Niemann (1987) and Zwart (1988) epitomized the debate in their point and counter-point conference articles about 'better information' resulting in 'better decisions' through modernized land information systems.

At the local level, 'just getting the job done' public agency practitioners generally have not been concerned about the role or impacts of spatial information. If they take time to consider these issues, it is likely that most would follow disciplinary training and assume that improvement in quality and availability of land information benefits the citizens and organizations they serve. If asked about 'public participation', they might also assume that improvements, particularly in data form and access, extend availability of information to audiences that otherwise may be excluded from decision-making processes.

Little empirical evidence has been reported to support or deny this belief in a positive role for land information in land use decision-making in the contemporary US local government context. Zwart (1991) defined indicators of the impacts on decision-making, though noted operational difficulties in using them. A theoretical model of the role of information in local land use planning was developed by Knapp *et al.* (1998). They explicitly looked at the effect of information about local government infrastructure investments and land use regulations on the timing of development decisions. Their study did show that information modifies development decisions. However, the models were limited to interaction between local government and developers; it did not account for all the other actors in development decisions, particularly actors who could be more influential if empowered by information.

In the land-use debate, the difficulty of ascertaining who may be affected by land-use decisions (and how this might change with differences in access to information) is exacerbated by the diffuse nature of the decision-making process. Many citizens are affected by land-use decisions but may not be directly involved in the decision-making process. The optimistic view is that of a Jeffersonian democracy, where well-informed citizens exert an indirect influence on the process, through elections, meetings, surveys, and even through consumer choices. A cynical view suggests that various elite groups, particularly those that benefit economically from development, control the decision-making process. From an information standpoint, a question critical to understanding which view prevails may be 'whose information?'...whose worldviews are represented in a data base and in analytic tools to understand the data, and do these representations exclude the views of segments of society?

Questions about the role of land information in local government decision-making have been difficult to resolve because spatial technologies are just maturing and because characterizing the decision-making process has been and continues to be difficult. In particular, it is difficult to determine what role land information plays in local land-use decisions because the process is influenced by so many other factors, including political, economic, legal, bureaucratic, personal and social pressures. And, the actors involved may not always be entirely open, knowledgeable or forthright about what has influenced decisions. Moreover, research must be done *in situ*; we don't have the luxury of controlled experiments in which we can suffuse a jurisdiction with information to observe the result while controlling or accounting for this host of other factors.

Our project contributes to the discussion about the role of data and land information in land-use decision-making by purposefully improving the type, quality, and availability of land information and analysis in a jurisdiction with an on-going and highly charged land-use decision-making scene. We will attempt to gauge the influence and impact this has on land-use decision-making processes and outcomes through first-hand observation, post-decision reconstruction, surveys, and other methods. Key questions include:

1 is new information being used?
   - in what form?
   - in what parts of the land-use planning process?
   - how is it used (to support decision-making or to justify decisions)?
   - does it or can it represent groups not traditionally empowered in decision-making?
2 who is using it?
   - do some groups use it more than others?

- are there technical barriers to fuller use by some groups?
- what would users be doing without it?

3 has the improved accuracy, specificity, and availability resulted in different decisions?
   - which of these information attributes are particularly important?
   - do the 'using it' groups have a real or perceived advantage in land use debates?
   - do any groups believe that information is missing or biased?

9.2 BACKGROUND

Dane County, Wisconsin (the County) is one of the fastest growing counties in the Midwest. It also continues to be one of the most productive agricultural counties in the state, typically ranking first in the state and in the top 50 nationwide in gross agricultural sales. The City of Madison is centrally located in the county, and contains half the population (about 200,000 in 1998). Madison is a regional employment centre, including a major state university, the state capital, and a rapidly growing high-tech industry. It consistently ranks high in various liveability indexes, including designation as the 'Best City' by *Money Magazine* in 1998.

For at least a couple of decades, the conversion of farmland to residential and commercial purposes has been contentious. The County Board frequently splits along rural/urban lines on land-use issues, with pivotal votes coming from fringe suburban areas. More effective land-use planning was a major theme in the campaign of the current County Executive. She followed her election with an effort to shift control from a regional planning commission that was regarded by some as ineffective, and incorporating this function in the County's more technologically sophisticated planning department. The *Design Dane* vision document (Falk 1998) embraces geospatial information and visualization technologies as part of a suite of tools to more effectively involve the public in land-use planning and management.

A cooperative relation has existed between the County and the University of Wisconsin-Madison Land Information and Computer Graphics Facility (LICGF) for almost two decades. LICGF has conducted research and development on land information technologies and applications in this 'real world' context with the County, essentially reducing the County's risk in adopting innovative technologies. As a result, the County has a sophisticated automated land information system used primarily for real property listing, tax assessment, deeds recording, and soil and water conservation (e.g. Tulloch *et al.* 1998; Miskowiak *et al.* 2000). They have recently begun to use it for land-use planning as well.

A recent rejuvenation of the LICGF–Dane County relation represents initial evidence that the 'whose reality is represented' question must be considered in understanding how information is used. The County Executive's interests were piqued when she was shown a very different picture of how much land could be considered 'open space and farmland' than that depicted in a 25-year land-use plan done by the autonomous Dane County Regional Planning Commission (DCRPC 1997), and shown in front-page graphics of a local daily newspaper as 'Room to Grow' (Hall 1995). We provided a different interpretation, from the County's own databases, that countered the RPC's suggestion of almost completely unencumbered open space beyond city boundaries. Our GIS-based analysis of land use classification from tax assessment roles indicated that less than 50 per cent of the county was developable farm and open space; in contrast, RPC used data for their land-use plan that showed 85 per cent of the county in this category, based on air photo interpretation. The tax assessors view includes all residences, including vacant lots slated for development, farm houses now used primarily as residences, and residences obscured by tree cover (Carlson 2000). It is arguably closer to the land owners view of what its use is or could be (Heinzel *et al.* 1996).

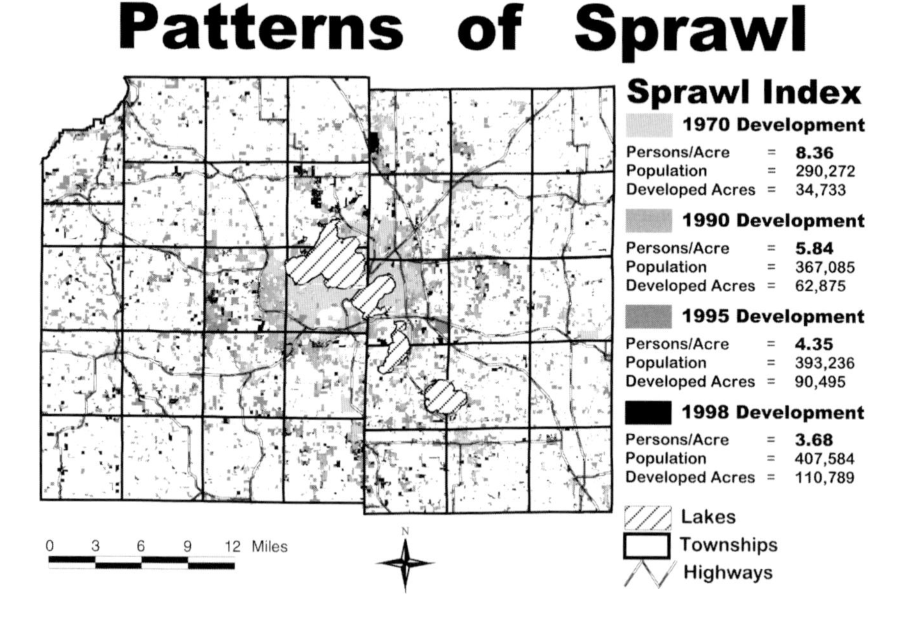

Figure 9.1 Patterns of Sprawl. This map displays patterns of development over three decades in Dane County, Wisconsin. It alerted citizens to the idea that development has become more land consuming and less dense with population over time.

In Spring of 1998, the newly elected County Executive reviewed the comparison of 'open/undeveloped land' in the RPC's 25-year land-use plan with our tax-assessment-based version (Figure 9.1). She immediately grasped the significance of the difference and the implications for where and how the county could grow. It was apparent that development was scattered throughout rural areas, generally following amenities such as prime vistas, forested lands, and water resources, as well as other factors traditionally thought to influence land use patterns such as proximity to good schools, jobs, and transportation systems. Another study indicated that enforcement of local subdivision ordinance and related land-use controls substantially affected farmland conversion (Bukovac 1999). The County Executive thought the differing land-use interpretations were significant. As a result, we were asked to participate in County-led forums on land use and in subsequent activities.

We have assisted the County in developing and disseminating land use information using a variety of venues and events. We have guided the County Executive's staff in the analysis, display and dissemination of their own geospatial data, particularly information related to land-use, ownership, assessment, and resources. We have attempted to make high quality geospatial data and information readily accessible to anyone interested in using it in local land-use planning, in several forms and through several venues.

In the Fall of 1998, Dane County was selected by the Federal Geographic Data Committee (FGDC) and Vice President Al Gore as the site of one of six Community Demonstration Projects. This led to a project that became known as 'Shaping Dane's Future', a collaboration of the University, the County, FGDC, the Natural Resources Conservation Service, and ESRI, Inc. The City and Town of Verona (two local units of government, adjacent to Madison) were selected as the project site because of significant land-use issues and interest from local officials in helping evaluate information technologies.

9.3 EVALUATING THE ROLE OF INFORMATION TECHNOLOGIES IN LAND-USE PLANNING

To begin answering some of our questions about the efficacy and impact of various information technologies, we have been providing information and analysis tools while observing how decision-makers and interested citizens and organizations react to and use geospatial technologies and information products. These can be thought of as experiments about form of and access to information. As part of collaboration with Dane County and the Shaping Dane's Future pilot project, we have conducted the following activities to learn about the role of information technologies in land-use decision-making.

- *Land-use forums* – Map-based and statistical overviews of County conditions, resources and trends based on geospatial analysis were presented in several venues, including four County-sponsored 'land-use forums', civic groups meetings, and a University-based seminar series on the role of geospatial technologies in land-use planning. The County land-use forum series included 'listening sessions', which provided an opportunity to observe directly how participants were thinking about and using geospatial information. In the largest County land-use forum (over 300 participants), about 25 large format (48 × 36 inches) maps were displayed. Participants were given a brief survey after an 'open-house' period of observation, with questions about which products they found useful, what other products might be useful for land-use planning, and how they saw themselves interacting with geospatial technologies. As an incentive to complete the survey, participants were promised a copy of the map of their choice. This experiment provided information both directly from the survey, and indirectly from tabulations of which maps participants selected as their reward for survey completion.
- *Allocation experiment* – In another County land-use forum session, participants were divided into small groups and asked to place dots on large format maps to designate areas for future residential development. The maps portrayed factors related to growth opportunities and constraints. Different dots represented different numbers of residences

Figure 9.2 Citizens participating in land-use allocation exercise.

(by dot colour) and development density (by dot size). Each group was responsible for allocating 75,000 new residents, the expected growth in Dane County by 2020 (Figure 9.2).

- *Land information bulletin series* – We developed a series of bulletins that guide land planning professionals, citizens, and decision-makers through components of land-use planning that involve spatial assessment and analysis. The series corresponds to the web-based 'planning tool kit'. The bulletins are distributed in hardcopy form to a variety of audiences, depending on subject.
- *Website development* – We developed a website (www.lic.wisc.edu/shapingdane), the 'Planning Resource Center', to assist citizens and planners in using geospatial technologies and the County's data for land-use planning (Figure 9.3). It includes WebGIS (based on ESRI's ArcIMS) to view data and create maps over the Internet. Advanced capabilities allow users to perform queries, download data, and post notes on the data layers and maps. Digital *post-it* notes create a mechanism by which users interact and share ideas via the Internet. The website also includes a portal to additional local land-use planning information and a link to the CyberCivic opinion registration and voting tools (see below). The website has been evaluated by local participants and modified to accommodate their needs and interests. We are monitoring domains of computers accessing the web-based material and asking viewers to provide feedback on the material and comments on the general approach with an open-ended *mailto*.

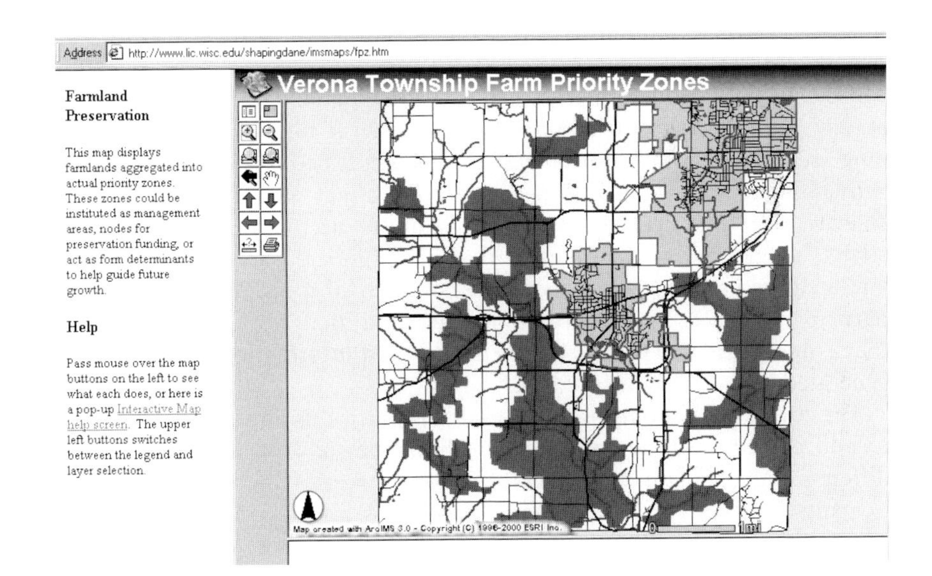

Figure 9.3 Planning Resource Center website (www.lic.wisc.edu/shapingdane).

- *Training experiment* – We provided free training in commercial GIS and visualization software, and their use in planning using the County data sets. Initially, we offered five 1-day sessions on five different aspects of land-use planning. This experiment provided us with data on who in the community were interested in improving their aptitude for GIS-based planning and for which aspects of planning they consider GIS to be an appropriate tool. The courses were so popular that we have continued to offer them on a cost-recovery basis. Additional courses have been set up specifically for the needs of the Verona project participants.
- *Software module development* – We assembled existing software and developed new software for five land-use planning elements (exploration, analysis, allocation, impact, public access). Selection of existing packages and development of new software were based on citizen and County staff participation during module development, who provided feedback on the usefulness, ease of use, and ability of modules to engage and involve citizens in land-use issues.
- *Electronic 'Town Hall'* – In collaboration with a local e-business (CyberCivic.com), we have linked to a web-based tool for citizens to register opinions with elected officials, vote their preferences on controversial land-use issues, and post comments in a chat-room and a bulletin board. This website was used as the hub of a live electronic town meeting, in which several elected officials responded to questions from a live audience, from phone lines, and from e-mail. Responses were simultaneously broadcast on a local cable television station, a local radio station, and the website.

The overt goal for our activities is to make geospatial data and information readily accessible to anyone interested in using it for local land-use planning and related applications. This was accomplished through the litany of activities listed above, as well as continual meetings with local officials and citizens. We tried to de-mystify and simplify user interfaces and other aspects of geospatial technologies that have hindered access to databases and analysis using currently available commercial tools. We also recorded participants' ideas and feedback on data and analyses, so that they felt they could guide what kind of information and information products were generated.

The covert goal of our project is to determine if this unprecedented access to and education about geospatial technologies and products makes any difference in the planning process. We have surveyed and interviewed participants to directly find out how individuals and factions/organizations perceive the impact and utility of geospatial technologies. We also gathered indirect evidence, such as where and how GIS-derived products and facts were used in documents and meetings, based on what kinds of participant-generated questions, whether these are used to help make decisions or used to justify decisions, and which factions or organizations seem to be able to make most effective use of them.

Two general strategies are being used to answer the three key questions raised in the introduction.

First, behavioural measures were used to observe land-use decisions and the products that were used in the process. Behavioural measures built into software programs tracked individual decisions. These were supplemented by direct observation of how selected officials and citizen-planners navigated and used web resources, and how they responded to open-ended suggestions for activities in training sessions. Behavioural measures were also obtained from observations of group meetings; notes were later evaluated using content analysis procedures.

Second, self-reported behaviour using questionnaires were used where direct measures of behaviour were not possible or inefficient. Questionnaires have the advantage of being able to explore beliefs and attitudes that underlie the behaviours in question. As part of the Shaping Dane's Future project, a survey was sent to all residents in the Town of Verona (about 700) as well as non-resident landowners (about 50). This survey followed a modified Dillman (1978) method of survey research, and resulted in over 70 per cent response rate. In addition to questions pertaining to land-use attitudes and beliefs, we asked how citizens learned about land-use issues and their degree of access to Internet and other computer resources. We will conduct a similar survey next year (2001) to directly measure if there has been a change in how people access information about land-use and to determine if there are changes in perceptions about the issues.

While true experiments comparing the influence and impact of land information analysis and visualization between different communities of users may not always be possible, the careful use of this quasi-experimental time-series design should be useful in isolating factors that influence which information technologies are adopted and how these are used.

9.4 RESULTS AND DISCUSSION

Most evidence to date indicates a high degree of interest in improved land information analysis and visualization from County staff, local (City and Town of Verona) staff and officials, and other factors typically involved in land-use decision-making. Requests for additional information products from new analyses continue. The requests are increasingly specific about the type and form of analyses and products, indicating a more sophisticated understanding of the spatial database and what is feasible in analyses and products. The county continues to invest more than 1/2 million dollars annually in land information systems and staff. The use of spatial information technologies is touted by County staff and officials as a key component in resolving some of the County's vexing land-use issues. The following is a brief description of what we have learned from the various 'experiments', as well as our meeting observations, interviews, and survey.

Based on the land-use forum and on the Verona survey, citizens come into the process with a relatively high level of computer acumen. Ninety-two per cent of the participants in land-use forums had home computers and 70 per cent had Internet access. Though these were presumably citizens motivated to become involved in county-wide land-use issues, approximately the same percentages of access to computers and Internet were recorded in the survey of *all* households in the Town of Verona.

Participants are also map-literate and readily learn spatial analysis. Participants in the forums, electronic town halls, and web-browsing consistently study or select complicated composite maps depicting several factors related to land-use and conditions, as opposed to simpler single-theme products. In sessions with GIS technicians, participants quickly grasped the information potential of thematic overlay, asking questions such as 'how many land parcels taxed as residential land but without improvements (e.g. vacant lots) exist at elevations about 900 feet.'

The allocation experiment provided explicit evidence about preferred development patterns and strategies of different societal factions (e.g. developers, farmers, rural or urban elected officials, environmentalists, etc., at least to the extent that the small groups were identifiable as particular factions). Preferred strategies conformed with expectations. For example, one group placed almost all their dots in and near existing urban areas, a 'compact growth' strategy favoured by environmentalists; another group scattered dots through a more rugged section of the county, with none in a region of highly productive farms, suggesting a farmland preservation strategy. All groups understood the spatial analysis involved in the exercise and how the underlying data contributed to their ability to make allocation decisions.

Almost all workshop participants found the land-use planning training very useful. Over 280 people applied for 100 available slots, indicating a high level of interest in becoming better able to use spatial technologies and data for land-use planning. Shaping Dane's Future project participants have eagerly participated in additional training and served as 'guinea pigs' as we develop and test new software. It appears that in a day of training, general planning concepts and 'hands-on' use of GIS software adapted to this domain can be conveyed to lay audiences.

It is too early in our experiments and observation to say definitively whether our infusion of better land information has engaged more people in the decision-making process or influenced land-use decisions. Clearly though, it has been an important component of the County's process. Land information and spatial analysis were prominent components of the public land-use forums. A land-use vision statement by the County – *Design Dane* (Falk 1998) – included many maps that were clearly the products of a GIS. A content analysis of that document revealed at least 38 different calls for information products or spatial analyses to support the County's proposed land-use goals. In it, the County Executive called for 'improving the way we do business by developing new information technologies to make more informed growth decisions'.

Verona participants have essentially re-designed their local land-use planning efforts to incorporate information technologies. GIS would not have been used if it had not been made accessible and understandable through the Shaping Dane's Future project. The citizen Land-use Planning Task Force was clearly excited by the information generated from using live, interactive GIS as part of their deliberations. They also quickly noted some key information (wetlands and floodplains) missing from the GIS data base, evidence that data availability is a key information attribute. The final plan, due out within a few months, will contain numerous GIS-generated products, including sophisticated analyses such as farmland preservation zones based on bio-physical and socio-economic criteria.

The extent to which other actors in land-use decision-making adopt and use the products and technologies remains to be determined. Though 'hands-on' use of GIS seems to be primarily by technologically inclined participants, it is apparent that most involved in the process grasp the mapping and spatial analysis concepts, and many use the web as an information resource, including the WebGIS component.

In the Verona situation, we have observed that many factions, including short- and long-term residents, farmers (though characterizing them as a single group on land-use issues is fallacious), environmentalists, and developers have embraced the use of spatial technologies in land-use planning and decision-making. This is not to suggest that other groups affected by land-use decisions will or will not recognize GIS's role in decision-making. We will learn more about this with the follow up survey next year. Similarly, we will only be able to determine if particular groups have significantly benefited or been disadvantaged by the Verona land-use after it is finalized and all groups have a chance to react to it. If we can surmount the difficult methodological hurdle of determining who wins and who loses in the overall land-use process, then we will determine whether GIS was among the causal factors in these outcomes.

At this point, practitioners promoting more accessible land records can be comforted that we have no evidence to suggest that this has disadvantaged any groups or individuals. Their influence appears to be positive, though of course this is an ongoing drama with many layers and perspectives that await more comprehensive evaluation.

REFERENCES

Bukovac, J. (1999) *Town Government's Role in Explaining the Spatial Variation of the Rate of Farmland Loss*, MS Thesis, Land Resources, University of Wisconsin-Madison.

Carlson, J. (2000) *Evaluating a Residential Land use Map Generated from Tax Assessment Data*, MS Thesis, Environmental Monitoring, University of Wisconsin-Madison.

DCRPC (Dane County Regional Planning Commission) (1997) *Vision 2020: Dane County Land use and Transportation Plan*, DCRPC, Dane County, WI.

Dillman, D. A. (1978) *Mail and Telephone Surveys: the Total Design Method*, New York: Wiley and Sons.

Falk, K. M. (1998) *Design Dane! Creating a Diverse Environment Through Sensible, Intelligent Growth Now*, Dane County Executive's Office, Dane County, WI.

Hall, D. J. (1995) 'Room to grow – the questions are: how and where?' *Wisconsin State Journal* (July 30): 1A.

Heinzel, W. M., Niemann, B. J. Jr. and Ventura, S. J. (1996) 'Integration of spatial technology: land tenure and land use change', *GIS '96 Conference Proceedings*, Fort Collins, CO. CD number B, track 1–3, GIS World, Inc.

Knaap, G. J., Hopkins, L. D. and Donaghy, K. P. (1998) 'Do plans matter? A framework for examining the logic and effects of land use planning', *Journal of Planning Education and Research* 18(1): 25–34.

Miskowiak, D., Ventura, S. J. and Sutphin, T. (2000) 'Farmland Preservation Zones', *Land Information Bulletin*, Number 2 (Spring 2000) Land Information and Computer Graphics Facility, University of Wisconsin-Madison.

Niemann, B. J. Jr. (1987) 'Better information for better decisions: no question about it', *Proceedings, URPIS 15*, pp. 187–194.

Tulloch, D., Moyer, D. D. and Niemann, B. J. (1998) 'Modernizing Dane County's register of deeds: saving property owners $6 million', *Land Information Bulletin*, No. 1, (Spring 1998), Land Information and Computer Graphics Facility, University of Wisconsin-Madison.

Zwart, P. R. (1988) 'Some observations on the real impact of integrated land information systems upon public decision making in Australia', *Papers from the 1988 Annual Conference of the Urban and Regional Information Systems Association*, 1: 68–79.

Zwart, P. R. (1991) 'Some indicators to measure the impact of land information systems in decision-making', *Proceedings of the Urban and Regional Information Systems Association*, vol. 2.

Chapter 10

Portland Metro's dream for public involvement

Mark Bosworth, John Donovan and Paul Couey

10.1 INTRODUCTION

This chapter examines the public involvement efforts of Metro, the regional government for the Portland metropolitan area, in using GIS technology to engage residents and policy-makers in making informed decisions about issues related to growth management, land-use and transportation. The first part of this chapter describes current programmes and the use of GIS technology. Section 10.4 proposes a new platform for public involvement that would allow public participation in planning efforts in 'real time'. Using Internet-based technologies, it is possible to create a new channel for public participation in the planning process.

GIS and public involvement professionals at Metro are exploring the potential for creating spatial representations of traditionally intangible information, such as what residents value about their homes and communities, what they hope to protect or pass on to future generations and what makes this region special and unique. By capturing this information, Metro could begin to use this 'value-based' information to help shape future policy initiatives as well as to illustrate the value systems that residents share. During the course of an intensive public outreach and long-range planning process known as the 2040 Framework, Metro has used GIS technology to enable residents in the decision-making process. These applications illustrate the potential of GIS technology as a platform for public involvement, and as a channel for accessing information about land-use policy decisions.

10.2 METRO AND ITS ROLE

Metro, the nation's only directly elected regional government, serves more than 1.3 million residents in the three counties and 24 cities of the Portland metropolitan region. In 1978, voters in Multnomah, Washington, and Clackamas counties approved the idea of a regional government to oversee

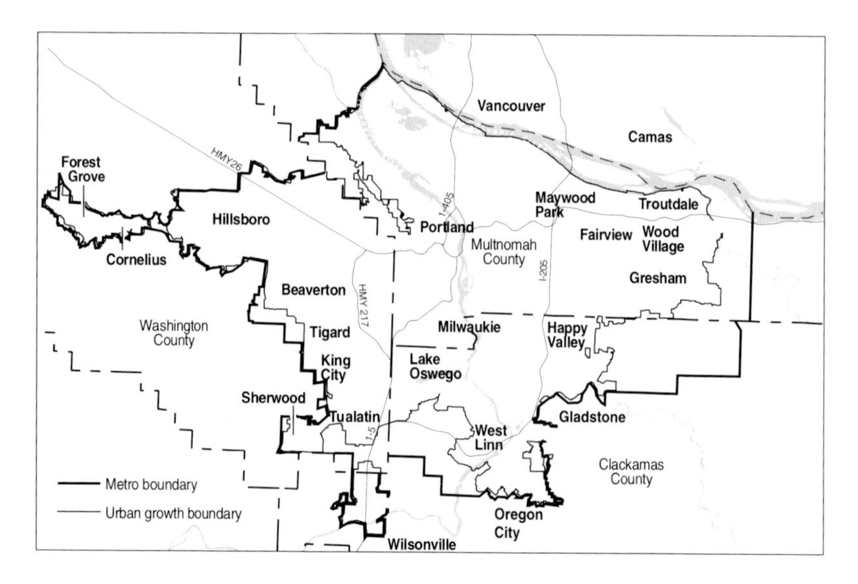

Figure 10.1 The Portland Metro area comprises the urbanized portion of three counties.

issues that transcend traditional city and county boundaries (Figure 10.1). Metro is responsible for transportation and land-use planning, solid waste management, regional parks and greenspaces, and technical services to local governments. Metro also manages the urban growth boundary, a planning tool that defines the limits of where urban growth can extend and where rural lands begin (Metro 1991).

The Metro Charter, approved by voters in 1992, calls for the creation of the Future Vision (Metro 1992). During the deliberations prior to the adoption of the Future Vision Report, advisory committee members and staff were able to use Metro's GIS system to provide interested citizens with a new view of the region. Through the use of shaded relief, full colour mapping images of the geography of the area within and surrounding the Metro service boundary, residents were able to understand the physical constraints of the area, how growth would likely occur and where critical natural resources still existed within our metropolitan urban form. This information helped shape the scope and final conclusions of the Future Vision Report.

> The bi-state metropolitan area has effects on, and is affected by, a much bigger region than the land inside Metro's boundaries. Our ecologic and economic region stretches from the Cascades to the Coastal Range, from Longview to Salem. Any vision for a territory this large must be regarded as both ambitious and a work-in-progress.
>
> Metro, Future Vision Commission 1995

10.2.1 The Region 2040 programme

When voters approved the new charter, they established growth management as Metro's primary mission and granted the agency authority to implement policies. Metro then began an intensive public outreach effort intended to involve residents in a regional planning process looking at how the region should grow for the next 50 years. Surveys designed to get answers about some basic livability questions were mailed to every household in the region (more than 500,000 homes). More than 17,000 people expressed their opinions. Metro learned that the survey respondents:

- value a sense of community,
- favour the preservation of natural areas, farm and forest lands,
- desire quiet neighbourhoods and accessibility to shopping, schools, jobs and recreational opportunities,
- value the 'feel' of this region, with open spaces, scenic beauty and the small town atmosphere, and
- favour a balanced transportation system that provides a range of travel opportunities including transit, walking, biking and autos.

At the same time, Metro employed cutting-edge urban analysis and forecasting technologies to study the ramifications of different growth management strategies. A wide range of possible approaches were identified and analysed for both positive and negative impacts to the region's neighbourhoods, transportation system, natural resources and key urban services. The results of this intensive study allowed Metro to focus on a smaller number of possible options to pursue and prepare for public review.

The Region 2040 planning programme culminated in the 2040 Growth Concept, a 50-year strategy for how the region will grow until the year 2040. The concept took four years to develop, including extensive public involvement outreach. The growth concept was adopted by ordinance in December 1995 by the Metro Council. This programme has been recognized as a national model of sustainable growth management planning (Metro 1995).

10.2.2 Evolution of a regional GIS

Early on, such comprehensive analysis of land-use and demographic patterns required a high level of detail. In 1989, the idea of a seamless, parcel-specific database began its evolution into a regional land base information system. This development has progressed from a computer-assisted drawing file into a mature GIS environment that has grown more 'intelligent' through substantial data conversion. It has also become widely accessible through a desktop version on CD and an online interactive mapping application that offers layers of geographic information individually or in combination to anyone who has Internet access and a web browser.

These interactive tools have been built on the framework of the Regional Land Information System (RLIS), an internationally acclaimed GIS programme created by Metro's Data Resource Center (DRC). RLIS is a parcel-based GIS, with data derived from assessment and taxation records from the three counties in Metro's boundaries as its base. Additional layers have been built in reference to the parcel base including street centrelines, digital ortho-photography, vacant lands, topography, soils, natural hazards, etc.

Metro data and map coverages are seamless across the region, eliminating problems that arise from data gaps and overlaps at city and county boundaries. This characteristic alone contributes greatly to the power of GIS to bring diverse groups together on issues; the debate focuses on the issues, not on the data or methodology used to arrive at a particular position.

10.3 PUBLIC INVOLVEMENT IN THE PLANNING PROCESS

Residents are the customers of Metro, with information and policy decisions as its product. Following this business orientation, Metro's public outreach policies interpret involvement according to a communications pyramid that divides the general public into four audience groups (see Figure 10.2). The pyramid illustrates how the size of the groups and the complexity of the

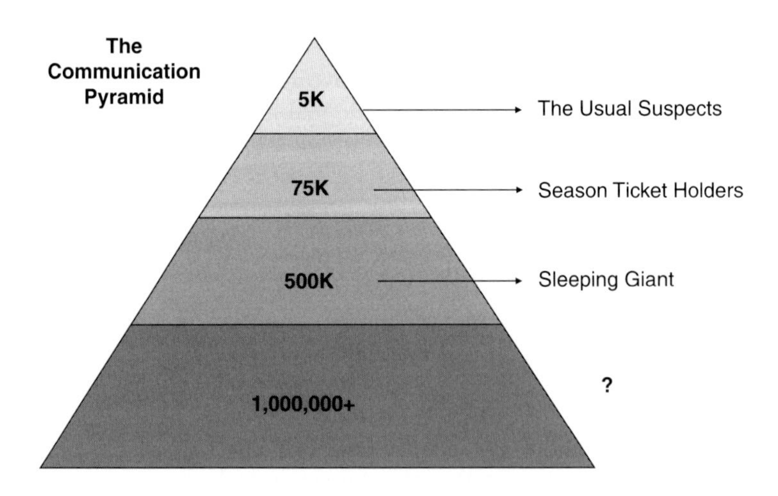

Figure 10.2 The communications pyramid showing the division of target populations for public involvement strategies.

messages interrelate. As interest/activity increases, number of people decrease. As interest/activity decreases, information gets simpler.

At the top of the pyramid, the *usual suspects* represent the smallest yet most involved group of people Metro reaches. Conversely, they require the most detailed information. There are approximately 5,000 people in this group and the entrance threshold is attending Metro meetings.

In the middle of the pyramid, we have the *season ticket holders*. This group is generally positively disposed toward our efforts and simply wants to be kept informed. They are 'in their seats', but only sporadically paying attention. They require less information than the activists, but must be kept regularly updated. There are about 75,000 people in this group.

The third group is *the sleeping giant*, all the people who are minimally active civically (registered to vote) but who are not paying much attention to Metro messages or programmes. This group requires simple messages packaged in a convenient form. There are approximately 500,000 people in this category.

Finally, we have the rest of the population that must be reached in new and creative ways in order to move them into the higher levels of our communications hierarchy. This makes up the last and largest portion of the pyramid.

Residents in the region have a wide variety of methods to communicate back to Metro. These methods can be organized by the communications pyramid structure. For all three pyramid levels of involved citizens, the most accessible means of expressing their preferences is through voting in elections and on ballot measures that relate to planning and resource protection. For the top two levels, the 'usual suspects' and the 'season ticket holders', responding to printed, electronic, phone or faxed surveys/questionnaires via one of Metro's outreach tools influences policy decisions. The most involved group, the 'usual suspects', influence policy by attending public meetings and engaging directly in dialogue with policy-makers and planners.

10.3.1 Existing GIS applications

RLIS Lite – A CD-ROM product created by the Data Resource Center for distribution of GIS data in a convenient format for desktop mapping. Simplified data themes, non-normalized data tables and 'human-readable' data elements – streams are labeled as 'Streams' instead of 'RIV' with an attribute of less than 4 in a related table, etc. This single product has opened up the distribution of GIS to a much greater audience than was previously possible. Created as a commercial product, the structure and format of the GIS data in this product have become the common language for data exchange and data usage in the region. A GIS distributed in this format made the following public involvement programmes at Metro possible.

'URSA-matic' – Metro employed a multi-criteria Excel program linked to a GIS to do comparative analysis of the potential areas for urbanization based on state-required factors for selection of urban reserves. This tool allowed planners, elected officials and residents to compare the suitability of these areas by weighting the different factors in different ways. 'URSA-matic' could be loaded onto a laptop computer and projected through an LCD projector so that groups of officials and residents could test various scenarios 'on the fly' in the course of meetings or hearings. The specificity of the application allowed residents to learn whether individual taxlots were 'in' or 'out' of the potential areas of development and whether they were likely to be included or not in the final selection of urban reserves. Residents could understand and be part of the decision making process, breaking out of the traditional 'black box' technical environment involving planners, lawyers and elected officials. More than 750 participants attending seven different workshops were given the opportunity to see the application and ask for specific information.

Natural resource protection workshops – Metro is working on new regional land-use policies on protection of areas along rivers, streams, floodplains and wetlands. Metro created new data layers for RLIS representing protected areas where the new policies would be applied throughout the region. The public outreach effort included workshops conducted around the Portland metropolitan area. Metro staff loaded the desktop version of RLIS, called RLIS-Lite, onto laptop computers and set up 'one-on-one' stations at the workshops. More than 600 participants attended two sets of workshops and sat down with planners to look at specific sites and examine the proposed overlay zones. They were also given a hard copy map to take home. Additionally, both planners and residents could fill out forms to request changes, with suitable documentation, to the maps if there were errors or omissions in the Metro data.

Metropolitan area disaster GIS – Since 1993, the federal government has invested in a Portland-area partnership of local, regional and state government agencies to identify earthquake hazard and the seismic risk posed by potential earthquakes to buildings and other structures. The distribution of the *Metro Area Disaster Geographic Information System (MAD GIS) CD-ROM* to risk managers and emergency managers ensures that these natural hazard and risk data will be used for risk assessment and disaster management planning to benefit communities and businesses in the region. The software can highlight and analyse essential elements of information related to hazard, risk and vulnerability. MAD GIS exploits the spatial analysis capability of GIS technology to paint a picture illustrating the vulnerability of people, structures and natural resources to damages that could be caused by an earthquake or other natural disaster.

MetroMap – a web-based view of the RLIS database provided to the public via the Internet. The application allows access to multiple layers of geographic information with some limited spatial analysis tools. Anyone with an Internet connection and a browser can have access to this application. The major 'analysis' function provided is a point-in-polygon tool, giving users the ability to determine what geography a location is 'in'. Boundary information can be generated in a list form and includes the following major categories:

- political (such as county, municipality, urban growth boundary and voting precinct boundaries),
- community (such as neighbourhood and school district boundaries),
- environmental (such as 100-year flood plain, wetlands, steep slopes and watershed basin boundaries), and
- infrastructure (detailing garbage hauler information).

MetroMap helps Metro communicate to constituents a better understanding of 'their geography' including landform, urban form, jurisdictional boundaries and critical services such as solid waste disposal and growth management. The impetus for building this application was to assist Metro staff in determining which of various jurisdictions and special service districts a

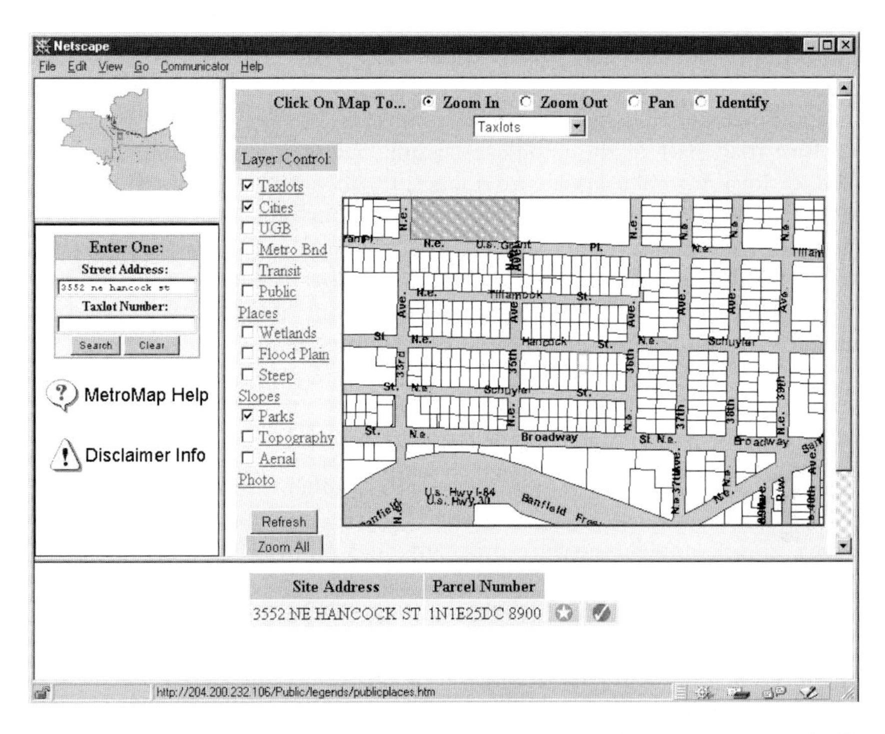

Figure 10.3 MetroMap is an interactive web-based application for accessing Metro's GIS data layers.

resident was in. Most recently, natural resources policy layers were added which mirrored the information available at the natural resource protection workshops (see above). The URL for MetroMap is http://www.metro-region.org/metromap (Figure 10.3).

10.3.2 Community impact

These tools offer alternatives to traditional ways of accessing spatial information about the region. Rather than searching through data at the census tract or other aggregate level in the various archives of local jurisdictions, individuals may find regional information already integrated into one seamless database in one place. This solution helps overcome two significant barriers to access, namely, lack of knowledge and time. Many residents are unaware of the various resources available through their local jurisdictions, and the time that is required to retrieve this data may discourage people from taking on such projects. Those who are impeded by these barriers may be empowered by the accessibility of integrated data in a central location. Individual residents, struggling non-profit organizations, and others who otherwise lack the resources to present their case through comprehensive GIS applications may convey their messages with the same precision and clarity as those who can afford these tools. The empowerment of such groups is consistent with Metro's record of making policy decisions under the guidance of public involvement.

More than half of the people accessing our web-based tools are using AOL or local Internet service providers (ISP) according to an analysis of log files. This tends to indicate that consumers of our Internet tools are accessing them from home. Other significant user communities are the education domain (edu) and local jurisdictions. A significant indicator of the perceived usefulness of these tools is that one in four of the users accessed the site more than once during a period of two weeks.

10.3.3 Technology environment

GIS technology has followed a similar path as other information-based systems over the last few decades: faster, cheaper hardware, combined with larger, more complex software programs, has allowed for a greater diversity of applications available to users. Desktop mapping tools have moved cartography away from the priesthood of highly specialized technicians and into a vocabulary that is available to uninitiated communicators.

The most significant technology trend to effect the development of a PPGIS is, of course, the rapid adaptation and growth of the WWW. Internet technology and infrastructure have been in place for years, but the metaphor of the browser – 'surfing' the web for information by following links – has captured popular attention and can be used as a solid base for application development for some time.

Web-based form input dialogue screens mimic the traditional survey form utilized by public involvement programmes. New technologies, such as JAVA, XML, database-driven pages, and Internet map servers, allow the addition of interactive spatial information to be gathered in the context of the traditional form. Now a user can use the 'point-and-click' simplicity of web-based forms to do a 'show and tell' about a particular location in space.

10.4 THE MAPMAKER'S DREAM

> In James Cowan's 1996 national bestseller, *A Mapmaker's Dream*, a fictional 16th century Viennese cartographer chronicles his attempt to create a 'mappa mundi' – a perfect map of the world – without ever leaving his monk's cell (Cowan 1996). Through traveler's stories and items that were brought back from unknown lands, the cartographer, Fra Mauro, finds that mapping perfection lies in capturing more than the physical geography of the world. He wrestles with how to map the essence of the places that made up the world, the cultures, the wonders, the 'feel' of the locations that visitors had experienced.

The goal of Metro's PPGIS long-range programme is to fulfil the 'mapmaker's dream' by providing meaningful information to residents about their neighbourhoods, communities and the region. Additionally, the programme will engage them on key issues/values, gather specific feedback and remind participants how these values relate to regional policies/strategies.

Through its previous experience with PPGIS programmes and investments in database and Internet technology, Metro is now in a position to turn the dream into reality. Metro's RLIS Lite has hundreds of data layers that empower users to become informed about the region. A simplified form has become available on the Internet. E-commerce functions and other interactive applications have been added recently. Staff on Metro's Growth Management Services Department have successfully supported use of GIS information at public events and will serve as a technical resource for broader availability of these tools in the future.

Besides being a powerful educational tool, Metro's 'Mapmaker's Dream' programme will create new 'value' data layers based on feedback that can help Metro, its partners, and residents understand what people care about and what shared beliefs we have as neighbourhoods, communities, and even as an entire region. The 'value' data will also help residents and policy makers better understand the unique characteristics and differences in beliefs that make up the Portland region.

By making the connection between GIS mapping capability and eliciting, sorting and reporting input to key values and tradeoffs questions, Metro will have a new way of helping regional officials and local partners identify residents' values and define generally vague concepts like 'livability'. Hence, the process that was initiated by residents in 1992 to develop a long-term plan for the region will find a suitable forum that may better accommodate the diversity and depth of individual goals. This prospect continues Metro's innovative tradition of citizen-directed sustainable growth management planning. The programme will use a web-based interface to RLIS data that will take the user through a short series of questions online – or in a form that can be quickly transferred to online outlets – such as newspaper ads or mailers with a short version of the questions.

By basing the programme on an Internet platform, we can automatically capture feedback, sort for various groups/jurisdiction boundaries and provide instant feedback on individual, neighbourhood, city, county and regional summaries. Further, this approach allows us to establish a schedule for more comprehensive reports of feedback to neighbourhood, city, county and regional officials.

The types of questions we could ask through the programme:

(All of these questions could be handled through an interactive, map-based interface plus pull down follow-ups to expedite the experience for the user.)

- Where do you live? (Point and click)
- Where do you work? (Point and click) How do you get there? (Pull down menu options) How long is your commute? (Pull down menu options)
- Where is your favourite park? (Point and click) Nearest park? (Point and click)
- When you do errands or shopping, where do you go? (Point and click) How do you get there? (Pull down menu options)
- When you have visitors from out of town, where do you take them? (Point and click)
- When you want to be out in a natural environment, where do you go? (Point and click)
- Where do you go to dispose of hazardous waste (paint, motor oil or old car batteries, for example)? (Point and click) Do you recycle? If so, what? (Pull down menu options)

Another meaningful set of questions that are quickly supplied and easily managed as data would ask for a form of a 'Livability haiku' (five words or less) to each of the following questions:

- What do you like best about your home?
- What do you like best about your neighbourhood or community?
- What do you like best about this region?

This 'show and tell' interface allows residents to provide general spatial information quickly on maps without having to provide name, address or other information that many feel may infringe on their privacy. The 'where do you live' question allows Metro and its local partners to generally track where the input is coming from and build data layers that have a spatial orientation for reporting purposes. Based on the first question, we can report to the user from where the cumulative input is taken from – their neighbourhood, community or the entire region. Specific internet/GIS tools might include: an interactive Internet site to provide 'ground-truth' information on urban habitat in and around an individual's home and neighbourhood, a real-time photo library of shots from around the region of habitat and open space areas, and educational programmes for children involving field survey and restoration work.

10.5 CONCLUSION

As Fra Mauro, the central character in *A Mapmaker's Dream*, finally achieved his goal of mapping the *essence* of the world, Metro hopes to map the intangible, the values and dreams of the region's residents. Though Fra Mauro's 'mappa mundi' was lost, we hope to share this vital information with those who participate and the community leaders who represent them.

The primary effect of providing these tools is to enable residents who have been previously involved in some level to have a more *active voice* in regional policy decisions. Policy-makers are given a more detailed picture of their customers'/constituents' feelings about an issue, a place, an idea. Any community of interest that has basic access to the tools is given the opportunity to move 'up the pyramid' both in terms of awareness as well as involvement.

This makes for a more efficient communication model, using fewer resources than more intense face-to-face communication models. A public workshop is considered a success if 60 people attend, while a website on the topic can reach 6,000 people a week. Properly designed, the site can provide a mechanism for feedback and may remove the perception of policy decisions being done in a 'black box'.

We recognize that this approach does not necessarily engage those communities comprising the larger foundation of the communications pyramid model. This group, by definition, is unengaged. In order to reach the broader audience, this tool must be tied to a major communications effort. Perhaps such a campaign would capture the attention of uninvolved groups. A well-designed experience with the web tools may motivate residents of this community to become more engaged. However, the strength of this programme lies in its potential to empower active residents with more powerful tools.

REFERENCES

Cowan, J. (1996) *A Mapmaker's Dream: The Meditations of Fra Mauro, Cartographer to the Court of Venice*, New York: Warner Books.
Metro (1991) *Regional Urban Growth Goals & Objectives*, Portland, Oregon, pp. 5–35.
Metro (1992) *Metro Charter*, Portland, Oregon, pp. 1–3.
Metro (1995) *2040 Growth Concept Report, Exhibit A*, Portland, Oregon, pp. 1–17.
Metro Future Vision Commission (1995) *Report of Metro's Future Vision Commission: Values, Vision Statements and Action Steps, Preamble*, Portland, Oregon, p. i.

Chapter 11

A community-based and collaborative GIS joint venture in rural Australia

Daniel H. Walker, Anne M. Leitch, Raymond de Lai, Alison Cottrell, Andrew K. L. Johnson and David Pullar

11.1 INTRODUCTION

Traditionally, the power to make decisions for natural resource use planning and management in Australia has been vested with regulatory authorities. However, sustainable resource use and participative democracy have emerged as increasingly influential paradigms since the 1950s. More recently, significant changes have occurred to involve the community in the decision-making process (e.g. McKenna 1995) that have challenged assumptions about requirements for sustainable resource use and, in particular, about the role of technocrats, resource users, and the broader community.

In Australia, natural resource management and rural development policy over the past decade has been underpinned by a rhetorical move toward participatory resource use planning (Dale and Bellamy 1998). This puts Australia at the forefront of international experience. The key feature of a participatory approach to planning is control of the information, evaluation, and decision-making process. In this type of approach, the community is responsible for developing a planning strategy and must have the capacity to undertake environmental analysis and evaluation.

Community-based decision-making represents a change in the organization and operation of information systems. To participate effectively, stakeholders must have:

- access to information pertinent to resource use planning,
- access to analytical tools required to make effective use of that information,
- a capacity to use the analytical tools and data sets, and
- a legislative and institutional environment that fosters effective participation.

Recent advances in information technology such as GIS have brought new opportunities for improving local capacity and participation in planning. As a result, community groups (rather than special interest groups) across Australia have driven a number of initiatives to create community resource information centres. Fostering effective use of GIS amongst a broad range of stakeholder groups and in the broader community requires investment in people as well as in data integration and provision. Community-based collaborative joint ventures can achieve both these objectives. This chapter reports the evaluation of a community-based, collaborative joint venture in tropical Australia and, on the basis of this experience, presents a set of principles for similar ventures elsewhere.

11.2 CASE STUDY: THE HERBERT RESOURCE INFORMATION CENTRE (HRIC)

11.2.1 The region

The Herbert River catchment drains a 10,000 km^2 area in Australia's tropical northeast into the Coral Sea (Figure 11.1). Large areas of the catchment contain natural vegetation, although approximately 35–40% of the coastal lowland has been cleared for crop production or pastures. The catchment has a population of approximately 21,000 people and is bounded by two World Heritage areas: the rainforests of the Wet Tropics on the steeper slopes of the central catchment, and the Great Barrier Reef immediately adjacent to the catchment. A plethora of government and statutory industry agencies claim, or are assigned, responsibility for managing different aspects of the catchment and a number of agencies provide research and development outputs.

This area has experienced strong economic growth in the agricultural and tourist sectors. The sugar industry dominates the local economy, having produced A$235 million worth of sugar from 1996 to 1997. However, the sugar industry may have significant environmental impacts on the Herbert catchment (Johnson *et al.* 1997). Riparian vegetation on stream banks and large areas of riverine rainforest have been removed in cane growing regions. Coastal wetlands, which provide important wildlife habitat and form an integral part of the hydrological regime, have also been cleared; soil erosion is a potential threat to long-term productivity. Diffuse source pollution may generate water quality problems in both ground and surface waters, including the area around the Great Barrier Reef.

Growing concern about potential environmental impact is balanced by a recognition of the regional and national importance of an economically vibrant sugar industry that is internationally competitive. To achieve ecological and economic sustainability within the Herbert catchment, effective means are required to manage and reconcile industry imperatives with the requirements of other users of the catchment (including conservation and environmental services). Recognition of such issues has led government agencies in Queensland to implement integrated approaches to resource management to avoid the environmental and social damage sustained by land-use conflicts. In practice, the effectiveness of these initiatives is often constrained by:

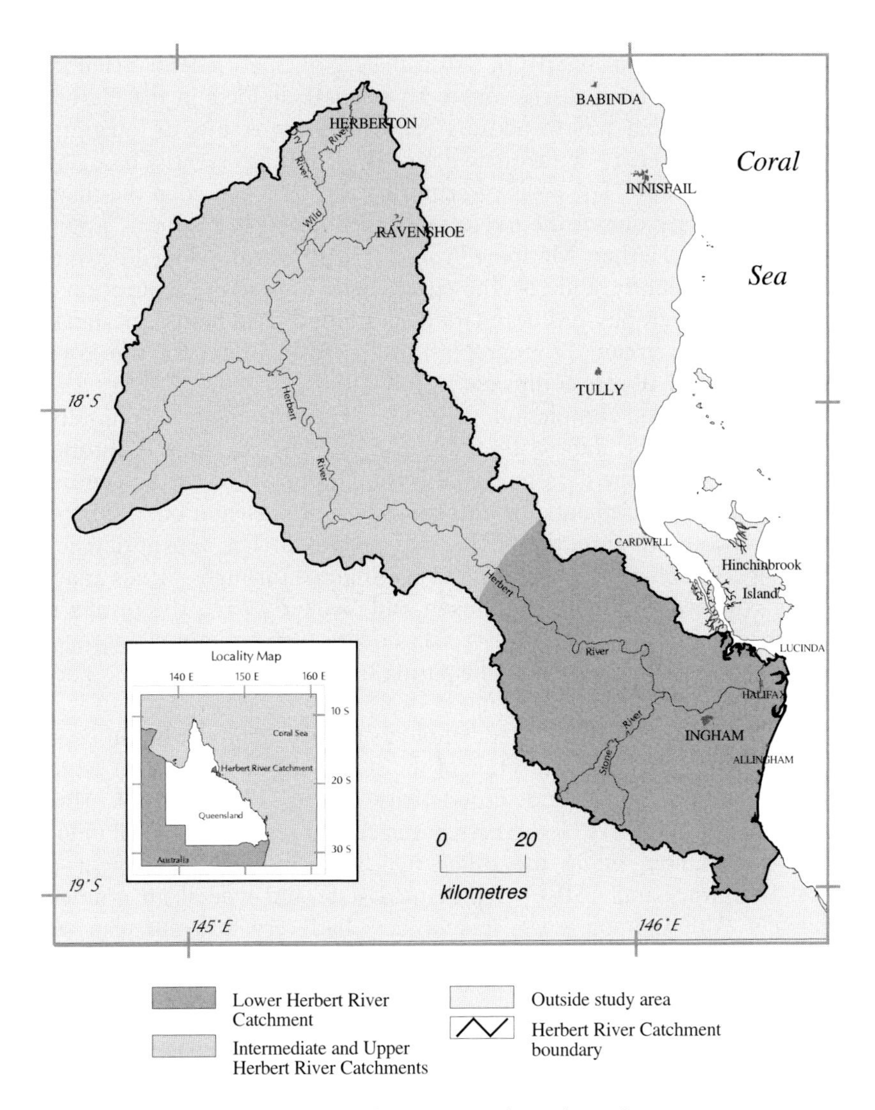

Figure 11.1 The Herbert River catchment in northern Australia.

- the paucity of data at spatial and temporal scales relevant to decision-making,
- poor coordination or communication between participating stakeholders,
- limits to the data processing and analytical capabilities of participants in the decision-making process, and
- poor understanding of key issues in sustainable resource use.

11.2.2 Creation of the HRIC

In mid-1993, scientists from the Commonwealth Scientific and Industrial Research Organisation (CSIRO), Australia's principal federal scientific research agency, initiated discussions with key stakeholders in the Herbert catchment. Their goal was to address one of the constraints to integrated catchment management – inadequate data – by acquiring essential base data at a scale of 1:10,000. The costs of acquiring this data exceeded the financial capacity of any one of the interested stakeholders. In response, a joint venture called the Herbert Mapping Project (HMP) was developed between 11 industry, community, and government agencies to fund the acquisition of digital orthophotography, cultural data (e.g. utilities, farm boundaries), natural features (e.g. streams, topography) and cadastral data for the lower catchment. The HMP was completed in July 1996.

As the HMP neared completion, it became evident to many stakeholders that the utility of the data collected could only be maximized through advanced analysis of the data in digital form. GIS provided the best means of satisfying the requirements for data analysis and presentation. A further collaborative joint venture, the HRIC, was proposed. The appropriateness and viability of such a joint venture was investigated through a needs analysis and a cost-benefit analysis (Johnson and Walker 1997), the results of which suggested that a collaborative GIS facility suited the organizational characteristics of the potential participating organizations and was a good public and private investment.

Based on this information, six stakeholders in the catchment began negotiations of a formal agreement. Four of the stakeholders (CSR Sugar Mills, Herbert Cane Protection and Productivity Board, Hinchinbrook Shire Council and Canegrowers Herbert River Executive) represented local industry and community, while the other two (Queensland Department of Natural Resources and CSIRO) represented state and federal government respectively. Although these six stakeholders had very different charters, organizational structures and cultures (some were even engaged in legal disputes with each other at the time) they were brought together by a desire and need to improve their business through better management of resources. In August 1996, a 10-year collaborative agreement was signed by the stakeholders to formally establish HRIC. The agreement secures the support of the stakeholders and binds them to uphold HRIC's status as a non-profit, community-based, collaborative GIS facility designed to support both economic and ecologically sustainable development in the Herbert catchment.

11.2.3 The nature of HRIC

HRIC is a catchment-based GIS facility that supports management of natural resources in the Herbert River catchment by providing and allowing access to geographic information, GIS tools, and expertise. The organization is intended to facilitate a common geographic view of the catchment and enable synergistic planning amongst the six HRIC stakeholders and the community. The HRIC also acts as a conduit for delivering research products to local decision-makers.

The four HRIC community stakeholders provide funding for HRIC. The two external stakeholders (CSIRO and the Queensland Department of Natural Resources) provide matching in-kind contributions such as data and technical and professional support. Two full-time GIS specialists staff HRIC, providing expertise and skills to facilitate the collection, storage, maintenance, and analysis of natural resource data. They ensure the products of these activities are delivered to HRIC stakeholders, provide consulting services and project management skills, and act as a conduit for the transfer of relevant research and development products. HRIC staff also build GIS capacity in the region by assisting stakeholders to implement GIS as part of their business operations, and promote improved communication and collaboration between HRIC stakeholders.

In addition to the active participation of community stakeholders, the community orientation of HRIC is demonstrated by a strong schools programme and documented use of HRIC's services by a range of community organizations including clubs and local Aboriginal representative bodies. In this sense, HRIC builds on rural Australia's strong history of active community and representative groups that play a key role in local politics and governance.

The objectives of HRIC are:

- improved quality of data available for the Herbert catchment,
- improved access to data,
- better-informed decisions in planning and implementing data collection and use projects,
- better-informed decisions in natural resource management, and
- improved collaboration.

11.2.4 HRIC structure

HRIC is a distributed cross-organizational corporate GIS (Figure 11.2). The organization offers a bureau service in its central office, and also provides

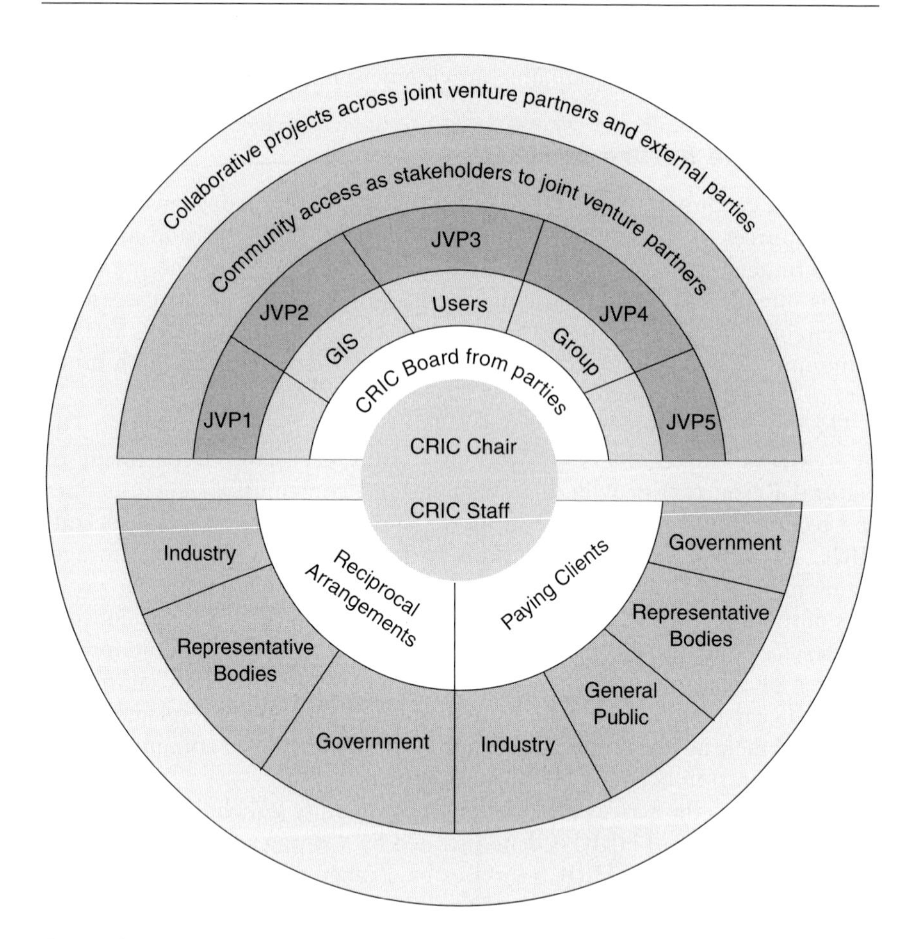

Figure 11.2 The structure of the Herbert Resource Information Centre.

a 'seat' at each of the partner sites to provide for their local requirements. In addition to the Centre staff, the collaboration involves 35 active GIS users who undertake project work under the coordination of HRIC. This structure enables small, project-based collaborations between individual joint venture partners and others, with HRIC staff providing project management and facilitation.

The Centre Manager plays a leadership and managerial role, and reports to the Board, which provides strategic direction to HRIC. Board members represent the range of GIS users in each of the joint venture partnerships. Within each of the partnerships there is a GIS group responsible for planning GIS work and implementing GIS as an enterprise system. The joint venture partners also represent a wide range of members of the community in their roles as ratepayers, taxpayers, farmers, etc.

There are other collaborators who, while not joint venture partners, have an ongoing relationship with HRIC through reciprocal data-sharing agreements. These collaborators include government departments, representative bodies, and local businesses.

11.3 EXAMPLES OF HRIC PROJECTS

The following projects illustrate the effectiveness of HRIC's collaborative approach.

11.3.1 Cane block mapping

The two sugar mills in the Herbert district were using inefficient (two years out of date) and inaccurate (by up to 80 m) land-based mapping methods to map farm blocks of sugar cane. HRIC and four of its joint venture partners collaborated to have the district photographed using highly accurate stereo-plotters. HRIC trained non-GIS specialists to subdivide the resulting blocks into cane blocks and to add relevant attributes to the map. This project saved A$1 million and nine years of work for the sugar mills. It also provided the four joint venture partners with a data set that met the needs of every partner, and provided the community with a core data set that has a wide range of uses including estimating cane crop, locating cane train sidings, positioning rubbish bins, valuing land, differential rates analysis, and mapping mosquito sites.

11.3.2 Use of spatial data by the Herbert Shire Council

Staff from HRIC facilitated a strategic GIS planning session with Herbert Shire Council staff from all levels and developed a three-year plan for the use of GIS by the Council. From this plan, an action list is developed annually with tasks and responsibilities clearly outlined. This structured approach to the application of GIS has enabled the Council to engage in many projects, including urban asset mapping.

11.4 EVALUATION

11.4.1 Objectives

A three-year evaluation programme was established at the commencement of HRIC in order to demonstrate rigorously the impacts of the initiative and derive lessons from its establishment. The objectives of HRIC were explicitly addressed during the evaluation, as were less tangible aspects of the project such as changed perceptions, attitudes, understanding and behaviour as a consequence of involvement in HRIC, particularly in relation to collaboration between groups.

11.4.2 Methods

The HRIC evaluation was conducted using qualitative research techniques (Denzin and Lincoln 1994; Patton 1987). Each year for three years, individual, face-to-face interviews were conducted with key participants in HRIC. Nineteen people were interviewed in February 1996 (six months before completion of the joint venture agreement), and follow-up interviews were conducted in February 1997 (19 interviewees) and March 1998 (17 interviewees). A total of 41 individuals were interviewed over the three-year period, with a core group of seven individuals who were interviewed all three times. Those interviewed included:

- HRIC staff,
- all key participants involved in the establishment of HRIC,
- all people from the partners who were involved in the operation of HRIC (this set evolved over the three years), and
- people external to HRIC – i.e. those who were not direct users of HRIC and those who had been involved originally but had subsequently reduced or terminated their involvement.

Interviews were conducted by two researchers from CSIRO who had not been involved in the establishment of HRIC. The interviews were semi-structured, with 30 topics used as a guide for discussion rather than as structured questions. The issues addressed in the survey are summarized in Table 11.1. In the first round of interviews, anticipated impacts were elicited. In the second and third rounds, anticipated and actual impacts to date were elicited. Each interview took approximately 90–120 minutes and was tape-recorded. After each set of interviews, interviewee responses were transcribed and collated. At the end of the three-year period, the entire data set was entered into the NUD*IST qualitative data analysis package (QSR 1997) and tagged against key evaluative criteria.

11.4.3 Outcomes

During the two and a half years of formal operation of HRIC covered by this evaluation, HRIC and its partners had collected, collated, and synthesized data from the catchment for a high quality spatial database. HRIC resources had been widely used by individual partners to plan infrastructure developments, assess resource bases and integrate monitoring activities. Direct (private) benefits had accrued to each of the joint venture partners. Specific outcomes corresponding to each HRIC objective are listed below.

- Improved quality of data available for the Herbert catchment and improved access to that data:

  'Totally replaced and enhanced previous data.'
  'A significant impact on data access...Not only have we been accessing data, but government agencies as well.'

Table 11.1 A summary of issues covered by the evaluation

Type of issues	*Description of issues*
Operational impacts	• Intentions in using new data sets that became available, implications for existing data, constraints to use • Impacts of involvement in HRIC on data availability, data collection, data storage, data access, complexity of decision-making, efficiency of decision-making, quality of decisions made, presentation of decisions • Impact of HRIC on the resolution of resource management issues within the catchment • Impacts of participation in HRIC on the types of activities in which the agency is involved
Evaluation of process	• Importance of HMP in triggering HRIC • Constraints to use of HRIC • Impact on other agencies • Use of HRIC by non-partners and impacts on those users • Interaction with other organizations: changes to general levels of frequency of interaction; understanding of the objectives of each agency; understanding of the constraints under which each agency operates; understanding of the data needs of each agency, willingness to work with the other agencies, nature and process of interactions, confidence in other groups • Evaluation of the dynamics of the process (key participants, positives, negatives)
Changes in understanding	• Awareness of the quality and availability of data • Credibility of data resources • Understanding of the limitations associated with spatial data • Understanding of data resources used by other groups in the project • Understanding of the data needs of other groups • Most important things learnt from involvement in HRIC • Understanding of the tractability of resource management issues • Understanding of the quality and limitations of data; awareness of the availability of data

Data access improved dramatically. Participants became more aware of the range of data available and had access to all data except confidential commercial data. There were still significant differences in perception between individuals regarding the general quality of data available in the Herbert, particularly between active and less active users. Nevertheless, many participants came to better understand the limitations of key data sets, including the implications of scale for the usefulness of data. Greater understanding of the data combined with a knowledge that all parties shared common data resulted in higher levels of confidence in using the data.

- Better-informed decisions in planning and implementing data collection and use:

  'Changed from pen/paper in drawers and files to digital form...'
  'The staff expertise really came through in the technical advice on how to go about our project.'

  Although processes for data collection were only moderately impacted for most parties, the need for data sharing and compatibility had a significant impact on data storage and management, both collectively and individually. For some activities, such as field surveys and orthophoto and satellite imagery, radical changes in data collection occurred. In general, although interviewees saw compatible data collection and storage as important, other factors – such as the opportunity to discuss differences in interpretation of shared data – were considered more important.
- Better informed decisions in resource management:

  'Efficiency and quality of decisions gets better...'
  'Without HRIC could not make decisions for (sugar) crushing agreement effectively because we did not know the exact area under cane.'

  Formal spatial analyses were used in planning decisions, often with a substantial cost savings, and resulted in a perception that decisions were as good as, and frequently better than, those achieved using traditional procedures. GIS-based products were increasingly used in negotiations regarding resource-use, although the inclusion of resource management issues that cut across sectors and stakeholders had not yet been achieved. Achieving this objective, however, was considered only a question of time rather than a function of more fundamental constraints.
- Improved collaboration:

  'HRIC has made me more aware of the...way people think and other people do business...drawn into a lot of projects.'
  'Everyone's willingness to share...it has changed attitude...not "what's mine is mine" but what's ours is ours'...to get a large public company and a shire council to work together is incredible...'

In terms of motivation for involvement and strategic direction, many participants initially saw HRIC as a data source and a means of cost sharing. Over time, however, HRIC was increasingly seen as a significant force for changing planning processes, sharing and developing skills, and brokering projects rather than as simply a technical service. The key assumption of the project – that a collaborative joint venture was an appropriate mechanism for fostering broader uptake of GIS technologies in the catchment – was perceived as having been borne out. For the people interviewed, involvement in HRIC has meant additional work, meetings and the need to quickly develop new skills in emerging fields. In some cases, it has also meant developing working relationships where none would have existed previously. All participants saw this as a positive experience for themselves and their community, and took pride in HRIC's achievements.

Equally significantly, the collaborative nature of the initiative had important impacts. Willingness to work together increased amongst the Centre's partners, and external use of HRIC by businesses and the broader community began to occur (although issues of user payment for HRIC services and third party data remained a complex challenge). Although improved collaboration between agencies was not reported for all the combinations of partners, in no case were relationships reported to have worsened. Some concern was expressed that groups that did not become formally involved in the Centre had become marginalized.

11.5 DEVELOPING BEST PRACTICE

The success of HRIC has attracted interest and enquiries from many parts of Australia. This prompted researchers from CSIRO to review the structure, history, strengths and weaknesses of HRIC in order to derive a generic model for a Collaborative Resource Information Centre (CRIC). It was believed this model could be used as a starting point by regional stakeholder groups interested in establishing their own resource information centres in a rural regional setting.

Funding from the Australian Surveying and Land Information Group enabled development of an information kit by staff from CSIRO and HRIC. The kit included a history and list of achievements of HRIC, as well as guidelines and resources for developing a CRIC. This kit was assessed at a workshop of representatives from ten potential CRICs and State and Federal agencies involved in facilitating the use of spatial data. The final information kit was used as the basis for a series of workshops including seven regional consortia from Queensland and Western Australia, policy and planning groups from private enterprise, local government, State agencies in Queensland, South Australia, and Western Australia. A complete version of the information kit is available at http://hric.tag.csiro.au. A summary of the CRIC model and the principles developed in these documents is presented below.

11.5.1 The model for a CRIC

A CRIC is a collaborative joint venture that seeks to provide access to information for use by partners of the collaboration and external groups or indi-

Table 11.2 Principles of a CRIC

Principle	*Description*
Joint venture	A CRIC is a joint venture that may include local, state, and federal agencies; businesses; and community and industry representatives. The CRIC model derives its advantages primarily from the data-sharing and cost-sharing that result from collaboration. A CRIC also improves linkages and working relationships between stakeholders, and thus it is beneficial to include a broad range of stakeholders within the partnership. A CRIC that is broadly representative and includes stakeholders with diverse (even conflicting) perspectives is more likely to become trusted and balanced than one with a restricted and partisan membership
A team approach	A CRIC is comprised of both the Centre staff and Management, and the GIS users within the stakeholder organizations. Staff within joint venture partners undertake much of the application of GIS to their core business. Investment in training and mentoring for staff within the joint venture partners as well as CRIC staff brings many disciplines and perspectives into a broad-based and expanding team responsible for implementing the CRIC's objectives
Independence	As collaboration is the life blood of a CRIC, it is important that the CRIC remain independent of joint venture partners. This might mean employing professional staff directly within the Centre, having an independent Chair to the Board, or being located independently of joint venture partners. The Centre staff face a fundamental challenge in maintaining equally effective working relationships with all the joint venture partners, while Board members need to find an appropriate balance between representing the interests of their organization and seeking to foster the best interests of the joint venture
Community ownership	The CRIC structure provides for a high level of community ownership. Community access and acceptance greatly increases community trust in and acceptance of the data sets generated, and removes a significant source of uncertainty with regard to analyses of resource use conflicts. If the joint venture partner membership of the CRIC is broad and represents a mix of organizations, a large proportion of the community should have effective involvement in the initiative through one or more of the joint venture partners

Private and community benefits	A CRIC meets the business needs of joint venture partners and has a charter for broader community benefits. A community-focussed approach that integrates private and public good by including government, private industry, small businesses, and local communities is highly desirable in rural Australia. Each sector of the community is dependant to a greater or lesser extent on the vitality of the other sectors. Thus, provided joint venture partners are getting enough out of their CRIC participation to justify their investment, benefits to other sectors in the community are to their ultimate advantage
Dual roles in data management and capacity building	A CRIC has a dual role: in managing and sharing data, and in building the capacity of joint venture partners and other groups to make effective use of that data. This approach can be usefully contrasted with a 'data warehousing' model, in which more data is made more widely available, but interest groups are not given assistance in using this new data. A CRIC that provides information analysis services rather than building capacity for analysis within partner organizations will fail to capitalize on the opportunities of the CRIC model
Linkages and roles	A CRIC fosters improved linkages between stakeholders and may change the roles traditionally assumed by the stakeholder organizations or the institutional arrangements within an area. For example, a CRIC may facilitate constructive relationships between groups where traditionally non-existed. This can be a positive development if it results in improved institutional roles and arrangements in an area, but might threaten some groups or individuals who are comfortable under current arrangements
Data exchange	A CRIC ensures data are exchanged between stakeholders which enables development of extensive spatial databases that meet a broad range of joint venture partner requirements and produce value-added data sets by bringing together existing complementary sets. The biggest threat to the sharing of data sets has come from the revenue potential of data. The market for data sets in rural areas is limited, however, and not sharing data because they may be sold by others is often a poor business decision
Best practice data management	A CRIC plays a central role in managing the aggregated data sets of the Centre's joint venture partners and, under some circumstances, those of other stakeholders through reciprocal data sharing arrangements. Data sets must be managed to high professional standards. A CRIC needs to maintain spatial data and meta data directories and comply with industry standards such as the Australian Surveying and Land Information Group (AUSLIG) Australian Spatial Data Infrastructure (ASDI) standards. A CRIC also has a role in helping others manage data better, allowing data to be used appropriately and future data collection exercises to be prioritized. Developing such an understanding among CRIC stakeholders is essential to best practice data management and use
Project brokering	A CRIC has a limited role in project work, but a key role in brokering projects within and between partners and external agencies. Overcommitment to specific long-term project work can

Table 11.2 (Continued)

Principle	*Description*
	undermine the ability of CRIC staff to support projects of partner agencies. CRIC staff should assume the role of project coordinator, facilitator and manager, while implementation should be undertaken by staff within the joint venture partners. This arrangement builds expertise and ensures that the skills of the Centre staff are as widely and effectively used as possible
Funding	A CRIC is self-funded. The financial input into a CRIC from the joint venture partners should be accompanied by expectations for specific results and accountability. A CRIC that is externally funded as a public service is unlikely to be dynamic and vibrant. This is not to say that there is no role for external seed funding to get a CRIC off the ground. Similarly, subsidation across joint venture partners is, in general, worth avoiding. Although apportioning benefits and working out appropriate levels of contribution is complex, the general principle of proportional contribution is important
Lifespan	A CRIC is a medium to long-term commitment; ideally planning should be on a ten-year timeframe (certainly not less than five). The task of collecting, synthesizing, and managing data and then building capacity within partner organizations to make better use of this data takes time. For this reason, it is hard to envision a CRIC realizing its full potential in less than five years. However, a CRIC need not be established as a permanent and ongoing organization. As the skills within the partner groups grow and technologies evolve, it maybe appropriate at some point to replace the CRIC with a simpler data-sharing structure.

viduals in order to improve resource use, planning, and management. Providing access means building the capacity to make effective use of the data as well as making data available. Thus a CRIC plays a role in data acquisition, synthesis, dissemination, and management. It also plays a key role in facilitating the use of that data by the joint venture partners and the broader community. An effective CRIC will meet the private needs of the joint venture partners while also promoting broader use by the community.

11.5.2 Principles of a CRIC

The CRIC model is based on a series of principles that outline its structure and function. These are described in Table 11.2.

11.6 CONCLUSION

Prior to the developments reported here, GIS had not been widely adopted in rural Australia. Perceptions existed that data are too expensive to collect and maintain, and that GIS required human and financial resources beyond the reach of many groups and communities. In addition, the business opportunities provided by GIS in a rural setting were not clear. HRIC provides an example of GIS in a rural community delivering clear financial and social benefits. The HRIC experience provides a valuable working model for other communities in Australia. This model has received widespread attention across Australia, and can be fine-tuned as it is applied and adapted in other settings.

ACKNOWLEDGEMENTS

The authors would like to thank for their participation in this evaluation, the members of the HRIC Board, HRIC staff, and HRIC users who took part in the interviews reported here. The authors would also like to thank Anne O'Brien and Stuart G. Cowell for their assistance in conducting the interviews in this chapter. The evaluation reported here was funded in part by the Sugar Research and Development Corporation and was supported by the Cooperative Research Centre for Sustainable Sugar Production. The Australian Surveying and Land Information Group, through the Spatial Data Infrastructure Partnership Program in 1999, provided financial support for the development of the Collaborative Resource Information Centre (CRIC) Model.

REFERENCES

Dale, A. and Bellamy, J. (eds) (1998) 'Regional resource use planning in rangelands: an Australian review', Land and Water Resources Research and Development Corporation Occasional Paper 06/98.

Denzin, N. K. and Lincoln, Y. S. (1994) *Handbook of Qualitative Research*, Thousand Oaks: Sage Publication.

Johnson, A. K. L., McDonald, G. T., Shrubsole, D. A. and Walker, D. H. (1997) 'Sharing the land – the sugar industry as a part of the wider landscape', in B. A. Keating and J. R. Wilson, *Intensive Sugar Cane Production: Meeting the challenges beyond 2000*, CAB International, Wallingford, New York, pp. 361–380.

Johnson, A. K. L. and Walker, D. H. (1997) 'Evaluating a corporate geographic information system (GIS): a case study in a coastal rural catchment', *Australian Journal of Environmental Management* 4(2): 112–129.

McKenna, B. (1995) 'Community participation in local government – a research report and critique', Report by the Australian Centre for Regional and Local Government Studies for the Office of Local Government, University of Canberra, Canberra, Australia.

Patton, M. Q. (1987) *How to Use Qualitative Methods in Evaluation*, Newbury Park: Sage Publication.

Qualitative Solutions Research (1997) *QSR NUD*IST User Guide*, Second Edition, Qualitative Solutions Research, La Trobe.

Chapter 12

Geographic information systems in the environmental movement

Renée E. Sieber

12.1 INTRODUCTION

Geographic Information Systems and related spatial technologies have become important tools for land management agencies to administer resources and protect the environment. Increasingly environmental and conservation non-profits use GIS in their own activities to better understand and advocate for their communities. Current applications range from inventories of spotted owl locations, thematic comparisons of toxic lead and poverty, and models of sustainable forest harvesting, to scenarios and solutions for urban sprawl. GIS, like many computing applications, holds great promise for environmental non-profits to maximize their traditionally limited resource base. Just as the word processor and desktop publishing have helped to publicize causes, and the Internet has provided an avenue for mass mobilizations, GIS could enable organizations to present a visually compelling image of an issue and quickly analyse data from disparate sources. Over time, GIS has become increasingly affordable spatial digital data seems ubiquitous. GIS could provide a critical implement to groups struggling to impact politics and empower environmentalist for social change.

Given their fragile resource base, particularly among the grassroots, environmentalists may struggle with system adoption and data acquisition. The GIS literature has long identified inadequate resources as an impediment to successful implementation (Croswell 1991). The PPGIS literature is even more pointed: GIS adoption might subvert the grassroots' *raison d'être*. This author presumes that, given sufficient information, non-profits have the right to make their own decisions about adoption. This chapter investigates the use and value of GIS from the vantage of activists within the environmental movement. It describes applications of GIS and computing technology by environmentalists. Their usage is compared to the literature on GIS diffusion, implementation, spatial data acquisition and sharing. How are groups applying GIS to their goals and missions is explored. Based on this information, I frame the use and value of GIS with recommendations on the appropriateness of GIS adoption by individual environmental organizations. Because of their tradition of scientific analysis and cartographic use, the environmental movement is uniquely positioned to take advantage of GIS capabilities and consequently provides a guide to GIS adoption in other social movement groups.

This chapter summarizes longitudinal research that began in 1996 (Sieber 1997a) and has continued with follow-ups in 1998 and 2000. The case study research assessed GIS applications, implementation, spatial data acquisition and sharing, and the contribution of GIS to organizational goals and missions. Work was conducted in California and the Pacific Northwest. All cases are in California.

12.2 APPLICATIONS OF GIS AND COMPUTING TECHNOLOGY

To understand how environmentalists and other non-profits might be applying GIS, a mail survey was sent to 100 environmental groups. The mail survey is more fully reported in Sieber 1997a. The mail survey found that groups are able to acquire and install GIS despite scarce or uneven resources. Twenty-seven per cent of respondents used GIS; 60 per cent of those had no paid staff (number of paid staff was used as a surrogate for financial resources (Snow 1992)). They need not purchase in-house systems, employ staff, or train workers; they can amass the initial hardware and software through grants or utilize a member's computer. In general, the survey results suggest that the environmental movement has laid the foundation for GIS use. Most know of GIS functionality and have seen it demonstrated. GIS-using groups build upon a movement-wide base of computer use, map use/creation, digital data acquisition, and scientific analysis.

Case study research revealed the varied applications of GIS. Below are examples from the five cases.

Case 1

The Greenbelt Alliance is an open space preservation organization in San Francisco. Greenbelt has applied GIS to capture the debate of where open space land is at risk for development and how it should be protected. Figure 12.1 details one such area at risk along US Highway 101 in Silicon Valley, California (dark areas along the highway in this black and white version indicate areas at the greatest risk). So clearly does its nine county version of 'Open Space and Farmland At Risk' GIS map express its goals that staff, members, and the press have labelled it the central metaphor of the organization. The 13 staff members believe GIS output provides them a powerful

Figure 12.1 Map prepared by the GreenInfo Network for the Greenbelt Alliance showing open space and farmland areas at risk for development along US Highway 101 in Silicon Valley, California.

persuasion tool, allowing the organization to more easily attract members, funding, and support from many area elected officials.

Greenbelt is one of the first users of GIS by a non-profit advocacy group in the United States; system implementation began in 1988. As gathered from expert interviews with other environmental GIS users and comments from vendors and environmental scientists, the organization is recognized as a pioneer in GIS usage, and representatives are frequently invited by foundations and other non-profits across the United States to lecture on the subject. Its expertise and belief in the technology's potential prompted the spin-off in 1996 of its GIS skills into a separate non-profit, the GreenInfo Network. At present, GreenInfo employs five people and has used its skills, data, and linkages to build GIS capacity in more than 90 other grassroots and non-profit organizations in the region. It currently maintains a GIS and produces the maps for Greenbelt.

Case 2

The Nature Conservancy at Lanphere Christensen Dunes Preserve, Arcata, was created as a small chapter of the national land trust organization, The Nature Conservancy, to preserve 450 acres of northern California coastal

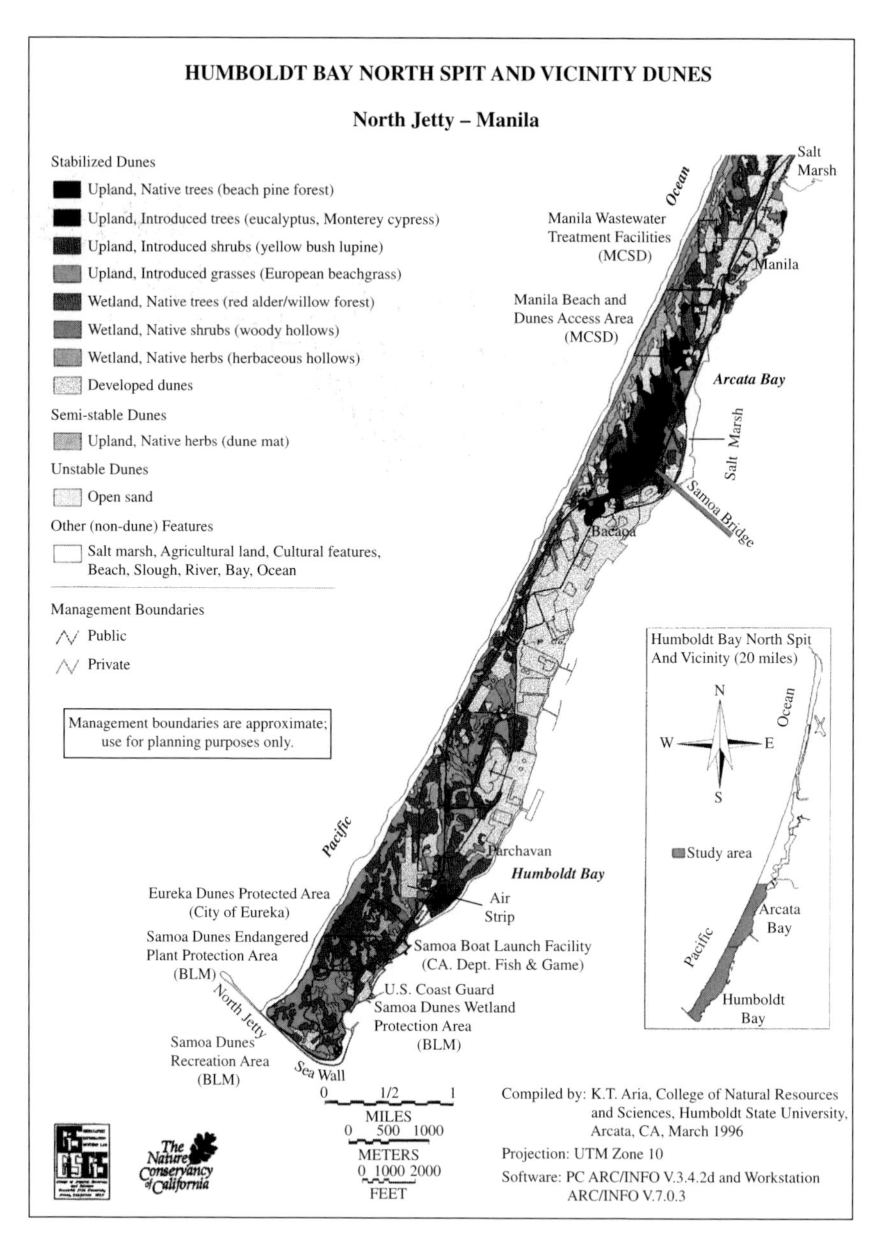

Figure 12.2 Map prepared for The Nature Conservancy – Lanphere Christensen Dunes Preserve showing dune vegetation on the Northern Spit, Humboldt Bay Dunes, California (courtesy: Travis Aria).

dunes. The one full-time director and two part-time staff have spent most of their time monitoring the spread of non-native vegetation and arranging for its eradication. Assistance on monitoring comes from various public agencies in the area and eradication is accomplished through another non-profit, the Friends of the Dunes. The director at the dunes preserve has outsourced for GIS services to the local university. Outsourcing reflects a strategy on the part of the national organizational to employ the assets of other agencies/institutions identified as better equipped to carry out an objective instead of acquiring the skills itself. In this manner, the dunes preserve can best leverage its resources (both natural and technical) to preserve habitats.

The director at the dunes preserve has utilized university students in GIS courses to track its mitigation of non-native vegetation. One application confirmed a long-held ecological growth model. With the creation of a bioregion-wide dunes database, the dunes preserve staff and university students have discovered a new way – a 'landscape level of analysis', according to its director – of comprehending the environment. The largest application is shown in Figure 12.2, and was prepared by a student as part of a masters degree. It shows both native and non-native dune vegetation in the preserve.

Case 3

Trinity Community GIS grew out of a community organizing effort in a town angrily split between loggers losing their livelihood and 'back to the land' environmentalists who had settled in the county because of the inexpensive land and expansive wilderness. This six-year-old organization has fashioned a unique approach to using GIS. Started by a Ph.D. in Landscape Ecology from University of California at Berkeley, it functions as a centre of GIS services under the umbrella of a small non-profit economic development corporation. Trinity conducts GIS research for area non-profits but also contracts for projects with local public agencies. One such project tracked 8,000 individual pesticide sprays for the California Basket Weavers' Association. Another project furnishes an economic development tool for the impoverished county; it trains unemployed loggers in global positioning systems (GPSs) and in spatial data collection.

Trinity's recent effort lies in fire safety; bringing residents together to identify locations of dense understory and brush in the forest, which create 'fuel', ladders and intensify the spread of fires. Figure 12.3 shows a map of proposed fuel breaks – reductions in brush – on Post Mountain. While simple in analysis, this represents a highly coordinated data collection and correction effort on information such as brush densities, water sources, culverts, and roads. It also serves to educate residents on the need to reduce fuel in specific areas. Work on fuel ladders, in turn, provides small diameter 'junk trees' to a refitted locally owned lumber mill.

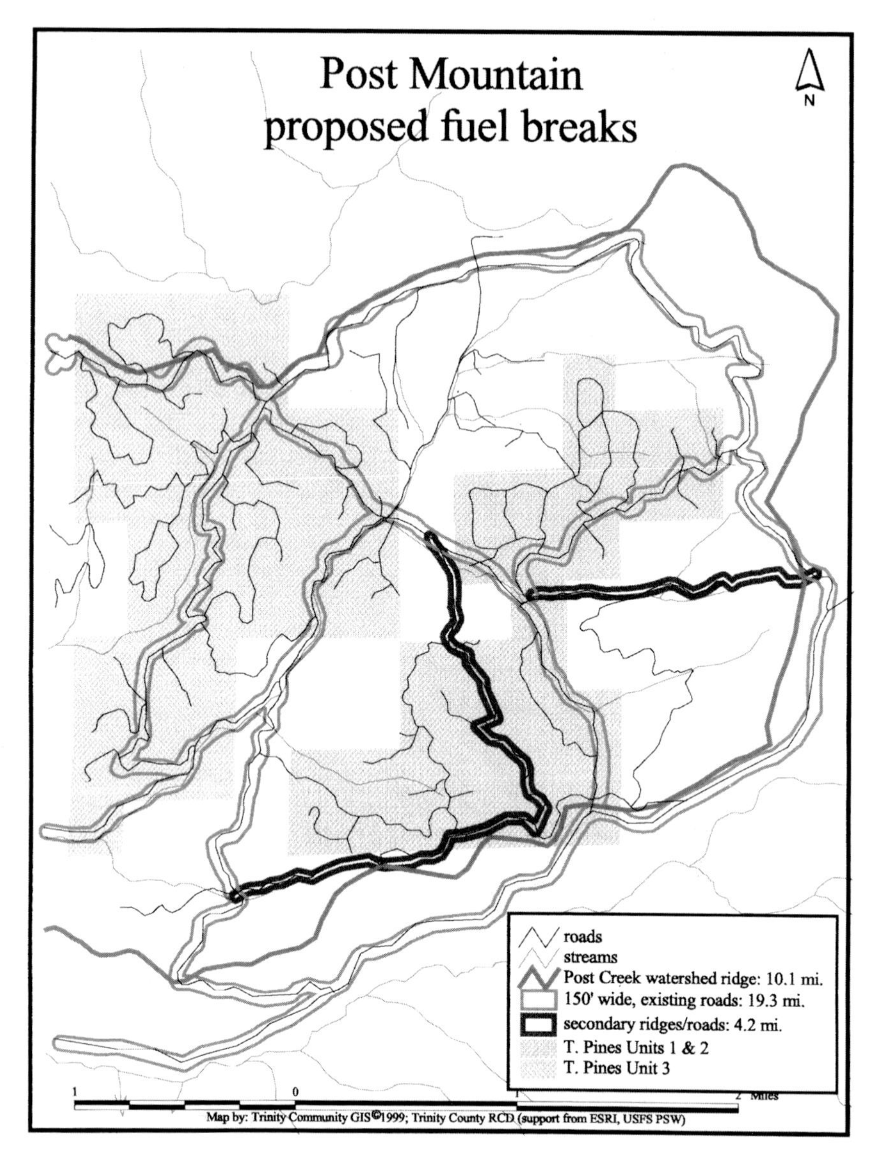

Figure 12.3 Map showing proposed fuel breaks (clearances of forest understory and brush to contain the spread of fire) on Post Mountain, Trinity County, California.

Case 4

The San Andreas Land Conservancy (SALC) is a one-person all-volunteer land trust that ambitiously covers the length of the San Andreas Fault Line and is headquartered in the director's home northeast of Santa Cruz.

Habitat protection and using GIS to enable that protection is not at the core of this individual's life; it is his life. To his endeavours, the director brings considerable knowledge: a Ph.D. in genetics, decades of activism, and intensive GIS training and experience. This is a case about a single individual with a singular ideal. That force of personality sets the agenda for GIS use and frames the agenda of the organization. As such he embodies the organization and, for him, the organization merely adds formality. He also wants to be paid for that expertise, to make both an avocation and a vocation out of his GIS expertise for environmentalism. Payment must come from the small non-profit sector, because his uncompromising belief impedes his entry into, for instance, a federal forestry career or an academic position. Far from unique, this individual's approach typifies several that I first met at the ESRI Users Conference in 1994. Theirs is an environmental entrepreneurial spirit that places them on the edges of the more institutionalized conservation organizations.

Previous to the creation of the trust, SALC's director had developed GIS applications for two organizations in southern California. To date, his GIS analyses have been strategic in preventing hunting in a wildlife refuge and protecting a key landscape linkage for area mountain lions. That mountain lion map has been instrumental in attracting new members. In SALC, the director has directed GIS toward management of the land trust and the monitoring of conservation easements.

Case 5

Located north of San Francisco in the Emerald Triangle, the *Environmental Protection Information Center* (EPIC) is dedicated to protecting the ecosystem of the old growth redwoods on both public and private lands. EPIC and its member organizations operate in a highly reactive and litigious environment. Some member organizations utilize direct action strategies and protect the trees with their bodies.[1] Most of EPIC's resources support legal fees to, for example, challenge lumber companies' timber harvest plans. Much of its activism has centered on the Headwaters Forest. Its most dramatic use of GIS was a courtroom demonstration that linked – and subsequently protected – a non-threatened species (redwoods) to an endangered species (the marbled murrelet).

EPIC employs four staff, but four of its five tries at GIS have been outsourced to other non-profits: two of which were all-volunteer. Figure 12.4 shows one of the maps done by one of these consultants – Legacy – The Landscape Connection (originally Legacy was all-volunteer but now it employs staff). EPIC continues to use GIS but GIS is not a primary concern. It currently pays one part-time staffer who manages the GIS but also must maintain the organization's website and manage the front office.

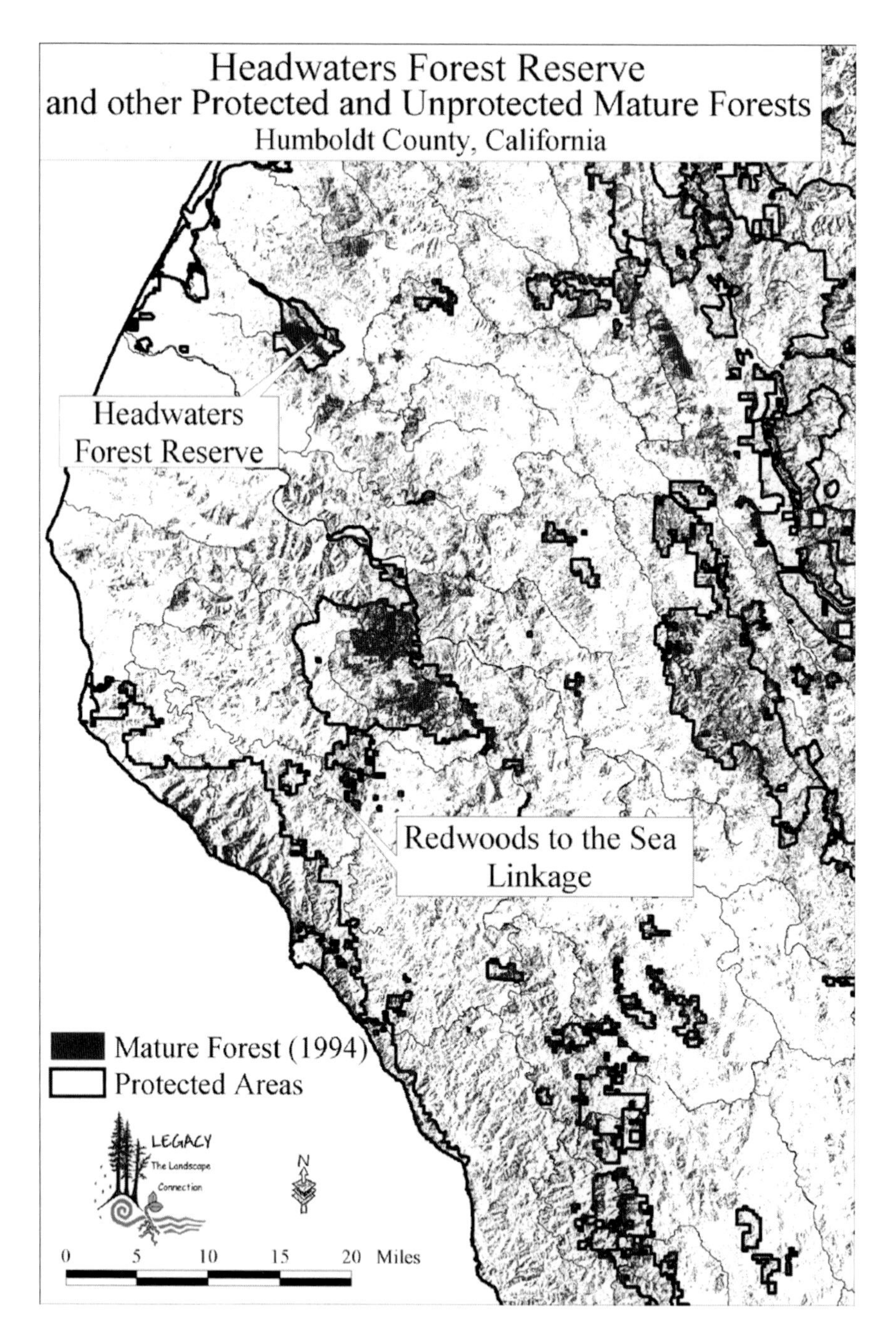

Figure 12.4 Map prepared by Legacy – the Landscape Connection showing the newly created Headwaters Forest Reserve as well as protected and unprotected mature/old growth forests in Humboldt County, California.

12.3 GIS IMPLEMENTATION

Given the demonstrated use of the technology, how was it accomplished? In general, organizations have discovered that GIS implementation encompasses far more than the purchasing and installation of hardware and software. Research in municipal, county, state governments demonstrates that successful implementation depends upon factors that include: (1) long-term upper management commitment to the project; (2) sufficient allocation of resources; (3) adequate staffing; (4) timely and sufficient training; (5) someone, called a GIS champion, who will shepherd the project from acquisition to use; and (6) organizational communication to smooth the transition to full utilization (Azad 1993; Croswell 1991; Onsrud and Pinto 1993).

The case study research demonstrated that non-profit implementation somewhat mirrors the technical and organizational issues found in governmental agencies, most significantly the need for an in-house GIS champion. Certainly, it helps to have other factors such as sufficient training and upper management commitment. For example, Trinity's director has ensured that all staffers are formally trained and it vigorously promotes GIS to other local non-profits. GIS receives considerable support in Greenbelt because its former and current directors spearheaded the GIS adoption.

However, the chief difference is that these factors, while important, are largely unnecessary to an environmental non-profit's successful implementation. The five cases were able to implement GIS because they improvised for a number of traditional resources, among them: (1) donations, primarily of GIS software (mostly from the ESRI Conservation Foundation); (2) universities (for hardware, software, data, and expertise); (3) volunteers/student interns; (4) a passionate 'hyper-champion' who may even sacrifice personal comfort for GIS; and (5) informal connections rather than formal policies. (More information on GIS implementation in these cases can be found in Sieber 2000.)

SALC exemplifies this non-profit implementation. Its director struggles to pay his rent but he would rather work part-time than compromise his activist and GIS work. GIS equipment, a SUN Sparc, was purchased second-hand. He received donated software, software support, and training from ESRI. He travels to a nearby university or to ESRI in Redlands when he needs large-format maps. Implementation has been slow but successful. Organizational issues have been minimized since the trust is small and board members are selected based on their agreement with his vision.

Notwithstanding the innovativeness of the improvisations and the passion of its champions, the capricious nature of these factors suggests that GIS remains a tenuous proposition for these organizations and their issues. All but one – SALC – arguably, the most vulnerable – has witnessed tumultuous changes, since this research began. In 1998–1999, Trinity lost two of its three full-time employees, including the founding director who possessed the strongest GIS skills. Greenbelt conducts no GIS in house now: 'they don't have the culture', according to GreenInfo's director. The Nature Conservancy sold the dunes preserve in 1999 to US Fish and Wildlife. Conversely Legacy, on the verge of collapse when this research began, has emerged as a GIS service centre with three full-time and several part-time employees. Whereas many non-profits might use GIS, instead of full implementation that could allow groups to concentrate on enhancing the sophistication of analysis and visualization, most might exist in a perpetual state of implementation.

12.4 DATA ACCESS AND SHARING

Considering the increasing availability of digital data, it might be far more efficient for organizations to acquire or share data instead of collecting it themselves. Similar to GIS implementation, sharing of spatial data can be impeded by factors such as: (1) variations in priorities and goals among participants; (2) differences in GIS resources and skills; (3) differences in the character of each organization, such as the level of bureaucratization and whether the participant is public or private; (4) differences in data quality, format, and reach; and (5) differences in organizational power stability (Azad and Wiggins 1995; Budic 1995; Campbell 1991; Pinto and Onsrud 1995). Overall, data sharing is optimized under a stable environment, with a history of prior positive relations among participants. In this complex environment of resource exchange, the arrangement policies cement these relations. Policies must establish data standards, responsibility/ownership, frequency of exchange, costs, and incentives (Calkins and Weatherbe 1995; Nedovic-Budic and Pinto 2000).

Unlike implementation, the organizational issues of data sharing appear heightened for the environmental groups. The dunes preserve has benefited from the local university's immense (40 georegistered layers, at three scales, of ten million acres) spatial database of the region's natural resources. However, it had to submit to the university regime, e.g. differing priorities of professors, cycle of the semester system, and varying degrees of student interest/work quality. Should non-profits acquire or share data with other institutions they can run risk of co-optation due to differences in power. Trinity inadvertently became a public service centre for a state agency, which itself was limited in resources or upper management directives, and was only too happy to off-load public inquiries about its data. Indeed, power differentials appear so large that all but Trinity were unable to enter into formal data sharing arrangements. Consequently, they built strength through informal arrangements and interpersonal linkages.

These variations can worsen under unstable political climates and a history of contentious relations. Trinity and the dunes preserve have interacted well with local public agencies and find data acquisition easy; SALC and EPIC are confrontational and uncompromising. As a result, SALC's director still enters much of his own data, so his data set has grown slowly and painfully. EPIC's GIS consultants have had to enter much of the necessary data since they are shut out even from county data. This makes GIS ill-suited to respond to the last minute reactionary environment in which EPIC operates.

Irrespective of the structure of data sharing, spatial data may not be designed to fit grassroots goals. Much computerized data may be available in formats incompatible to requestors' systems. Forestry data sets of Trinity County tend to consist of aggregated features (e.g. ecological units); however, Trinity needs the actual species data. Some data can be physically inaccessible; it may be, e.g. located in a distant site. EPIC's outsourcer often must travel hours north to Arcata for data. Rather than maximizing resources, non-profit attempts to acquire government data actually may consume more resources than entering the data from the beginning.

Increasingly local and county governments have recovered the cost of expensive GIS hardware through the sale of their GIS data (Dando 1992). Canada has mandated cost recovery for its government units (Roberts 1999). Often the non-digital version of this data is free or priced at the cost of its distribution medium; governments are selling the processed electronic version. For SALC, the high cost of accessing his county's cadaster presents a prohibitive obstacle to utilization.

Agencies may be disinclined to release data that can be privatized or, more importantly, be regarded as politically sensitive. The government retains the power to classify information as sensitive – for the protection of individual privacy or national security, or for its own protection in the uncertain status of liability for accuracy – or simply to withhold information it deems a threat to the status quo (Castells 2000; Epstein and Roitman 1987). The dunes preserve may easily obtain university data; however, the university was not always sanguine about distributing its database to environmentalists, since they include direct action groups. In another case, Orange County refused to release to the SALC director the digital locations of a threatened species that were part of a voluntary conservation programme between the government and the largest landowner. Refusal was based on the grounds that the director might do harm to the species. To obtain governmental computerized data, activists have engaged, and likely will continue to be forced to engage, in legal challenges (Archer and Croswell 1989). EPIC, e.g. used a motion for discovery to access digital data.

Over the years of this research, data issues have lessened as more data sets have become available. The real advances have arisen as non-profits have established their own data repositories. Legacy, in coordination with other north coast organizations, is creating a non-profit spatial data clearinghouse conforming to the standards of the Federal Geographic Data Committee. Other non-profits in the region are already referring to Legacy as a 'data center'. Peer-to-peer data sharing will likely increase as other non-profits adopt GIS or (more likely) collect spatial data to be used in database programs.

12.5 GOAL ATTAINMENT

Will GIS contribute to advancing the many and varied goals of the environmental movement? The literature presents a conflicting story on IT (detailed in Sieber 1997a) in which advocates and critics square off to paint IT alternately as political saviour to a participatory democracy or as an electronic gilded knife in the heart of social activism. GIS can improve analysis (Dangermond 1991; Parent 1989) but it might be the wrong tool (Convis 1995). It could help groups compete better in the political environment (Goldman 1991; Klosterman 1987), drawing particularly on its double ability to exploit the perception of neutrality of computers and objectivity of maps (Monmonier 1991; Wood 1992). In doing so, it may cripple groups' effective strategies and deny them their democratic foundation (Gittell 1980; Kweit and Kweit 1987; Piven and Cloward 1977; Rubinyi 1989). Environmental activists might gain methods to define meaning over their space (Castells 1983) because mapping technologies can harness the power of the image (Aberley 1993; Wood 1992). Conversely groups simply may be adopting an instrument to hide corporate power (Curry 1995; Harris and Weiner 1998; Pickles 1995). Drama aside, environmental organizations already utilize GIS and other sophisticated computing technology; their use alone places value on the diffusion of technical innovation into the movement. If authors agree on nothing else, this varied literature is unequivocal that these technologies exert a perceptible impact upon organizations.

Craig and Elwood (1998) have researched the different goals to which community-based organizations apply their GIS. They categorized the applications into four types: (1) administrative (e.g. evaluate programmes), (2) strategic (e.g. assess neighbourhood needs), (3) tactical (e.g. produce counter maps), and (4) organizing (e.g. recruit members). Case study participants were clear that GIS enhanced many of their existing goals. Greenbelt is using GIS to target new members and demonstrate their effectiveness to foundations through numbers and types of acreage saved. A map produced by GreenInfo (Figure 12.5) for the Packard Foundation shows how GIS can be used to monitor organizational overlap and highlight areas of potential coordination. GreenInfo has found GIS to be invaluable to manage the hundreds of data sets from governmental units and developers. In what one respondent at The Nature Conservancy's California office believed was a limited and highly competitive environmentalism 'marketplace', GIS helps environmental groups gain a larger piece of the public and foundation financial pie.[2]

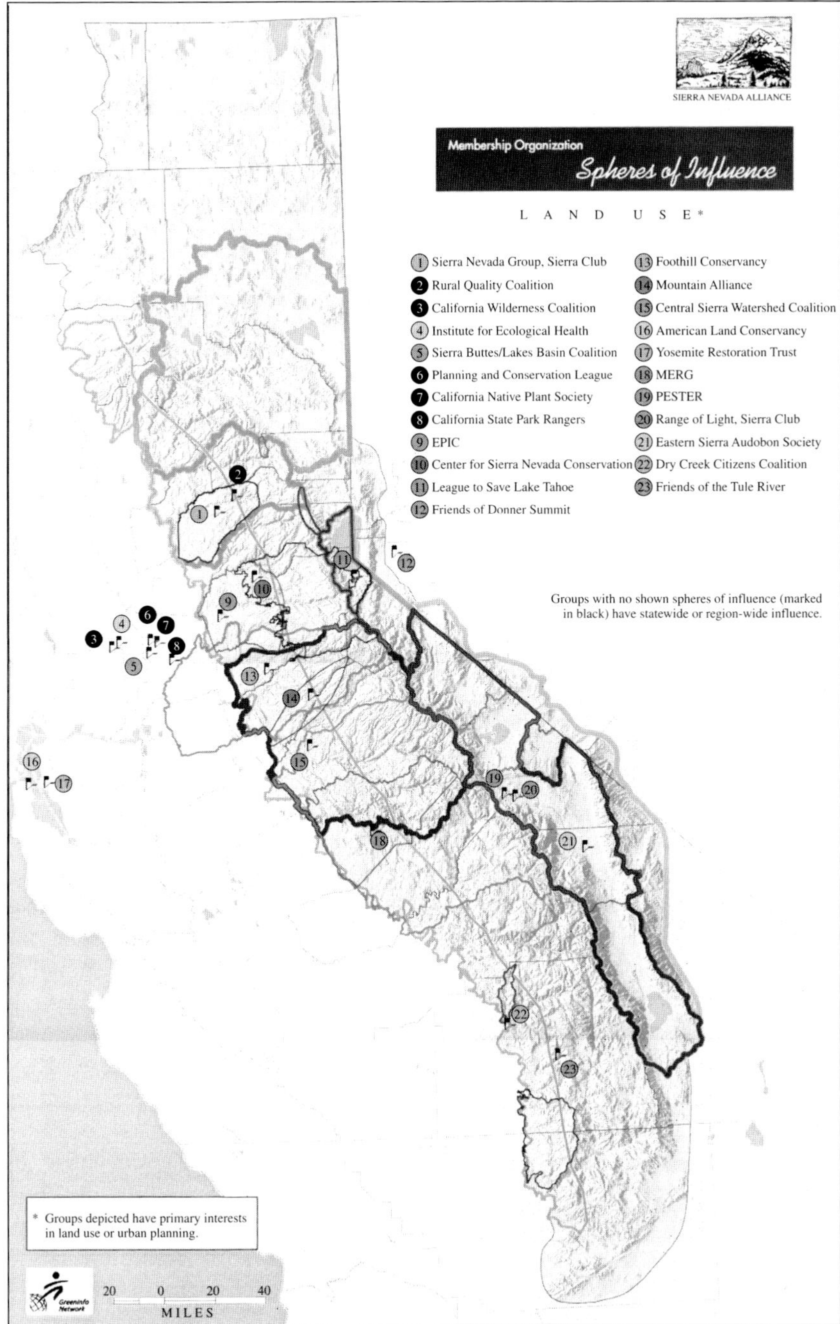

Figure 12.5 Map prepared by GreenInfo Network showing spheres of influence of non-profit organizations engaged in land-use or urban planning issues in north-eastern California.

In some instances, GIS was being used strategically to 'reverse engineer' the GIS of their adversaries or data suppliers. Case participants have discovered, much to their dismay (or delight), numerous errors in secondary spatial data, some unintentional[3] and some deliberate miscalculations. Three of five groups have found that GIS can be used to reveal those errors, omissions, or agendas in data collection/use. Groups also use GIS to dissect the resultant output and consequently understand the constituent data and decisions. The SALC director uses his GIS as a 'guerrilla ground truthing tactic' to discredit opponents' information before information is concretized into policy.

Respondents from most cases argued that, more than analysis, they have been assisted by GIS's strong visualization functionality – a shorthand of complex environmental problems. The dunes preserve director found that the management of information conferred greater clarification and identification of issues: 'where it makes sense to advocate for preservation'.

Goal enhancement exhibited an upper limit: groups would distance themselves from GIS rather than have it affect the organizational agenda. Two groups (Greenbelt and SALC's director's former organization) spun off their GIS capacity; a third, Trinity's umbrella non-profit, created Trinity as a separate organization. By outsourcing, the dunes preserve and EPIC stay true to their core goals. These GIS-only organizations choose the means and not the ends of GIS as their goals. This focus on means does not go uncriticized: the SALC director questions environmentalists' inordinate emphasis on GIS and the reification of the quantifiable to the exclusion of the qualitative.

In the larger scale, does GIS use exert an impact on policy and the environment (natural and man-made)? Most participants would say yes; however, they are more guarded about whether that impact is all positive. SALC's spatial analysis confirmed for policy-makers that hunting in the region would prove detrimental to wildlife as well as humans. Trinity's efforts put unemployed loggers back to work, through conducting GPS work, sustainable harvesting of special forest products and small diameter logging, although staffers are reflective about whether GIS offers an alternative to traditional resource extraction or optimizes operations such as clearcuts. The At Risk Map has proved a powerful asset to influencing the political agenda, but it elevates Greenbelt's already considerable status in the Bay Area.

EPIC would never argue that GIS was key to winning the court battle. In 1998, thousands of acres of land were saved through a state buyout. Were these accomplished because of GIS or because of the unending tree-sitting, barricading, and 'cat-and-mouse' games played with the chainsaws? Moreover, members are driven by a spiritual connection to the trees and many reject all forms of technology. They would rather chain themselves to trees instead of chaining themselves to a digitizing table.

Despite the argument on the direct impacts on a discrete policy, GIS allows groups to express their own version of ground-truth, combining experience, science, passion, and people. The cases are able to show their intimate knowledge of their environment, demonstrate their scientific understanding, and bring people and skills together to protect the land they love. In Trinity's case, GPS has enabled groups to exert a physical ground truth over their own area. Overall, GIS helps groups promote their vision in two ways. To the extent that public policy is based on science and the determination of accurate correlations, GIS helps activists by promoting the value of lay science and exposing weaknesses in institutional data. To the extent that policy in the political sphere is driven by other agendas, then GIS can be used to bypass policy-makers and reach out to the public using the visual sphere.

12.6 THE ROLE FOR SUPPORTERS

Elsewhere I have discussed the crucial role played by universities (Sieber 1997c), vendors, local government, professionals, and non-profits (Sieber 1997b) in diffusing GIS to the environmental movement. Every case depended upon these supportive institutions and individuals. Although it now has GIS in-house, Greenbelt began GIS by drawing land-use patterns onto USGS quads and driving them to two local universities for digitizing and overlay analysis. The dunes preserve staff have employed local university students who utilize university equipment and data. Trinity has made use of a university's SUN Sparcs for processing of large data sets, and tape drives for unusual tape formats (although the trip is eight hours).

As the reliance is tempered by the differing priorities of the institutions, so must the assistance. Supportive individuals can 'shore-up' the fragile structure with resources (peripherals, students, and expertise), continuity, and act as an intermediary to acquire and massage data. Vendors can build on their tradition of donations and remove the initial hardware and software barriers. Technology transfer is tempting because it gives the groups the skills to tailor GIS to their own needs. However, too many diffusion attempts have failed because needs are so much more prosaic: groups may just want the maps. Supporters first must allow groups to discover their own resources and needs before attempting to push GIS adoption. Supporters are not the only institutions to be confronted by this reality. GreenInfo has since changed its mission of technology transfer to other organizations and now produces high-quality maps for these organizations. In the 'Give a person a fish and he/she will eat for a day; teach a person how to fish and he/she will never be hungry' debate within GreenInfo, the giving of fish has won.

As time passed, cases have reduced their reliance on universities. They became increasingly independent and interdependent for their implementation (and, to a lesser extent, for data sharing). Peer-to-peer relations grew as more non-profits in the region obtained GIS; the likely reason is that other non-profits speak the same language because they share the same constraints in time and resources. For example, in exchange for access to

Table 12.1 Appropriateness matrix: who should or should not use GIS compared to who does and does not use GIS?

	SHOULD NOT USE GIS	SHOULD USE GIS
DOES NOT USE GIS	*Operation* Cannot spare resources to devote to GIS, Has largely computer-illiterate staff/board/volunteers Does not work with maps, Wants illustration only, May not know about GIS or have seen it demonstrated *Implementation/Data sharing* Employs one or less staff or has no GIS champion, May be involved in science/data collection but does not want to invest in data computerization *Goals/Mission* Is ideologically opposed to computing (and possibly nature's deconstruction)	*Operation* Creates and/or analyses maps, Is involved in scientific/public policy research and analysis Recognizes thematic as well as analytic functionality of GIS Shows interest in receiving technical support from other GIS-using institutions *Implementation/Data sharing* GIS champion emerges, Can get value back/see results in one year (prototype), Utilizes at least one level of government data and or other non-profits' data *Goals/Mission* Is involved in habitat/open space issues, Engages in proactive, negotiation-oriented strategies
DOES USE GIS	*Operation* Analysis is not part of organizational objective nor is it done in other computer applications Is not connected to the infrastructure of GIS users and resource substitution Uses GIS for one-off projects instead of planning for data/application reuse *Implementation/Data sharing* Remains unaware of or unwilling to acknowledge GIS implementation/data access costs *Goals/Mission* Goals and missions are better suited to non-GIS strategies Operates in crisis-oriented environment with a reactive and no compromise style	*Operation* Is involved in scientific research, publications, media campaigns, Uses, creates, and/or analyses maps Uses CAD, graphics software, Creates analysis and graphics/presentation applications *Implementation/Data sharing* Realizes implementation factors: commitment, computer literacy, grants of hardware and software, voluntarism, Has a GIS champion Recognizes costs of data collection, computer training Possesses good relations with institutional data providers *Goals/Mission* Employs GIS in long-range and proactive vision Does not allow GIS to divert goals, mission Shows interest in offering technical assistance

GIS services GreenInfo is subletting their space from the Trust for Public Lands.

Based on my analysis, I propose a set of characteristics of groups that should and should not adopt the innovation. The two-by-two matrix presented in Table 12.1 shows skills that must be acquired, groups that are incompatible with required skills and conditions, and the broader implications for all groups interested in GIS technology. One axis shows characteristics of groups that do and do not have GIS; that axis is crossed with another that displays groups that should have and should not have GIS. The profile should not be construed as an exhaustive list of criteria, nor does it imply that groups cannot surmount obstacles to adoption. However, it synthesizes the experiences of studied groups that can frame discussion on the role of supportive institutions.

12.7 CONCLUSION

Research demonstrated that environmental groups use and value GIS. Differences can certainly be observed in the applications of GIS but also in the objectives of GIS such as visualization, analysis, or skill-building. Implementation can also be quite varied, from traditional in-house implementation to complete outsourcing to resource-sharing within a consortium. Groups, however, do share an understanding of the complexities of and constant need for commitment to the technology. All groups exhibit an indigenous demand for GIS, backed by a history of scientific and technical knowledge. A leader committed to championing GIS innovation, so critical in government implementation, emerged in each of the cases. Improvisation demonstrated that resources did not represent a barrier to GIS implementation. This was not the case for acquiring digital data, however, which favoured groups engaged in proactive and non-confrontational agendas. All five groups also use GIS with passion and for advocacy. Research also emphasized the pivotal role that universities and other professionals played in groups' successful utilization of GIS. This support must focus on the infrastructure of GIS environmentalism before it begins GIS diffusion.

NOTES

1. A dozen activists were arranging for bail during my first field visit. In September of 1998, one activist was killed as a tree was felled on him (see http://www.headwatersforest.org/david.chain/index.html).
2. Several respondents also questioned the size of the pie.

3. Some would argue that information created in the context of power is never free from malign intent. By 'unintentional', I mean misplacing a datapoint or decimal point or, at worst, sloppiness in positioning.

REFERENCES

Aberley, D. (ed.) (1993) *Boundaries of Home: Mapping for Local Empowerment*, Philadelphia, PA: New Society Publishers.

Archer, H. and Croswell, P. (1989) 'Public access to geographic information systems: an emerging legal issue', *Photogrammetric Engineering and Remote Sensing* 55(11): 1575–1581.

Azad, B. (1993) 'Organizational aspects of GIS implementation: Preliminary results from a dozen cases', Paper presented at the Urban and Regional Information Systems Association Conference, Atlanta, GA, 13pp.

Azad, B. and Wiggins, L. L. (1995) 'Dynamics of inter-organizational geographic data sharing: A conceptual framework for research', in H. J. Onsrud and G. Rushton (eds) *Sharing Geographic Information*, New Brunswick, NJ: Center for Urban Policy Research (CUPR) Press, pp. 22–43.

Calkins, H. W. and Weatherbe, R. (1995) 'Taxonomy of spatial data sharing', in H. J. Onsrud and G. Rushton (eds) *Sharing Geographic Information*, New Brunswick, NJ: Center for Urban Policy Research (CUPR) Press, pp. 65–75.

Campbell, H. (1991) *Impact of Geographic Information Systems on Local Government*, TRP 101, Sheffield, UK: The University of Sheffield, Department of Town and Regional Planning.

Castells, M. (1983) *The City and the Grassroots*, Berkeley, CA: University of California Press.

Castells, M. (2000) *The Rise of the Network Society*, Oxford: Blackwell Publishers.

Convis, C. (1995) Keynote speech. *Specialist Meeting on Multiple Roles for GIS in U.S. Global Change Research*, Washington, DC, 5 March 1995.

Croswell, P. (1991) 'Obstacles to GIS implementation and guidelines to increase the opportunities for success', *Journal of the Urban and Regional Information Systems Association* 3(1) (Spring): 43–56.

Craig, W. J. and Elwood, S. (1998) 'How and why community groups use maps and geographic information', *Cartography and Geographic Information Systems* 25(2): 95–104.

Curry, M. R. (1995) 'GIS and the inevitability of ethical inconsistency', in J. Pickles (ed.) *Ground Truth: The Social Implications of Geographic Information Systems*, New York: Guilford Press, pp. 68–87.

Dando, L. P. (1992) 'Open records law, GIS and copyright protection: life after *Feist*', *Journal of the Urban and Regional Information Systems Association* 4(1) (Spring): 45–55.

Dangermond, J. (1991) 'Citizen use of GIS technology', *ArcNews*, Environmental Systems Research Institute, Redlands, CA (Summer): 16–18.

Epstein, E. F. and Roitman, H. (1987) 'Liability for information', *Proceedings of the Urban and Regional Information Systems Association* 4: 115–125.

Gittell, M. (1980) *Limits to Citizen Participation: The Decline of Community Organizations*, Beverly Hills, CA: Sage Publication.

Goldman, B. A. (1991) 'The environment and a community's right-to-know: information for participation', *Computers in Human Services* 8(1): 19–40.

Harris, T. and Weiner, D. (1998) 'Empowerment, marginalization and "community-integrated" GIS', *Cartography and Geographic Information Systems* 25(2): 67–76.

Klosterman, R. E. (1987) 'The politics of computer-aided planning', *Town Planning Review* 58(4): 444–451.

Kweit, M. G. and Kweit, R. W. (1987) 'The politics of policy analysis: the role of citizen participation in analytic decision making', in J. Desario and S. Langton (eds) *Citizen Participation in Public Decision Making*, New York: Greenwood Press, pp. 19–37.

Monmonier, M. (1991) *How to Lie with Maps*, Chicago, IL: University of Chicago Press.

Nedovic-Budic, Z. and Pinto, J. K. (2000) 'Information sharing in an interorganizational GIS environment', *Environment and Planning B* 27(3): 455–474.

Onsrud, H. J. and Pinto, J. K. (1993) 'Evaluating correlates of GIS adoption success and the decision process of GIS acquisition', *Journal of the Urban and Regional Information Systems Association* 5(1) (Spring): 18–39.

Parent, P. (1989) 'Issues arising from the proliferation of information', *Journal of the Urban and Regional Information Systems Association* 1(1): 17–26.

Pickles, J. (1995) 'Representations in an electronic age: geography, GIS and democracy', in J. Pickles (ed.) *Ground Truth: The Social Implications of Geographic Information Systems*, New York: Guilford Press, pp. 1–30.

Pinto, J. K. and Onsrud, H. J. (1995) 'Sharing geographic information across organization boundaries: a research framework', in H. J. Onsrud and G. Rushton (eds) *Sharing Geographic Information*, New Brunswick, NJ: Center for Urban Policy Research (CUPR) Press, pp. 44–64.

Piven, F. and Cloward, R. (1977) *Poor People's Movements*, New York: Pantheon.

Roberts, A. (1999) 'Closing the window: how public sector restructuring limits access to government information', Ottawa, Canada: Government Information in Canada/Information gouvernementale au Canada, Number/Numéro 17. Also available at http://www.usask.ca/library/gic/17/roberts.html.

Rubinyi, R. M. (1989) 'Computers and community: the organizational impact', *Journal of Communication* 39(3): 110–123.

Sieber, R. E. (1997a) 'Computers in the grassroots: environmentalists, geographic information systems, and public policy', Ph.D. dissertation, Rutgers University, New Brunswick, NJ.

Sieber, R. E. (1997b) 'GIS and the grassroots: the role for everyone else', presentation at the ESRI Users' Conference, San Diego, CA.

Sieber, R. E. (1997c) 'GIS in the grassroots: the role for universities', Paper presented at the University Consortium for Geographic Information Science, Bar Harbor, ME.

Sieber, R. E. (2000) 'GIS implementation in the grassroots', *Urban and Regional Information Systems Association Journal* 12(1): 15–29.
Snow, D. (1992) *Inside the Environmental Movement*, Covelo, CA: Island Press.
Wood, D. (1992) *The Power of Maps*, New York: The Guilford Press.

Chapter 13

There must be a catch: participatory GIS in a Newfoundland fishing community

Paul Macnab

> While the land has been seen by cultural geographers and others as layered with proprietary rights, use rights and cultural symbols, the water has been seen as empty.
>
> Jackson 1995

> That's a good idea to get the fishing grounds down on the charts. You know, its like I've got a map of the grounds in my head.
>
> Newfoundland fisherman 1995

13.1 INTRODUCTION

Five hundred years ago when John Cabot explored the coast of present day Atlantic Canada, he lowered a basket into the sea and pulled it out full of fish. Today, there are hardly enough codfish left to grace the dinner table in Newfoundland, Canada's easternmost province. Eight years have passed since the Atlantic Groundfish Moratorium was declared in 1992 and there are still too few cod in much of the region to permit commercial extraction. Beyond the environmental degradation that this stock collapse represents, the social impact has been devastating for fisheries-dependent communities, particularly those reliant on the traditional small-boat inshore harvest. Confronted by the ominous spectre of rotting skiffs, closing hospitals and massive out migration, many groups are working diligently to conserve remaining fisheries, such as lobster, and the traditional way of life that now depends on them. Before the crisis, the knowledge and concerns of fishers and their families were often disregarded – indeed marginalized – by biologists and ocean-related agencies. Now, communities expect to participate actively in every facet of fisheries science and management, especially where spatial and temporal limitations to harvesting may be implemented. This chapter describes a GIS project that evolved to link harvesters and government organizations in central Bonavista Bay, a historically strong fishing area on the northeast coast of Newfoundland. I discuss a collaborative project intended to capture local fisheries knowledge through participatory mapping aided by emerging geographic information technologies, principally, GIS.

13.2 CASE STUDY OVERVIEW

The research described here occurred over a three-year period (1994–1997) when I worked at Terra Nova National Park (see Figure 13.1) to explore conservation measures and related information needs for Bonavista Bay. Through the course of my research and employment with Parks Canada, I was invited to participate in small-boat fishing activities with local harvesters. I also facilitated a series of community meetings to discuss conservation measures. As a reaction to industry demands that government managers and conservation agencies acknowledge and incorporate local knowledge, I began organizing a GIS project to capture traditional fishing patterns. The

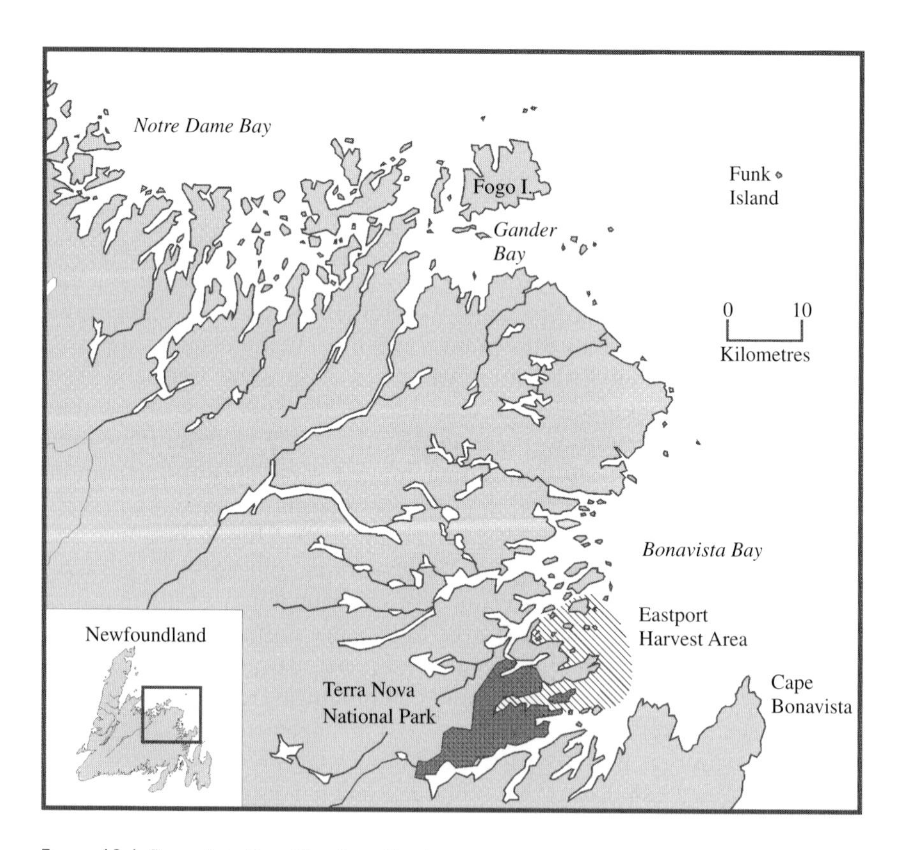

Figure 13.1 Bonavista Bay, Newfoundland.

project evolved as a collaborative effort with input from several government agencies, a local fishermen's committee, a GIS training programme and a software firm. Using digital topographic maps and newly collected hydrographic data, a prototype chart was customized for use in participatory mapping sessions where harvesters delineated fishing grounds, spatial management controls and local toponyms. Annotated charts were digitally rendered to produce composite maps that have since been used to help communicate fishing patterns.

13.3 BACKGROUND

13.3.1 Coastal Newfoundland and the collapse of a fishery

Typical of northeast Newfoundland, Bonavista Bay encompasses shoals and deep troughs, exposed shorelines, archipelagos and sheltered fjords. The cold waters of the Labrador Current support a wide variety of fish species as well as populations of North Atlantic seabirds, seals and whales. These resources have supported humans for over 7,000 years as evidenced by numerous archaeological sites. Europeans arrived for a seasonal fishery in the 1500s and settled permanently in the 1600s. Cod, the primary species harvested, was salted and dried for export markets by family enterprises until well into this century. Over time, larger fibreglass vessels replaced home-built wooden boats while monofilament nets supplanted hook and line gear. The intensification and expansion of the inshore sector was also accompanied by the imposition of an increasingly centralized management regime, new regulatory measures and scientific stock assessments. After Canada declared a 200-mile fishing zone in 1977, stern trawlers harvesting on the offshore banks delivered a welcome bounty to land-based processing plants.

All seemed fine until the early 1980s when fishers from the small boat inshore sector started to express concerns about declining catch rates and decreasing fish size (Neis 1992; Finlayson 1994). A considerable drop in biomass was finally detected in the offshore stocks towards the end of the 1980s (see Hutchings and Myers 1994; Finlayson and McCay 1998) and by 1992, the Atlantic Groundfish Moratorium was declared leaving close to 40,000 harvesters and plant workers without a livelihood. Life in post-moratorium Bonavista Bay carries on, but coastal communities' modern day dependence on the fishery has become painfully evident (e.g. see Woodrow 1998). The strengthening of other sectors such as aquaculture and tourism has been promoted, but many assert that coastal communities will survive only with a renewed fishery. Were it not for the lucrative lobster and crab fisheries that remain open, an entire way of life would be much eroded.

13.3.2 Dialogue on conservation

In the years immediately preceding the moratorium, Bonavista Bay was short-listed by Parks Canada as a candidate site for a national marine conservation area. Following some resource mapping and an 'experts workshop' the Bay was selected over three others to best represent the natural and cultural heritage of northeast Newfoundland (Mercier 1995). How would fish harvesters, the dominant stakeholder group in Bonavista Bay, react to such a proposal in a time of crisis? Would Newfoundland communities respond to participatory approaches successfully employed in other countries (e.g. Wells and White 1995)? Could local needs and priorities be reconciled with federal conservation goals? It became the responsibility of field staff to initiate local dialogue in an effort to answer these questions (see Macnab 1996; 1997).

From early discussions on the range of precautionary approaches available for marine resource management, no-take areas attracted considerable attention from harvesters, especially for the potential conservation of spawning fish, juveniles, sedentary species and supporting habitats. Instructive lessons from New Zealand and the tropics were conveyed by Parks Canada planning staff: resident species in areas set aside from harvesting will grow in size, increase egg production and replenish the surrounding fishery. The possibility that reserves could act as 'insurance policies' against overfishing (Ballantine 1995) received very little argument from fishers; however, where to establish such harvest refugia and how to make up for lost fishing space were questions not easily answered. Meanwhile, an assessment of marine resource data for the Bay showed that existing scientific knowledge was inadequate for a purely ecological approach to identifying and planning reserves. Information on human activities was also shown to be lacking. In particular, areas fished by small boats remained uncharted and unknown to those outside the fishery. To complicate matters, the existing nautical chart for the Bay, produced by the British Admiralty in 1869, was inaccurate, small-scaled and largely unsuitable for inventory purposes. Modern hydrographic surveys were in progress, but finished charts were estimated to be many years from publication.

Over time, it became evident that locally supported reserves would emerge through dialogue about conservation measures as they related to specific locations and fishing activities. On many occasions, fishers pointed to a spot on the chart explaining both the need for special protection and the likely displacement of fishing effort that would result. With very little scientific guidance available in the way of depth, bottom type or optimal placement, a group of fishers active in the waters adjacent to Terra Nova National Park began to discuss seriously the establishment of closed areas for lobster. Members of the Eastport Peninsula Inshore Fishermen's Committee eventually agreed that their fishery might benefit from trial closures. Harvesters started to discuss potential refugia based on local harvest patterns, observed oceanographic circulation and long-term knowledge of the lobster stocks.

13.3.3 Local marine knowledge

The rich knowledge base of resource users has been recognized as an important complement to scientific modes of inquiry for environmental management and protected area planning (e.g. Sadler and Boothroyd 1994; Pimbert and Pretty 1997). Mailhot (1993: 11) characterizes this knowledge as 'the sum of the data and ideas acquired by a human group on its environment as a result of the group's use and occupation of a region over many generations'. Johnson (1992) extends the definition to include 'nonindigenous groups such as outport fishermen' and describes three categories of knowledge: (i) a system of classification; (ii) a set of empirical observations about the local environment; and (iii) a system of self-management that governs resource use. Known by many names including traditional ecological knowledge (e.g. Berkes 1999), common sense geography (e.g. Egenhofer and Mark 1995) and indigenous knowledge (e.g. Warren *et al.* 1994), 'local knowledge' avoids some of the semantic and conceptual problems associated with other labels and is adopted here after Ruddle (1994).

Research on local knowledge systems in marine settings has been undertaken by a range of investigators, many of whom see it as essential for effective fisheries and coastal management regimes (e.g. Dyer and McGoodwin 1994; Jackson 1995; Neis and Felt 2000). The demands from non-governmental organizations, communities and scientists in Newfoundland are captured in the *Report of the Partnership for Sustainable Coastal Communities and Marine Ecosystems*:

> There is a neglect of fishers' information and an absence of serious efforts to use this to supplement scientific research. Partnerships should be established and supported between federal and provincial governments to develop appropriate databases for integrating scientific and traditional knowledge.
>
> National Round Table 1995: 32

What often goes missing in such broad calls, however, are the challenges of collection, veracity, analysis, application and ownership of local knowledge. Many researchers have suggested that local knowledge should be integrated or somehow blended with scientific forms of knowledge after collection and careful evaluation by 'outsiders' (e.g. DeWalt 1994; Murdoch and Clark 1994). Others argue that local knowledge is developed and transmitted *in situ*, and therefore must be captured and applied by people who live 'inside' the socio-cultural setting where it has evolved (e.g. Agrawal 1995; Heyd 1995; Chambers 1997). Is it really a 'black and white' case of scientific extraction versus community empowerment? Is there not some middle ground that could accommodate both of these perspectives? What if, as Fox (1990) argues for social forestry programmes, participatory research is conducted to help communities and outsiders 'learn about each other, develop a foundation for cooperation, and begin negotiating on the design and implementation of [resource] management plans' (120)?

13.3.4 Facilitated community inventories

Few would disagree that fishers and other customary users of marine resources have a substantial body of knowledge that could be useful for science and management, but if the information flow is only in one direction – knowledge extracted for use by outsiders – communities will most certainly be reluctant to contribute. If an inventory of local marine knowledge is to stimulate participant concern for resources and lead to stewardship activities, it must be community-based, and ideally, it should be community-driven: 'experience in Canada tells us that it is at the community level where the required actions to maintain coastal resources are implemented; it is from this level that the true effort springs' (Norrena 1994: 160). It is fine to have a conceptual notion of a community-driven inventory, but it is quite another thing to enable one. Unless such a plan originates at the community level, how is a community to become interested? There are also structural considerations. Communities should conduct their own studies, but with limited access to government information and cartographic production techniques, manual or digital, how can community groups best capture and display their own geographic knowledge?

Here, there is a definite role for collaborators, especially when it comes to technical assistance, project funding and linkages with scientific authorities. Where government participation is regarded with suspicion at the local level, academic researchers and NGOs have helped to gather and organize information with and for interested communities, often to support and reinforce traditional stewardship activities (e.g. see Fox 1990; Sirait *et al.* 1994; Berkes *et al.* 1995; Nietschmann 1995). A common element in many of these projects is the degree of control maintained by participating communities; coordination is provided by existing organizations (e.g. First Nation Elder Councils) and knowledge is often protected by some form of copyright. Problems of cross-cultural communication are lessened when local people collect knowledge and work as facilitators in their own communities (Brice-Bennett 1977). Outsiders might provide elicitation skills and technical support, but ideally, the knowledge is captured, held and applied by the community.

13.3.5 A role for geographic information technologies?

Local knowledge is often dismissed as being qualitative and unscientific, particularly within a positivist conservation paradigm that only considers opinion when it is stated in scientific terms (Pimbert and Pretty 1997). Does this hold true for the 'art, science and language' of cartography? Consider two case studies in which maps were used to depict local people's understanding of natural resources. Peluso (1995) describes constructive meetings between government mappers and Indonesian 'peasant groups' possessing legitimate and technically acceptable maps. Contrast the ready acceptance of these digitally enhanced forestry maps with the government rejection of sketch maps 'prepared by peasants' in an effort to claim lake portions of the Titicaca National Reserve in Peru (Orlove 1993).

When defined orally, or drawn without scale, orientation and formal grid reference, local knowledge remains anecdotal. Geographic information technologies provide a more technical and precise, if not more 'scientific', means of capturing the spatial components of local knowledge. When cognitive landscapes are inscribed and georeferenced in the field with affordable GPS, or merged with government maps and remotely sensed digital imagery, local knowledge assumes far more authority than possible with oral descriptions and simple sketch maps (e.g. see Bronsveld 1994; Conant 1994; Thomas 1994; Poole 1995; Dunn *et al.* 1997). Decreasing costs have permitted these technologies to be applied in ethnographic surveys and local knowledge documentation projects around the planet. Published applications include studies in forestry (Fox 1990; Cornett 1994; Sirait *et al.* 1994; Sussman *et al.* 1994; Peluso 1995), agriculture (Tabor and Hutchinson 1994; Gonzalez 1995; Harris *et al.* 1995; Lawas and Luning 1996) and indigenous land use (Duerden and Keller 1991; Poole 1995; Harmsworth 1998). In the marine realm, applications have been described for coral reef habitats (Stoffle *et al.* 1994; Nietschmann 1995; Calamia 1996), spawning fish (Ames 1997) and management regions (Clay 1996; Pederson and Hall-Arber 1999; St Martin 1999).

Suggesting that 'low quantitative salience' has prevented broad acceptance of social scientific data in fisheries, McGoodwin (1990) recommends that practitioners 'develop more rigorous techniques and the kind of data that will permit comparability, as well as integration, with other already formalized means of analysis' (187). GIS offers considerable promise in this regard. Information that was once dismissed by biologists as anecdotal (e.g. experiential knowledge of spawning sites) can be made more compatible with accepted 'scientific' forms of spatial knowledge (e.g. depth, temperature and salinity) through proper georeferencing.

13.3.6 The data challenge for coastal fisheries

Scientific mapping of the world's oceans and coasts has progressed remarkably in the last decade with the introduction of multi-beam hydrography, better remote sensing devices, enhanced digital processing equipment, GPS enabled navigation systems and GIS (Wright and Bartlett 2000). Generally though, our oceanic knowledge still pales by comparison with that of terrestrial environments. There are many reasons for this, not least of which are the challenges and expenses posed by a mobile ecosystem that demands mapping in four dimensions and a management regime that is administered by numerous agencies, each with distinct and at times redundant, conflicting and incompatible data collection programmes (Ricketts 1992; Furness 1994). Despite the limits to marine data collection and analysis, Bonavista Bay was subject to extensive surveying in the mid-1990s. Beyond the aforementioned hydrographic exercise, the Bay received a digital shoreline classification, hydro-acoustic and airborne stock assessments, visits by navy submersibles and telemetry tracking of fish implanted with acoustic devices. Still, with all of this ocean research and the proliferation of digital data that followed, there was minimal scientific knowledge of inshore fishing locations.

Fisheries scientists have adopted GIS for stock assessment and spatial analysis (e.g. Meaden and Chi 1996), but much of the newer work in fisheries GIS, particularly in Atlantic Canada, has been directed towards offshore areas where catch statistics and survey data are recorded with precise geographic coordinates (e.g. Mahon *et al.* 1998). Closer to shore, where small-boat fishers ply their trade over bottoms too rough for offshore sampling gear, GIS and related tools remain limited for the analysis of local fishing patterns. To begin with, harvesters report their catch by port of landing; logbook data recorded at this scale reveals little of fishing locations. Remote sensing instruments may help indicate fish stocks, important habitat (e.g. Simpson 1994) or boat locations, but they cannot detect how people are fishing or what they are catching. Similarly, land-use mapping, which relies upon the correspondence between land cover and land use (e.g. a field indicates agriculture), is not of much use for delineating fishing grounds – especially grounds which have not been fished since the moratorium was declared. Generally speaking, mapping human use of the world's oceans remains little practiced. Why? Activities on land are relatively fixed and basically two dimensional; by comparison, fishing activities are mobile and four dimensional (i.e. occurring at different times and levels in the water column). Furthermore, unlike a cut boundary or fence on land, or even a natural boundary, fishing territories cannot generally be detected, photographed or visited, and thus mapped, without some kind of local interpretation (e.g. Acheson 1979; Clay 1996). To collect such knowledge, two workable options appear to be available: (i) visit fishing locations and map the grounds with GPS and sounders (e.g. Nietschmann 1995); or (ii) map harvest areas from memory onto suitable hydrographic charts. The case study presented here details a project designed to work through the second option.

13.4 THE EASTPORT MAPPING PROJECT

13.4.1 Initiating the project

The idea for a fishing grounds inventory was discussed initially with the Chair of the Eastport Peninsula Inshore Fishermen's Committee. I had been investigating marine mapping for some time and had regularly communicated my findings to the Chair, so he was aware of recent hydrographic surveys and local mapping initiatives in other areas. While reviewing various charts with the Chair, his wide knowledge and local perspective were demonstrated with reference to specific locations. For example, while discussing some of the features that he had pointed out on an earlier lobster fishing trip, the Chair motioned to an inlet far too small for annotation on a government map. The inlet was known locally as 'Hospital Cove', named for a past fishers' practice of leaving sick and injured lobsters there to recover without the threat of capture. I suggested that we could relabel the maps with local names and add fishing patterns. My function, I explained, would be to provide the cartographic support necessary for such an undertaking; fishers would provide the information to be mapped.

The Committee Chair could see the value in documenting local knowledge, but would other fishers share his interest? To find out, the idea was presented at a committee meeting with a display of sample inventory maps from other jurisdictions. New hydrographic fieldsheets (1:20,000), which many fishers knew existed, but few had ever seen, were demonstrated alongside the familiar British Admiralty chart of the Bay. The inventory was presented not as an extractive government exercise or an impersonal academic survey, but as way for fishers to communicate their knowledge. Visualization by way of graphic display, I suggested, could demonstrate local concerns and help to identify conservation priorities to outside agencies. Attention was drawn to the copyright statement included on maps drawn by harvesters in Nova Scotia: 'This mapping series was compiled under the direction of the Guysborough County Community Futures Fisheries Sub-Committee and is now the property of the Guysborough County Inshore Fisherman's Association. The information and basemaps can only be duplicated or altered with permission of the Association.' The message was simple: fishers' knowledge leads to fishers' maps. The Chair borrowed these sample maps for the next committee meeting to gauge whether or not the larger membership agreed that harvest area mapping was a desirable undertaking. At that session, the committee discussed and endorsed the project. Afterwards, the Chair indicated formal acceptance of the inventory project and invited me to proceed.

13.4.2 Collaboration in GIS

The harvesters' proximity to Terra Nova National Park, a committee structure and keen interest, coupled with existing relationships and an established rapport made the Eastport group a strong candidate for collaboration. Initially, I believed that fishers could provide valuable information about sensitive areas and thus help to guide further scientific investigations and conservation planning efforts. Before long, the project focus shifted towards the committee's objective: harvest area maps for use in their own deliberations and in dealings with outside agencies. Parks Canada provided funding, computers, data and in-kind support for the project. The federal Department of Fisheries and Oceans, a central coordinating agency for coastal inventories, grew interested in the project and committed financial assistance; officials also wished to add the collected information to a Province-wide database. The research continued to evolve with digital contributions from several bodies including the Canadian Hydrographic Service and the Newfoundland Department of Natural Resources. Universal Systems Limited of Fredericton, New Brunswick, made available a complementary version of their CARIS software (Computer Aided Resource Information System), a GIS package that is installed and used widely in hydrographic offices and Canadian government organizations. Finally, instructors and displaced fisheries workers training for a GIS diploma provided technical assistance and plotting services.

13.4.3 Methods and procedures

As outlined earlier, I worked from Terra Nova National Park and met with fishers to explore their ideas for marine conservation. Participation in lobster and crab trips enabled me to see fishing patterns up close; it also demonstrated that I was genuinely willing to learn from harvesters. Spending time in boats with fishers also helped me become familiar with a substantial part of the seascape that was to be charted. Honesty, and perhaps my own experience as a commercial fisherman, led to an open exchange of ideas and information. In dry land map discussions involving digitally produced hydrographic data, which I was able to access easily through government sources, I was the specialist with something to contribute, but on the water, fishers were clearly the specialists possessed of their own unique brand of expertise.

Technical support was provided to the Eastport Fishermen's Committee in an interactive and adaptive fashion. It seemed opportune to take advantage of recent sounding data, digital topography and the possibilities enabled by GIS to create custom maps. Meetings were held with Committee members to review data sources, to demarcate the Eastport fishing territory and to determine basemap features. CARIS was then utilized to combine topographic and hydrographic data for the area. The intent was to build a geographic database that would reflect the members' worldview, a view that still relied on terrestrial features for navigation (e.g. Butler 1983) and experiential knowledge of water depths for fish detection and gear placement. By using the tools available within CARIS, it was possible to customize data according to the harvesters' wishes. For example, from metric soundings, depth contours were interpolated in fathoms, still the standard measure in the fishing industry. Successive topo-bathy maps were generated, plotted, reviewed by fishers and reworked to produce a 1:25,000 basemap depicting the Eastport harvest area.

To capture information about fishing grounds, individuals and small groups used Mylar to create thematic overlays. Knowledge elicitation and documentation methods were inspired by research in several fields including marine resource mapping (Butler *et al.* 1986), indigenous land-use and occupancy studies (Elias 1989; Usher *et al.* 1992; Robinson *et al.* 1994; Poole 1995; Huntington 1998), participatory rural appraisal (Chambers 1997; Townsley *et al.* 1997), toponomy (Canadian Permanent Committee on Geographical Names 1992; Gaffin 1994) and the bioregional movement (Aberley 1993). Many practitioners in these fields stress the importance of relaxed rapport and informal checklists of potential items to be mapped. As the outside 'specialist' in the Eastport project, I facilitated the mapping sessions, occasionally prompting for categories of information, but participants did the actual sketching and map delineation of features and activities. In most cases, fishers had a clear idea of what information they wished to capture. Mylar sheets were compiled for digitization and thematic entry. Draft place name and composite harvest area maps were then generated and laser-printed on 11″ × 17″ paper to enable low-cost reproduction and wide distribution. A set of these maps was returned to each participant for review and corrections.

13.4.4 Results and outcomes

Fishers were generally interested in the new hydrographic data and the potential of GIS, but for the most part, they were after printed maps that would portray traditional harvesting activities. Individuals and small-groups demonstrated tremendous above and below water environmental recall as they documented the harvest in water surrounding Eastport. Clearly, local knowledge – spatial, biological, technical, ecological and historical – continues to inform the cognitive basis of inshore fishing. There was a form of built-in peer review when mapping sessions were conducted by groups of fishers; as the information was filled in, the group automatically performed checks to make sure that the map was 'complete'. Group work also permitted those less comfortable with map reading to sit back and describe the fishing grounds while others charted the information. Longstanding fisheries such as those conducted for cod, lobster, squid and capelin received a considerable amount of attention. Amongst newer fisheries, skate, crab and lumpfish were easily charted. Emerging fisheries such as urchin and shrimp remain experimental and somewhat competitive. As a result, knowledge of these grounds was not shared. Women's impressions of fishing space and coastal environments were not captured in Eastport, though they have been elsewhere (e.g. Pocius 1992) and methods for gendered resource mapping are documented (e.g. Rocheleau *et al.* 1995).

Annotated maps showed that committee members continue to regulate fishing space within their communities by means of informal local boundaries, lottery-like draws for prime trap berths, individual tenure for lobster bottom and acceptance of local customs for net spacing. Much of this local area management is accomplished with toponyms used to denote bays, grounds, rocks, islands and landforms. That many of these smaller features are left unnamed on published maps came as no surprise to participants; however, that 24 names on the official topographic map were locally unrecognizable revealed as much about the cultural landscape as it did about government cartography. In many ways, the mapping process was far more valuable than the actual maps produced. The process helped government officials and harvesters move beyond concepts and theories to discuss real locations and pressing issues in the fishery. Combining information in an atmosphere of trust and openness helped to build common understandings of a shared marine environment. In the final analysis, maps and mapping were a catalyst for learning and action. A small number of government staff came to appreciate the complex psychological sea claim that fishers had in an area previously depicted as a series of crude ecological overlays (e.g. Mercier 1995). For harvesters, a certain pride evolved as the collective local knowledge base was revealed through mapping. The project maps were eventually used in community discussions and in meetings with scientists and managers to help establish lobster closures and to explain community-defined boundaries. Government agencies identified potential applications in coastal zone management such as oil spill planning and aquaculture siting.

13.5 LESSONS LEARNED

Collaboration, interaction and adaptation enabled people, knowledge and data to be assembled in this undertaking for far greater efficacy than would have been possible with individual efforts. Regrettably though, funding shortfalls, academic commitments, reporting deadlines, technical glitches and a variety of other factors limited the final outcomes of the exercise. There was a perception that mapping with digital data would somehow be quick and easy – this simply is not the case with multi-participant GIS projects. Government, community and educational collaborators had high hopes for the project. However, as with many GIS undertakings, the amount of lead-time in the Eastport project remained invisible. Participants asked the predictable question: 'We keep spending all of this time and money on GIS – why haven't we seen any useful maps yet?' Our collaboration with displaced fishery workers enrolled in a GIS training programme created additional problems. An informal partnership with the educational firm seemed cost effective and entirely appropriate at the outset, but when the company running the programme went bankrupt, staff and students dispersed without finishing the maps. A formal agreement requiring delivery of the maps might have prevented this unfortunate outcome. In summary, project champions must secure senior-level interest, funding support and staff commitments from one or more organizations if collaborative and participatory GIS projects are to succeed.

GIS provided for the adaptive improvement of basemaps, and in that fashion, it did assist in the documentation of local knowledge. We had the digital data and the right tools; it would have been a shame not to, as Tortell (1992) suggests, 'tailor-make' the printed map to meet the user's needs. Knowledge capture by and with fishers was faithful, but the filtering required to transfer the information into a GIS necessitated compilation and some interpretation. Generalization helped to produce a series of maps, but the subtleties of local context were inevitably lost as years of experience and layers of meaning were reduced to points, lines and polygons. Was the technical experimentation worth the effort? Yes, but a 'low-tech' approach utilizing existing paper charts would have freed up more time for participatory mapping and learning in the community. By drawing directly onto published basemaps and using manual compilation methods (e.g. Butler *et al.* 1986; Harrington 1999), an acceptable set of preliminary maps could have been generated quite quickly. Compilation sheets would have reproduced well on a blueprint machine and they could have been digitized at a later date for GIS treatment.

13.6 FUTURE OPPORTUNITIES

Now that the Eastport Fishermen's Committee has reviewed and corrected draft maps, additions and editing of the database can take place. Ideally, this would be followed by full-size colour plots annotated with appropriate copyright statements. Digital versions of the database are being considered for distribution on a CD. A growing number of harvesters operate home computers, so if the database is bundled with some form of shareware for viewing and simple queries, many more participants could access the collected knowledge. Several distribution issues remain, in particular, user agreements for electronic versions of the contributed local knowledge and the licensed government data. Given the shift towards new technology in the fishing industry (e.g. electronic navigation charts, GPS units, sounders) the potential for field truthing and continued documentation is unlimited. With due respect for potential conflicts, the project could also be expanded to include other user groups such as scuba divers and recreational boaters. Federal funding has been secured to undertake a larger inventory project in Bonavista Bay; if the agencies involved collaborate in an open and honest fashion, GIS and computer assisted visualization will continue to benefit inshore fishing communities.

POSTSCRIPT (JANUARY 2001)

Data access remains a challenge for inshore fishers in Eastport. The 500th anniversary of Cabot's arrival in Newfoundland accelerated the production of navigation charts for Bonavista Bay, but, unfortunately for resident fishers, the new charts (1:60,000) contain only a fraction of the information portrayed on the source-data fieldsheets (1:20,000). As it stands, the Parks Canada license to use hydrographic data does not permit further distribution of digital fieldsheets. A paper fieldsheet that cost approximately $16 in 1994 has recently jumped in price to $150, thereby making the set of five for Eastport prohibitively expensive and impractical for fishers. Some time after this GIS project was completed, Parks Canada launched a full study to assess the feasibility of a national marine conservation area in the waters of Bonavista Bay. The genuine two-way learning described here was difficult to continue at a community level once a formal advisory committee was established. The conservation area proposal met with growing opposition as locals grew suspicious of government agendas and in 1999, the feasibility assessment was terminated by the advisory committee. Eastport, however, has become a model for successful community-based fisheries management in Newfoundland (Rowe and Feltham 2000). Voluntary lobster reserves were eventually supported in regulations by the Department of Fisheries and Oceans. It is difficult to evaluate the role that mapping and GIS played in this process, but it is safe to conclude that information exchange and dialogue helped to create an environment where government could support community-driven conservation initiatives.

ACKNOWLEDGEMENTS

The Eastport Peninsula Inshore Fishermen's Committee contributed their time, interest, consent and wealth of knowledge to this project. I was humbled on many occasions and the learning has been permanently imbedded in my psyche. The generous financial support of Parks Canada and the Department of Fisheries and Oceans enabled the project to realize its present life. The views and opinions expressed here come as result of extensive reading, interaction with hundreds of individuals and through my employment with the Government of Canada, but in no way should the content be construed as representative of those agencies and people with whom I have collaborated. A detailed report of this undertaking is available in my Master's thesis, a piece of work that never would have been completed were it not for the patient encouragement of Dr Gordon Nelson, my advisor at the University of Waterloo. An NCGIA Seed Grant shared with Barbara Walker permitted me to travel to the First GIS in Fisheries Science Symposium where my presentation, *The Data Collection Challenge for Inshore Fisheries: Atlantic Canada's Experience*, permitted some refinement of the current paper. Much of the original material was first presented in Santa Barbara, California, at *Empowerment, Marginalization and Public Participation GIS.*

REFERENCES

Aberley, D. (ed.) (1993) *Boundaries of Home: Mapping for Local Empowerment*, Gabriola Island: New Society Publishers.

Acheson, J. M. (1979) 'Variations in traditional inshore fishing rights in maine lobstering communities', in R. Anderson (ed.) *North Atlantic Maritime Cultures: Anthropological Essays on Changing Adaptations*, The Hague: Mouton Publishers, pp. 253–276.

Agrawal, A. (1995) 'Dismantling the divide between indigenous and scientific knowledge', *Development and Change* 26: 413–439.

Ames, E. P. (1997) *Cod and Haddock Spawning Grounds in the Gulf of Maine*, Island Institute, Rockland, Maine.

Ballantine, W. J. (1995) 'Networks of "No-Take" Marine Reserves are Practical and Necessary', in N. L. Shackell and J. H. M. Willison (eds) *Marine Protected Areas and Sustainable Fisheries*, Wolfville: Science and the Management of Protected Areas Association, pp. 13–20.

Berkes, F. (1999) *Sacred Ecology: traditional ecological knowledge and resource management*, Philadelphia: Taylor and Francis.

Berkes, F., Hughes, A., George, P. J., Preston, R. J., Cummins, B. D. and Turner, J. (1995) 'The persistence of aboriginal land-use: fish and wildlife harvest areas in the Hudson and James Bay lowland, Ontario', *Arctic* 48(1): 81–93.

Brice-Bennett, C. (ed.) (1977) *Our Footprints are Everywhere: Inuit Land Use and Occupancy in Labrador*, Nain: Labrador Inuit Association.

Bronsveld, K. (1994) 'The use of local knowledge in land use/land cover mapping from satellite images', *ITC Journal* 4: 349–358.

Butler, G. R. (1983) 'Culture, cognition, and communication: fishermen's location-finding in L'Anse-a-Canards, Newfoundland', *Canadian Folklore Canadien* 5(1–2): 7–21.

Butler, M. J. A., LeBlanc, C., Belbin, J. A. and MacNeill, J. L. (1986) *Marine Resource Mapping: An Introductory Manual*, Fisheries Technical Paper 274, Rome: Food and Agriculture Organization of the United Nations.

Calamia, M. A. (1996) 'Traditional ecological knowledge and geographic information systems in the use and management of Hawaii's coral reefs and fishponds', *High Plains Applied Anthropologist* 16(2): 144–164.

Canadian Permanent Committee on Geographical Names (1992) *Guide to the field collection of native geographical names*, Natural Resources Canada (http://geonames.nrcan.gc.ca).

Chambers, R. (1997) *Whose Reality Counts? Putting the first last*, London: Intermediate Technology Publications.

Clay, P. M. (1996) 'Management regions, statistical areas and fishing grounds: criteria for dividing up the sea', *Journal of Northwest Atlantic Fishery Science* 19: 103–126.

Conant, F. P. (1994) 'Human ecology and space age technology: some predictions', *Human Ecology* 22(3): 405–413.

Cornett, Z. J. (1994) 'GIS as a catalyst for effective public involvement in ecosystem management decision-making', in V. A. Sample (ed.) *Remote Sensing and GIS in Ecosystem Management*, Washington, DC: Island Press, pp. 337–345.

DeWalt, B. R. (1994) 'Using indigenous knowledge to improve agriculture and natural resource management', *Human Organization* 53(2): 123–131.

Duerden, F. and Keller, C. P. (1991) 'GIS and land selection for native land claims', *Operational Geographer* 10(4): 11–14.

Dunn, C. E., Atkins, P. J. and Townsend, J. G. (1997) 'GIS for development: a contradiction in terms?' *Area* 29(2): 151–159.

Dyer C. L. and McGoodwin, J. R. (eds) (1994) *Folk Management in the World's Fisheries: Lessons for Modern Fisheries Management*, Colorado: University Press of Colorado.

Egenhofer, M. J. and Mark, D. M. (1995) 'Naive Geography', in A. U. Frank and W. Kuhn (eds) *Spatial Information Theory: A Theoretical Basis for GIS*, Lecture Notes in Computer Sciences No. 988, Berlin: Springer-Verlag, pp. 1–15.

Elias, P. D. (1989) 'Rights and research: the role of social sciences in the legal and political resolution of land claims and question of aboriginal rights', *Canadian Native Law Reporter* 1: 1–43.

Finlayson, A. C. (1994) *Fishing for Truth: A Sociological Analysis of Northern Cod Stock Assessments from 1977–1990*, St John's: ISER Press.

Finlayson, A. C. and McCay, B. J. (1998) 'Crossing the threshold of ecosystem resilience: the commercial extinction of northern cod', in F. Berkes and C. Folke (eds) *Linking Social and Ecological Systems*, Cambridge University Press, pp. 311–337.

Fox, J. (1990) 'Diagnostic tools for social forestry', in M. Poffenberger (ed.) *Keepers of the Forest: Land Mangement Alternatives in Southeast Asia*, Quezon City: Ateneo De Manila University Press, pp. 119–133.

Furness, R. A. (1994) 'Data access for effective coastal zone management: a *cri du coeur* for openness', *Cartography* 23: 11–18.

Gaffin, D. (1994) 'The geographic identities of Faeroe islanders', *Landscape* 32(2): 20–27.

Gonzalez, R. M. (1995) 'KBS, GIS and documenting indigenous knowledge', *Indigenous Knowledge and Development Monitor* 3(1). Internet edition: http:// www. nuffic.nl/ciran/ikdm/3-1/articles/gonzalez.html.

Harmsworth, G. (1998) 'Indigenous values and GIS: a method and a framework', *Indigenous Knowledge and Development Monitor* 6(3). Internet edition: http://www.nuffic.nl/ciran/ikdm/6-3/harmsw.html.

Harrington, S. (ed.) (1999) *Giving the Voice a Land: Mapping Our Home Places*, Land Trust Alliance of British Columbia, Salt Spring Island, British Columbia.

Harris, T. M., Weiner, D., Warner, T. A. and Levin, R. (1995) 'Pursuing social goals through participatory geographic information systems: redressing South Africa's historical political ecology', in J. Pickles (ed.) *Ground Truth: The Social Implications of Geographic Information Systems*, New York: The Guildford Press, pp. 196–222.

Heyd, T. (1995) 'Indigenous knowledge, emancipation and alienation', *Knowledge and Policy* 8(1): 63–73.

Huntington, H. P. (1998) 'Observations on the utility of the semi-directive interview for documenting traditional ecological knowledge', *Arctic* 51(3): 237–242.

Hutchings, J. A. and Myers, R. A. (1994) 'What can be learned from the collapse of a renewable resource? Atlantic cod, *Gadus morhua*, of Newfoundland and Labrador', *Canadian Journal of Fisheries and Aquatic Sciences* 51: 2126–2146.

Jackson, S. E. (1995) 'The water is not empty: cross-cultural issues in conceptualising sea space', *Australian Geographer* 26(1): 87–96.

Johnson, M. (1992) *Lore: Capturing Traditional Environmental Knowledge*, Ottawa: IDRC.

Lawas, C. M. and Luning, H. A. (1996) 'Farmers' knowledge and GIS', *Indigenous Knowledge and Development Monitor* 4(1). Internet edition: http://www. nuffic.nl/ciran/ikdm/4-1/articles/lawas.html.

Macnab, P. A. (1996) 'Fisheries resources and marine heritage in Newfoundland: crisis, conservation and conflict', *Environments* 24(1): 106–115.

Macnab, P. A. (1997) 'Exploratory planning for a proposed national marine conservation area in Northeast Newfoundland', in G. Nelson and R. Serafin (eds) *National Parks and Protected Areas: Keystones to Conservation and Sustainable Development*, NATO ASI Series G, vol. 40, New York: Springer, pp. 133–143.

Mahon, R., Brown, S., Zwanenburg, K., Atkinson, B., Buja, K., Claflin, L., Howell, G., Monaco, M., O'Boyle, R. and Sinclair, M. (1998) 'Assemblages and biogeography of demersal fishes of the east coast of North America', *Canadian Journal of Fisheries and Aquatic Sciences* 55: 1704–1738.

Mailhot, J. (1993) 'Traditional ecological knowledge: the diversity of knowledge systems and their study', Background Paper No. 4, Montreal: Great Whale Public Review Support Office.

McGoodwin, J. R. (1990) *Crisis in the World's Fisheries: People, Problems and Policies*, Stanford: Stanford university Press.

Meadon, G. J. and Chi, T. D. (1996) 'Geographical information systems: applications to marine fisheries', Fisheries Technical Paper No. 356, Rome: Food and Agriculture Organization of the United Nations.

Mercier, F. (1995) 'Report of a workshop to identify a potential national marine conservation area on the NE coast of Newfoundland', in N. L. Shackell and J. H. M. Willison (eds) *Marine Protected Areas and Sustainable Fisheries*, Wolfville: Science and the Management of Protected Areas Association, pp. 240–248.

Murdoch, J. and Clark, J. (1994) 'Sustainable Knowledge', *Geoforum* 25(2): 115–132.

National Round Table on the Environment and the Economy (1995) *The Report of the Partnership for Sustainable Coastal Communities and Marine Ecosystems in Newfoundland and Labrador*, Ottawa: NRTEE.

Neis, B. (1992) 'Fishers' ecological knowledge and stock assessment in Newfoundland', *Newfoundland Studies* 8(2): 155–178.

Neis, B. and Felt, L. (eds) (2000) *Finding Our Sea Legs: Linking Fishery People and Their Knowledge with Science and Management*, St John's: ISER Press.

Newfoundland fisherman (1995) Comment from an inshore fisherman at a meeting convened by the author on Fogo Island, Newfoundland, Winter 1995.

Nietschmann, B. (1995) 'Defending the miskito reefs with maps and GPS: mapping with sail, scuba, and satellite', *Cultural Survival Quarterly* 18(4): 34–37.

Norrena, E. J. (1994) 'Stewardship of coastal waters and protected spaces: Canada's approach', *Marine Policy* 18(2): 153–160.

Orlove, B. (1993) 'The ethnography of maps: the cultutral and social contexts of cartographic representation in Peru', *Cartographica* 30(1): 29–46.

Pederson, J. and Hall-Arber, M. (1999) 'Fish habitat: a focus on New England fishermen's perspectives', *American Fisheries Society Symposium* 22: 188–211.

Peluso, N. L. (1995) 'Whose woods are these? Counter-mapping forest territories in Kalimantan, Indonesia', *Antipode* 27(4): 383–406.

Pimbert, M. P. and Pretty, J. N. (1997) 'Parks, people and professionals: putting "participation" into protected-area management', in K. B. Ghimire and M. P. Pimbert (eds) *Social Change and Conservation: Environmental Politics and Impacts of National Parks and Protected Areas*, London: Earthscan Publications Ltd, pp. 297–330.

Pocius, G. L. (1992). *A Place to Belong: Community Order and Everyday Space in Calvert, Newfoundland*. Montreal: McGill-Queen's University Press.

Poole, P. (1995) 'Indigenous peoples, mapping & biodiversity conservation: an analysis of current activities and opportunities for applying geomatics technologies', Biodiversity Support Program Discussion Paper Series, Washington: WWF; The Nature Conservancy; World Resources Institute.

Ricketts, P. J. (1992) 'Current approaches in geographic information systems for coastal management', *Marine Pollution Bulletin* 25(1–4): 82–87.

Robinson, M., Garvin, T. and Hodgson, G. (1994) *Mapping How We Use Our Land: Using Participatory Action Research*, Calgary: Arctic Institute of North America, University of Calgary.

Rocheleau, D., Thomas-Slayter, B. and Edmunds, D. (1995) 'Gendered resource mapping: focusing on women's spaces in the landscape', *Cultural Survival Quarterly* 18(4): 62–68.

Rowe, S. and Feltham, G. (2000) 'The eastport lobster project', in B. Neis and L. Felt (eds) *Finding Our Sea Legs: Linking Fishery People and Their Knowledge with Science and Management*, St John's: ISER Press, pp. 236–245.

Ruddle, K. (1994) 'Local knowledge in the folk management of fisheries and coastal marine environments', in C. L. Dyer and J. R. McGoodwin (eds) *Folk Management in the World's Fisheries: Lessons for Modern Fisheries Management*, Colorado: University Press of Colorado, pp. 161–206.

Sadler, B. and Boothroyd, P. (eds) (1994) *Traditional Ecological Knowledge and Environmental Impact Assessment*, Vancouver: University of British Columbia.

Simpson, J. J. (1994) 'Remote sensing in fisheries: a tool for better management in the utilization of a renewable resource', *Canadian Journal of Fisheries and Aquatic Sciences* 51: 743–771.

Sirait, M., Prasodjo, S., Podger, N., Flavelle, A. and Fox, J. (1994) 'Mapping customary land in East Kalimatan, Indonesia: a tool for forest management', *Ambio* 23(7): 411–417.

St Martin, K. (1999) 'From models to maps: the discourse of fisheries and the potential for community management in New England', Unpublished dissertation, Clark University.

Stoffle, R. W., Halmo, D. B., Wagner, T. W. and Luczkovich, J. J. (1994) 'Reefs from space: satellite imagery, marine ecology, and ethnography in the Dominican Republic', *Human Ecology* 22(3): 355–378.

Sussman, R. W., Green, G. M. and Sussmen, L. K. (1994) 'Satellite imagery, human ecology, anthropology, and deforestation in Madagascar', *Human Ecology* 22(3): 333–354

Tabor, J. A. and Hutchinson, C. F. (1994) 'Using indigenous knowledge, remote sensing and GIS for sustainable development', *Indigenous Knowledge and Development Monitor* 2(1). Internet edition: http://www.nuffic.nl/ciran/ikdm/2-1/articles/tabor.html

Thomas, G. (1994) 'Traditional ecological knowledge and the promise of emerging information technology', *Nature and Resources* 30(2): 17–21.

Tortell, P. (1992) 'Coastal zone sensitivity mapping and its role in marine environmental management', *Marine Pollution Bulletin* 25: 88–93.

Townsley, P., Anderson, J. and Mees, C. (1997) 'Customary marine tenure in the South Pacific: the uses and challenges of mapping', *PLA Notes* 30.

Usher, P. J., Tough, F. and Galois, R. (1992) 'Reclaiming the land: Aboriginal title, treaty rights and land claims in Canada', *Applied Geography* 12: 109–132.

Warren, D. M., Brokensha, D. and Slikerver, L. J. (eds) (1994) *Indigenous Knowledge Systems: The Cultural Dimension of Development*, London: Intermediate Technology Publications.

Wells, S. and White, A. T. (1995) 'Involving the community', in S. Gubbay (ed.) *Marine Protected Areas: Principles and techniques for management*, London: Chapman and Hall, pp. 61–84.

Woodrow, M. (1998) 'A case study of fisheries reduction programs during the Northern Cod Moratorium', *Ocean and Coastal Management* 39: 105–118.

Wright, D. J. and Bartlett, D. J. (2000) *Marine and Coastal Geographical Information Systems*, Philadelphia: Taylor and Francis.

Chapter 14

Environmental NGOs and community access to technology as a force for change

David L. Tulloch

14.1 INTRODUCTION

Environmental NGOs are finding themselves, and as a result their constituencies, increasingly empowered as users of geospatial technologies in New Jersey. A common concern regarding geospatial technologies is that the systems require significant technical knowledge in order to be properly applied to a problem. The average citizen lacks the requisite basic technical skills, thus limiting the opportunities for PPGIS. Finding a way in which these citizens can participate in the application of a community-based or community-oriented system is a challenge. Special interest groups purporting to represent various segments of their larger community can serve as the interface between citizens and government by operating, evaluating, or opposing public systems.

A basic assumption of this chapter is that NGOs can either

1. interface with an otherwise inaccessible public system, thus rendering it a PPGIS despite the system's initial failings, or
2. develop on behalf of members of the community a system that can serve as a PPGIS, despite parallel local government efforts.

14.2 FACTORS SUPPORTING NGO ACTIVITY

With this assumption in mind, this chapter will highlight four factors responsible for accelerating NGO activity in New Jersey and empowering citizens through a series of state-level NGO-PPGIS. These factors include:

1. prominent environmental and land-use issues that require attention
2. a traditional local government political structure that has limited development of public geospatial data and systems
3. a state government 'champion' that has assisted NGOs with data and software

4 a state-wide NGO 'champion' that has provided technical assistance and assisted with communication and coordination between groups.

The role of any individual factor in promoting or inhibiting PPGIS development is hard to identify; rather, these factors have acted in concert to promote or inhibit the development of geospatial systems (Tulloch 1999). This chapter will address each factor and describe how they have interacted to promote or inhibit PPGIS development in New Jersey.

14.2.1 Factor 1: physical and social conditions affecting the New Jersey land use puzzle

New Jersey has unique physical and social conditions that have accelerated the need for environmental response in the state. As the most densely populated state in the nation (over 8 million residents living in less than 8,000 square miles), New Jersey is home to dense urban areas (e.g. Newark, Camden, and Paterson), extensive sprawl, large industries (e.g. pharmaceuticals and petrochemicals), and significant transportation systems (e.g. the Port of Newark, Newark International Airport, the New Jersey Turnpike, and Amtrak's Northeast corridor). This intense development exists alongside some impressive natural areas, including the Pinelands (the largest body of open space on the mid-Atlantic seaboard between Richmond and Boston), the Hackensack Meadowlands, the Delaware Water Gap, and the New Jersey Highlands. In addition, New Jersey's extensive agricultural areas provide seasonal produce for Philadelphia and New York City, and place it

Table 14.1 1997 surface area of land-cover/land-use in New Jersey, based on the National Resources Inventory (Natural Resources Conservation Service 1999)

Land-cover/use classification category	*Acres*	*Percentage of NJ*
Developed (includes urban, built-up land, and rural transportation zones)	1,848,900	35
Forestland	1,624,700	31
Agricultural land (includes cropland, pastureland, and Conservation Reserve Programme land)	682,500	13
Water areas	530,200	10
Other rural land (includes barren land and marshland)	380,500	7
Federal lands (includes military bases, National Wildlife Refuges and National Park Service properties)	148,300	3

Source: Natural Resources Conservation Service 1999.

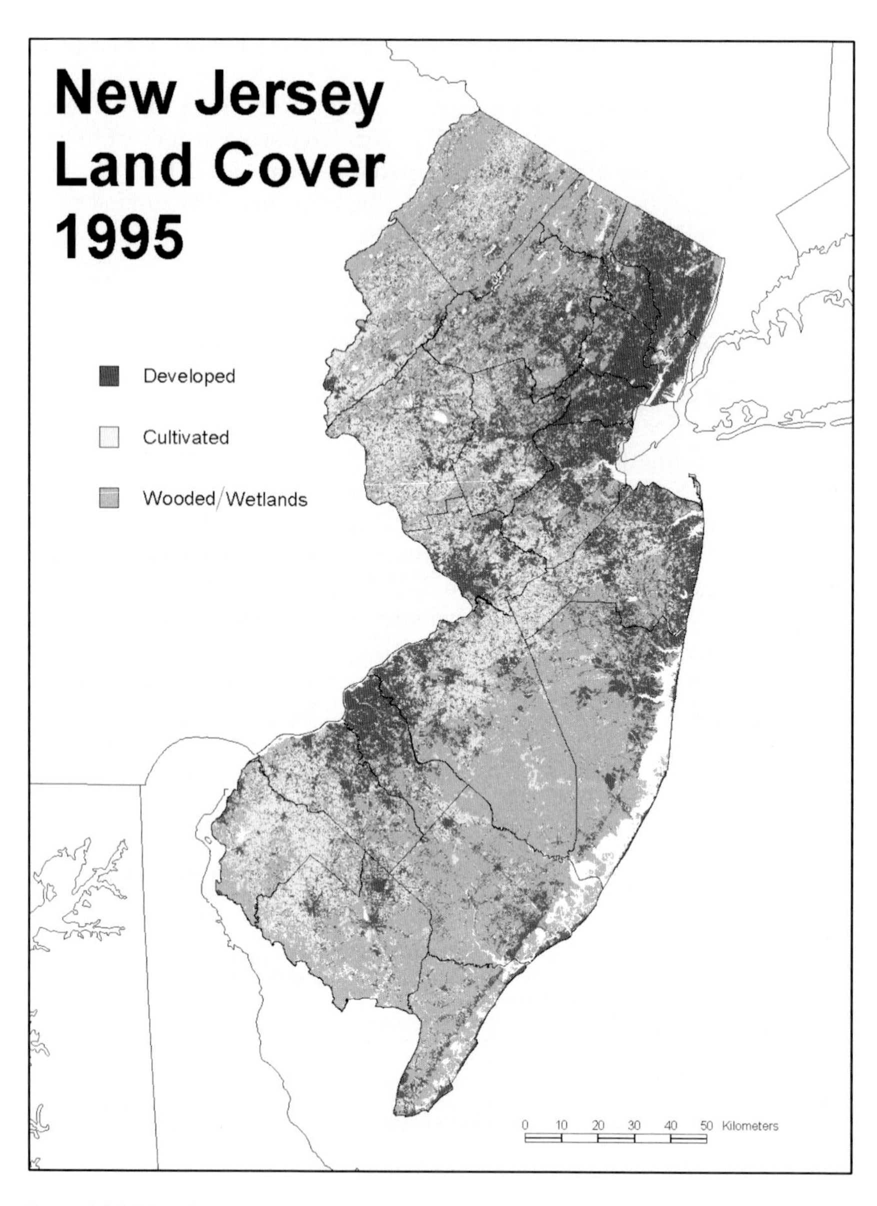

Figure 14.1 New Jersey land-cover, 1995.

among the nation's top ten producers of bell peppers, spinach, lettuce, cucumbers, sweet corn, tomatoes, snap beans, cabbage, escarole/endive, and eggplant, as well as a number of specialty crops including cranberries, blueberries, peaches, and asparagus.

What is unique about New Jersey is the cheek-by-jowl relationship between these diverse land-uses (see Figure 14.1 and Table 14.1). Since at least the 1950s, NGOs have formed in response to the conflicts that have emerged at the convergence of agricultural land-uses, natural areas, and urban development. One indicator of this complex relationship between urbanization and agriculture is that the average per acre-value of New Jersey farmland ($8,370) is the highest in the nation. The constant tension between these broad categories of land-use has caused the destruction of irreplaceable resources, and has contributed to the increased role of NGOs in providing solutions to competing land uses.

14.2.2 Factor 2: strong home rule and limits on local technology development

An important force in New Jersey is the state's tradition of strong home rule. As a result, the state has 566 independent municipalities (shown in Figure 14.2) that control land-use and address development-related environmental issues, with only a few able to support local development of GIS. This creates a particularly difficult challenge for the development of NGO systems because local governments are an important source of foundational spatial data sets in other parts of the country.

With the state sliced into 566 independent municipalities, many communities find themselves without the tax base necessary to support the development of even a rudimentary geospatial system. Most are small communities: 63 per cent of the municipalities in New Jersey have less than 10,000 residents, while over 25 per cent have less than 3,000. It is almost inconceivable that accurate, detailed information could be compiled at any level other than the local level, particularly for data themes like parcels and land-use (as opposed to the more generalized land-cover data as described in Table 14.1). In other states, strong home rule could serve as a negative factor for NGOs who find themselves stymied by the lack of local data. However, in New Jersey this local data void has provided a rallying cry for NGOs; some are trying to produce their own complete local data sets, while others have focused on ways to encourage or assist the municipalities within their jurisdiction to develop databases.

It should also be noted that strong home rule has contributed to environmental and growth management problems in New Jersey (Mansnerus 1998). The state has been severely limited in its ability to address land use and environmental problems at the local level. A significant portion of New Jersey's sprawl has come as a result of the state's municipalities competing against one another for new development (and property taxes). Strong home rule has also had the unintentional outcome of promoting fragmented landscapes that are inefficient for providing community services, make farming

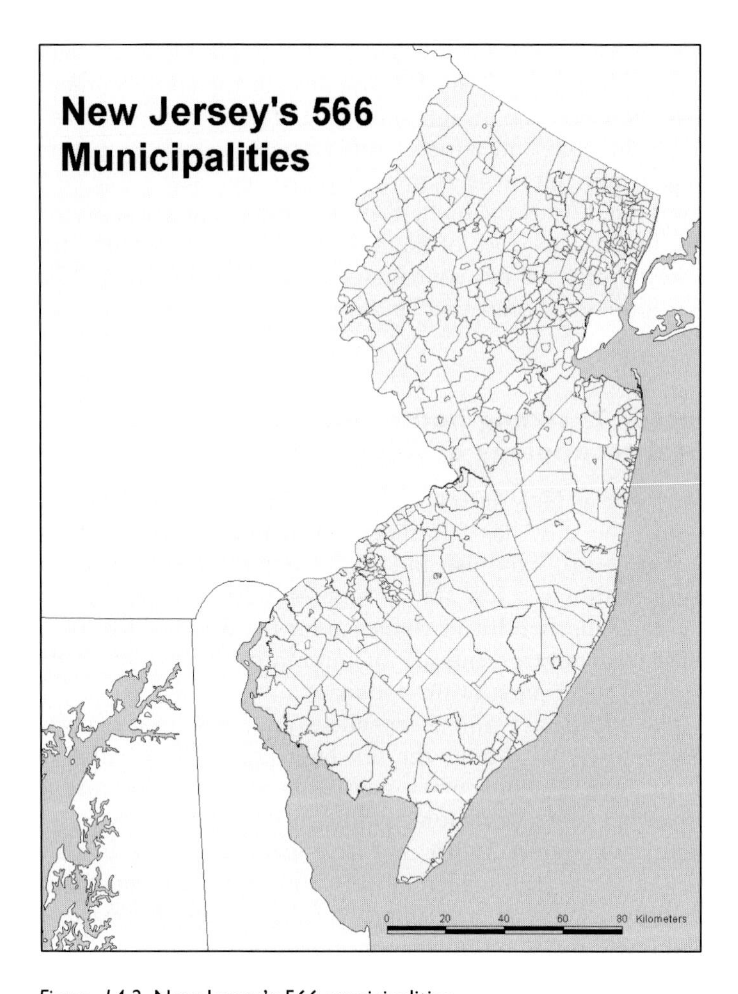

Figure 14.2 New Jersey's 566 municipalities.

difficult, and create landscapes ill-suited for ecologically desirable native species.

14.2.3 Factor 3: New Jersey Department of Environmental Protection and NGO-based GIS

The New Jersey Department of Environmental Protection (NJDEP) has recognized the fertile ground provided by factors 1 and 2, and sown the seeds for NGO-based GIS participation throughout the state. The NJDEP, acting through the New Jersey State Mapping Advisory Committee (SMAC), has published a series of CD-ROMs that provide a variety of statewide coverages (by county), including transportation, land-use/land-cover, soils, public lands, open spaces, coastal areas, wetlands, and floodplains, as well as legislative districts and state, municipal, and county boundaries. The NJDEP began disseminating its data as a CD-ROM series beginning in 1996, eventually distributing a total of five CDs (NJDEP 1996a,b,c,d; 1997).

In 1997, the NJDEP also began distributing specially attained 'free' licenses of ESRI's ArcView to local government agencies and environmentally oriented NGOs. The use of the license was conditional on an agreement by the receiving agency to produce suitable hardware, and assure that a reasonable number of its staff would be trained to use the software. So far, around 200 such licenses have been granted.

Financially challenged organizations have been able to convert this assistance into newly developed systems that better enable them to participate in public decision-making processes (Gibson 1998). Although it does not provide complex analysis of the issue, an article by Parrish and Patterson (1998) of the Great Swamp Watershed (GSWA) attests that graphic capabilities enabled by these basic data sets and desktop mapping software have played an important role in getting and keeping the attention of local environmental commissioners and planning board members. Perhaps the best application of these graphics programmes has been their production of a watershed open space and greenways plan (Parrish and Walmsley 1997) and a build-out analysis of the watershed (Patterson 1999).

14.2.4 Factor 4: New Jersey Non-Profit GIS Community and NGO-based GIS

The final source of support for PPGIS in New Jersey, particularly for smaller NGOs, is the New Jersey Non-Profit GIS Community (NGC) (http://www.princetonol.com/ngc). Founded in 1996 by Doug Schleifer, a GIS specialist at the Upper Raritan Watershed Association, the NGC offers environmentally oriented 501(c)(3) non-profit organizations 'facilities with technical and conceptual support for projects requiring the use of Geographic Information Systems technology' (New Jersey Non-Profit GIS Community 1997: 1).

The NGC did not become a reality until it was populated by a membership of various New Jersey NGOs and designed to provide support for NGOs struggling with GIS problems. Although the more sophisticated users in the state use the NGC as a GIS users group, less sophisticated users are able to go to this group for the actual hardware and software needed for geospatial analysis (Schleifer 1998).

The NGC's provision of training sessions for members has been pivotal for these NGOs. The free ArcView license through the NJDEP required

NGOs to get employees or members trained to use the software. As my experiences with the Lawrence Brook Watershed have proven, this training is neither cheap nor easily accessible. The NGC allowed its members to quickly and affordably become compliant with the NJDEP's requirements,

Table 14.2 Current membership of the New Jersey Non-Profit GIS Community and their preferred acronyms

Appalachian Mountain Club (AMC)	The Nature Conservancy of NJ (TNCNJ)
Association of NJ Environmental Commissions (ANJEC)	NJ Audubon Society (NJAS)
Bergen Save the Watershed Action Network (BSWAN)	NJ Conservation Foundation (NJCF)
Building Environmental Education Solutions, Inc. (BEES)	NJ Housing & Mortgage Finance Association (NJHMFA)
Center for Environmental Responsibility (CER)	NJ Marine Sciences Consortium (NJMSC)
Delaware & Raritan Greenway (DRG)	NJ RailTrails (NJRT)
East Coast Greenway Alliance (ECGA)	NJ ReLeaf (NJRL)
Friends of Hopewell Valley Open Space (FHVOS)	NJ Water Supply Authority (NJWSA)
Friends of Monmouth Battlefield (FMB)	NY/NJ Baykeeper (NJBAY)
Friends of Princeton Open Space (FOPOS)	NY/NJ Trail Conference (NYNJTC)
Friends of the Rockaway River (FORR)	Oldmans Creek Watershed Association (OCWA)
GeoEnvironmental Research (GER)	Passaic River Coalition (PRC)
Greater Mercer Transportation Management Association (GMTMA)	Paulinskill-Pequest Watershed Association (PPWA)
Great Swamp Watershed Association (GSWA)	Raccoon Creek Watershed Association (RCWA)
Green Pond Environmental Foundation (GPEF)	Rancocas Conservancy (RC)
Heritage Conservancy (Doylestown, PA) (HC)	Red Bank River Center (RBRC)
Highlands Iron Conservancy (HIC)	Ridge and Valley Conservancy (RVC)
Isles, Inc. (ISLES)	Skylands CLEAN (SCLEAN)
Keep Middlesex Moving (KMM)	Sierra Club Coalition of Rutgers University (SC)
Kingston Greenways Association (KGA)	Soil and Water Conservation Society-Firman E. Bear Chapter (SWCS)
Lawrence Brook Watershed Partnership (LBWP)	South Branch Watershed Association (SBWA)
Meadowlinks Meadowlands Transportation Brokerage Corporation (MLINKS)	South Jersey Land Trust (SJLT)
Morris Land Conservancy (MLC)	Stony Brook-Millstone Watershed Association (SBMWA)
MSM Regional Council (MSM)	Upper Raritan Watershed Association (URWA)
Musconetcong Watershed Association (MWA)	Washington Crossing Audubon Society (WCAS)

and thus these NGOs have been able to quickly start applying the technology to community problems in an appropriate manner.

For an NGO operating with a limited budget, the NGC's support (training, technical advice, and hardware/software use) has been attributed as the difference between successful GIS use and development and opting for other less technical projects (Gibson 1998). A crude but rather effective measure of the success of this group is that its membership has quickly swollen to 50 New Jersey NGOs (Table 14.2). It holds regular user-group style meetings in which the more advanced members present their successes and failures as lessons for others.

The integration of the technology into the activities of NGOs has played another significant role. It has brought about a change in cognitive and analytical processes. As explained by Kim Ball Kaiser of the Association of New Jersey Environmental Commissions, the technology has expanded the ability of NGOs to consider less traditional boundaries to problems: 'Before GIS, the world ended at the Township line' (Kaiser 1999). In particular, she cites the ability of technology to integrate data from many sources to facilitate more meaningful representations, such as watershed maps. In this sense, the technology is helping to circumvent some of the problems associated with strong home rule as explained above. Another change in thinking was described by Beth Davisson of the New Jersey Conservation Foundation and the Mendham Township Open Space Trust Committee, who felt that the technology was leading to more 'justifiable or defensible' decisions by changing the criteria used in decisions and allowing for a complete consideration of all properties in a study area 'rather than people bringing parcels to the committee that they just happened to know about (which was the pre-GIS method)' (Davisson 2000).

14.3 SO WHAT? ACCOMPLISHMENTS OF NGO SYSTEMS

As a result of these four factors, NGOs throughout the state have become very active in system development. The interplay of these factors is somewhat reminiscent of John Mayo's (1985) push of technology and pull of society thesis. The first two societal factors play the role of 'pulling' the NGOs into the state's many environmental conflicts. At the same time, the second two technological factors serve to 'push' the NGOs to develop solutions to address the conflicts present.

Simply accepting the free software and data does not assure progress, which makes assessment of system outcomes an important step. The relative newness of the process described in this chapter makes assessment difficult at this time. However, some anecdotal evidence demonstrates the benefits of these efforts. Some of these benefits are direct, such as altered outcomes of public meetings, while others are indirect, like the development of a state-wide parcel-mapping guide (Parrish 1999).

In many cases, NGOs are providing political and technical support for the development of systems at the municipal level. This was evident when SMAC produced a state guidebook for parcel mapping (Parrish 1999): the volunteer editor/coordinator and many of the contributors were NGO employees. The NGO contributors were individuals whose involvement is largely fueled by the combined efforts of the NJDEP and the NGC. Despite assistance from both groups, the NGOs still felt the parcel handbook was an important investment of their time and might encourage local governments throughout the state to become more involved in the automation of this important base layer. Karen Parrish is also working to equip environmental commissioners with data for land resource-related decisions (Parrish and Patterson 1998). This indirect benefit is one way that the NJDEP and NGC may have aided a broader set of geospatial system development efforts than was at first expected.

Direct benefits of the NGO systems are defined, in part, by the missions of the organizations. These organizations often are engaged in efforts to alter land-related resource allocation systems while using geospatial technologies as a tool in that process. For example, the GSWA reports that their ability to produce sophisticated map products (especially in circumstances where the municipality lacks similar resources) has earned them greater influence in local decisions (Parrish and Patterson 1998). When attending municipal planning board meetings and similar public forums, they report that the boards respond strongly to these map products, often treating them as if produced by the board's own staff. Although this benefit lacks the quantitative charm of reduced staff or faster response, it represents the benefit most valued by the NGO community: empowerment.

David Peifer, executive director of the Upper Raritan Watershed Association (URWA), has described a fairly concrete example of this empowerment. A developer had proposed an extensive condominium development on top of a ridge overlooking the township of Bedminster, NJ. Using a free copy of GRASS software and mostly publicly available data, the URWA was able to conduct a viewshed analysis and produce a map showing that the development would be visible from about three-quarters of the township. Although the developer employed an expensive legal defense, Peifer represented the URWA and the community using only inexpensive GIS map products. Still, the technology empowered Peifer to actively participate in the public hearing on the development and succeeded delaying the project and, eventually, altering the plan significantly by pushing back the line of development over 50 yards. Peifer realizes that GIS alone was not sufficient to empower his organization; 'It took a Board that was ready to see the evidence and prepared to act on it' (Peifer 1999).

14.4 A CHANGING LANDSCAPE

After helping many NGOs start using GIS, the NGC has encountered several new challenges in its effort to serve NGOs throughout New Jersey. As with so many non-profits, funding became a significant stumbling block. The NGC was conceived with support from the Victoria Foundation – an arm of the Chubb Insurance Company – which sponsors environmental activities in New Jersey. However, the foundation places emphasis on starting efforts rather than sustaining them. As a result, the NGC currently finds itself without ongoing funding.

Another challenge facing the NGC comes directly from its successes. The technical support that it originally offered other NGOs was of a relatively simple nature – fixing minor software glitches, offering printing assistance, helping applicants for free software, and distributing basic data sets. Having accelerated GIS use by so many NGOs, the NGC now finds itself experiencing an increased demand for advanced assistance, such as sophisticated analysis and more and better data. One solution to the problem has been to offer some advanced assistance on an at-cost consulting basis. This still helps the local non-profits, without taxing the NGC staff. This solution may soon develop into a distinct non-profit organization that offers NGOs low-cost GIS consulting assistance.

The NGC has been successful in providing new data to the NGO community. Even the more technologically sophisticated NGOs prefer to let the NGC collect significant data sets, reformat the files, and redistribute the data on CD-ROMs, thus reducing duplication of effort. The NGC has developed a working relationship with data-distributing agencies, allowing them to get early access to data when they become publicly available.

A new role for the NGC has also emerged: an organizing force for the NGO community. After waiting more than a year for updated land-use/land-cover data from the NJDEP, NGOs were informed in fall 1999 that the department had decided to release data only to municipal agencies. For NGOs who had initiated major projects that depended on these data, this situation was seen as a crisis. The NGC immediately began a letter writing campaign to the NJDEP, and within a matter of weeks the policy was changed to an Internet-based public release of the data. This quick, concerted response to a political problem demonstrates the potential advocacy role for the NGC (Parrish 2000).

One other significant external change may still impact the NGC and its future roles. In 1999, the governor of New Jersey and the state chief information officer formed a state Office of Geographic Information. Although the Office was formally designated to coordinate and direct state-level GIS activities, little is yet known about the long-term role that this office will play. If the office engages in data distribution, establishment of standards, or assistance in community GIS use, it could significantly change the future of the NGC.

14.5 IMPLICATIONS

For future development of PPGIS in other areas of the country, the New Jersey approach outlined in this chapter provides a general template for how to jumpstart groups otherwise impeded by financial limitations. However, the template is not one easily applied to all locations; finding a lead agency to provide such high levels of assistance as those provided in New Jersey and finding a central NGO to serve the others can be difficult. It is also hard to tell if the 'push' of hardware, software, and data provided by the NJDEP and NGC would be enough in a region lacking the strong social 'pull' of environmental problems.

The examples discussed here provide an important demonstration of the value of publicly accessible data as a possible antidote to communities that insist on charging exorbitant rates for access to public data in the name of cost recovery. Had the NJDEP chosen a less suitable cost recovery approach, not only would most of the NGOs have chosen a non-technological path, but the citizens of the state also would have been deprived of representation by the NGOs.

It seems likely that factors 1 and 2, although very specific to New Jersey, could easily be paralleled in other states or regions with similar conditions. The situation in New Jersey can be generalized as one in which external forces (the environment, home rule) created a condition where enough demand for action existed that NGOs could generate strong grassroots support. This situation might suggest that even a sophisticated PPGIS could be threatened by a tranquil situation in which few citizens feel compelled to participate or support their representatives (including NGOs). It also seems to suggest that under the conditions represented by factors 1 and 2, and without the help of the NJDEP and the NGC, these citizen groups could risk marginalization when competing with other groups for resources or attempting to sway decisions.

One of the lessons here is that the open nature of the NJDEP was the first step toward democratization. This begs the question: Could the first step toward broader public participation and citizen empowerment simply be encouraging more data producers to engage in the basic democratic act of free and open access?

What seems most clear is that the dynamics of participatory systems are enormously complex because they include both direct and indirect participation. This means that identifying the extent of participation may become increasingly difficult as citizens learn to support and rely upon these groups for the employment of sophisticated GIS technologies.

ACKNOWLEDGEMENTS

None of this research could have been performed without the help of the various members of the New Jersey Non-Profit GIS Community and its director, Doug Schleifer, to whom I am extremely grateful. This research was conducted with support from the New Jersey Agricultural Experiment Station (Hatch No. 84101).

REFERENCES

Davisson, B., New Jersey Conservation Foundation and the Mendham Township Open Space Trust Committee (2000) Personal email communication with author, 28 January.

Gibson, A., Former Project Coordinator for the Passaic River Coalition (1998) Interview with the author, 20 August.

Kaiser, K. B., Association of New Jersey Environmental Commissions (1999) Interview with the author, 1 December.

Mansnerus, L. (1998) 'Home rule: a history of defeat', *New York Times*, New Jersey Section, 27 September, p. 8.

Mayo, J. S. (1985) 'The evolution of information technologies', in B. R. Guile (ed.) *Information Technologies and Social Transformation*, Washington, DC: National Academy Press, pp. 7–33.

Natural Resources Conservation Service (1999) 'Summary Report 1997 National Resources Inventory', (http://www.nhq.nrcs.usda.gov/NRI/1997/summary_report/report.pdf) Washington, DC: Natural Resources Conservation Service, United States Department of Agriculture.

New Jersey Department of Environmental Protection (1996a) *GIS Resource Data: Southern New Jersey*, New Jersey Department of Environmental Protection (CD-ROM), Series 1, vol. 1.

New Jersey Department of Environmental Protection (1996b) *GIS Resource Data: Central New Jersey*, New Jersey Department of Environmental Protection (CD-ROM), Series 1, vol. 2.

New Jersey Department of Environmental Protection (1996c) *GIS Resource Data: Northern New Jersey*, New Jersey Department of Environmental Protection (CD-ROM), Series 1, vol. 3.

New Jersey Department of Environmental Protection (1996d) *GIS Resource Data: Tidelands Claim Maps and Integrated Freshwater Wetlands with Land Use/Land Cover*, New Jersey Department of Environmental Protection (CD-ROM), Series 1, vol. 4.

New Jersey Department of Environmental Protection (1997) *GIS Tools for Decision Making: Mapping the Present to Preserve New Jersey's Future*, New Jersey Department of Environmental Protection (CD-ROM), Series 2, vol. 1.

New Jersey Non-Profit GIS Community (1997) *NGC Newsletter*, August.

Parrish, K. (ed.) (1999) *Digital Parcel Mapping Handbook: Standards and Strategies for New Jersey's Parcel Mapping Communities*, Chicago, IL, Urban and Regional Information Systems Association (published previously as K. Parrish (ed.) (1999) *Digital Parcel Mapping: Standards and Strategies for New Jersey's Parcel Mapping Communities*, Trenton, NJ, New Jersey State Mapping Advisory Committee).

Parrish, K., Project Director for the Great Swamp Watershed Association (2000) Interview with the author, 21 January.

Parrish, K. and Patterson, K., Project Director and GIS Specialist for the Great Swamp Watershed Association (1998) Interview with the author, 26 August.

Parrish, K. and Walmsley, A. (1997) *Saving Space: The Great Swamp Watershed Greenway an Open Space Plan*, New Vernon, NJ: Great Swamp Watershed Association.

Patterson, K., GIS Specialist for the Great Swamp Watershed Association (1999) Interview with the author, 7 December 1999.

Peifer, D., Director of the Upper Raritan Watershed Association (1999) Interview with the author, 30 November 1999.

Schleifer, D., Director of the New Jersey Non-Profit GIS Community (1998) Interview with the author, 20 August.

Tulloch, D. L. (1999) 'Theoretical model of multipurpose land information systems development', *Transactions in Geographic Information Systems* 3(3): 259–283.

Chapter 15

Mexican and Canadian case studies of community-based spatial information management for biodiversity conservation

Thomas C. Meredith, Gregory G. Yetman and Gisela Frias

> How does one obtain reliable data within...a framework where nothing is constant and everything is on the move? [The] best one can do...is to accept that there is not any one desirable and sustainable state for society – only near continuous transition, often coupled with the impossibility to forecast even the near future. [Successful adaptation requires that] the system – whether an individual or a social system – collects information about its own functioning, which in turn can influence that functioning.
>
> Felix Geyer (1994: 18)

15.1 ADAPTATION, SUSTAINABILITY AND PPGIS

Sustainable development has come to summarize the acknowledged importance of non-destructive land-use. The idea has become widely accepted – perhaps because of its inherent constructive ambiguity, or perhaps because, like motherhood and apple pie, it is simply a notion that is hard to argue against. But unlike motherhood, it is not something to which an irrevocable commitment can arise from a moment of irrational passion and, unlike apple pie, it has no simple recipe. The challenge, as Geyer (1994) observes, is: How can dynamic communities with changing needs, aspirations and technologies maintain a non-destructive relationship with an environment that is itself dynamic and constantly changing? This clearly requires an adaptive process, and in the time frame that matters to us now, that adaptive process needs to be based on human intelligence and environmental information. Finding ways to optimize the use of available information and ensure that all providers and users of information have effective links to decision-making processes is an essential step towards sustainable development. GIS provides tools to discover, analyse and communicate the spatial relevance of data and information. A critical question still remains, however: How can high technology information management tools be brought into the public forum in a way that fosters fairness and increases decision-making competence (Webler 1995) rather than increasing polarization and marginalization?

This chapter describes community-based research intended to bring local spatial information into public consciousness and build local capability to manage and use that information. It focuses on two initiatives in mountain forest villages that are experiencing rapid environmental change. One of these initiatives is taking place in Invermere, British Colombia, Canada, located in the Upper Columbia Valley between the Rocky Mountains and the Purcell Mountains in an area that is ecologically diverse and largely unspoiled, but under competing land-use pressures. The other initiative is taking place in Huitzilac, Morelos, Mexico, in an area of spectacular mountain forests less than one hour's drive south of Mexico City. In both cases, groups that involve academic researchers and local citizens manage the projects.

This chapter explores two particular issues arising from the research initiatives: (1) barriers to information flow (Meredith 1997a); and (2) the impact of access to information on the dynamics of community adaptation (Meredith 1997b). The conclusions of the chapter are three: (1) PPGIS outcomes may be determined by data selection that is constrained or even arbitrary; (2) the best GIS technology will always, by definition, be ahead of the public's ability to participate; and (3) with PPGIS, the process is the product – that is, by the time the public has become involved in generating or understanding a system, the educational and analytical benefits of public participation may already have been achieved.

15.1.1 Rural communities' role in environmental protection: the socio-cybernetics of conservation

Anthropologist John Bennett (1993) wrote that the requisites for achieving sustainability (a dynamic balance between resources and sustenance) are nothing short of a 'restructuring of human purpose and a total reassessment of cultural, political and moral problems' (p. 79). Environmental management decision-making is an essential element of this restructuring; environmental management decisions are 'about human behaviour rather than physical things' (Grumbine 1997: 42). For these practical reasons alone (i.e. without invoking ethical and equity considerations at all), public participation in environmental planning is essential (Fisher 1996; Pepper 1996).

Rural communities are the custodians of many ecological resources. They are often economically dependent on those resources, but at the same time, their citizens have a great appreciation of the rural landscape. This sometimes leads to local conflicts – in the worst cases to a 'downward spiral' of environmental degradation that leaves 'habitats half protected, rural economies weakened and personal principles bargained away' (Johnson 1993: 16). More effective decision-making is needed for effective local adaptation. Public participation provides a promising option but requires radical changes in information management skills. Geyer (1994) notes that for successful adaptation, 'the minimum requirements... are self-observation, self-reflection and some degree of freedom of action' (p. 11). This sequence – perceive, interpret and respond – is the foundation of sound decision-making and it is contingent on effective information flow. Section 15.2 discusses some of the barriers that were observed in the Invermere case study.

A second issue relates to the conceptual framework for spatial decision-making. The concepts of systems theory, and in particular of cybernetics, provide an analytic paradigm for assessing the role of environmental decision-making. This requires a distinction between first- and second-order cybernetics (Geyer 1994). First-order cybernetic systems are those external to an observer; second-order are those of which the observer is part. In second-order systems, the observer's understanding of the system becomes part of the system. In environmental management, this is the difference between decision-makers who are part of the ecosystems they are managing (community-based) and those who are external to those systems (technocratic).

The concept of *rational expectations* in the field of economics recognized that the way systems function is based not just on externally measurable or quantifiable parameters of the economy, but also on what human members of the economic system know about those measures. Observers are seen to be part of the system, so their perception, interpretation and response are also part of the system. This concept radically altered economic research, and arguably, its relevance and impact. So might recognition of the role of community-level information users alter the theory and practice of environmental modeling and planning. For example, land cover change modeling based on Markov chains or on logit regression assumes that what has happened in the past will happen again. But as Scott Adam (1997) glibly puts it as 'any doom that can be predicted won't happen' (p. 6).

Viewing community–ecosystem interactions as cybernetic systems can shape our understanding of environmental problems and solutions. If exploring land-cover change at the community level alters the perception and awareness of the causes of change, the causes themselves may be altered. Geyer (1994) asked whether science should support concentrated technical capability and therefore centralized planning or, rather, 'strive to improve the competence of factors at the grassroots level so that these factors can steer themselves and their own environment with better results?' (p. 13). Geyer's own arguments strongly support the latter. PPGIS can make a contribution; surprisingly, the process itself might be more important than any concrete product it generates. This is explored in Section 15.3, which considers the Mexican case study.

15.2 BARRIERS TO INFORMATION FLOW: THE CASE OF INVERMERE, BC

The Upper Columbia River Valley is very diverse ecologically: within a few miles one can find vast permanent wetlands, semi-arid grassland benches, dry Douglas fir forests, montane spruce-fir forests and alpine tundra. Despite economic strategies that have included over the years from fruit production, mining, forestry, ranching and tourism, the valley has remained relatively undeveloped and has attracted residents who are drawn by, and appreciate, the generally unspoiled landscape. Invermere is the largest of several settlements.

The economic, recreational and aesthetic character of the community is bound up in the environmental quality and so, understandably, the range of perspectives on environmental issues is diverse. Local stakeholders are now involved in commercial activities such as forestry, ranching and nature-based tourism, and in personal activities such as hiking, hunting, fishing or simply nature appreciation. Clearly, differences in *values* will cause disagreements between stakeholder groups. For example, clearcut logging is simply seen in different ways by loggers – whose livelihood derives from the practice and who can point to healthy second growth forests as proof of the viability of the practice – and, say, amateur naturalists – who see nothing but ecological wasteland in the clearcuts and simplified artificial monocultures in the second growth. These value differences may be very difficult to overcome. But in addition to differences in values, differences in the perception of *facts* can also cause disagreements between stakeholder groups, and these differences can more readily be overcome through information management. This is the intention of the Invermere project.

There are two GIS-related facets to the project. The first is an effort to create an environmental atlas which will help present information about local environments and thereby help support community-based environmental decision-making (Figure 15.1). The procedures, in brief, were to involve members of the local community in identifying: (a) priority issues; (b) data needs; (c) data sources; (d) information 'targets'; and (e) communication strategies. The second facet was an effort to produce a dynamic land-cover change map for the region based on satellite imagery (Figure 15.2). These exercises led to the discovery of a number of disempowering realities which can be considered as 'barriers to information flow'. These are discussed in the order they would typically be encountered. The technical and communication barriers are of most interest to PPGIS concerns.

Dispersion of data sources The most obvious barrier is ignorance of the fact that specific information, or even of a class of information, exists. In the case of the environmental atlas, this proved to be one of the most challenging obstacles. Amassing an inventory of reliable, current and relevant

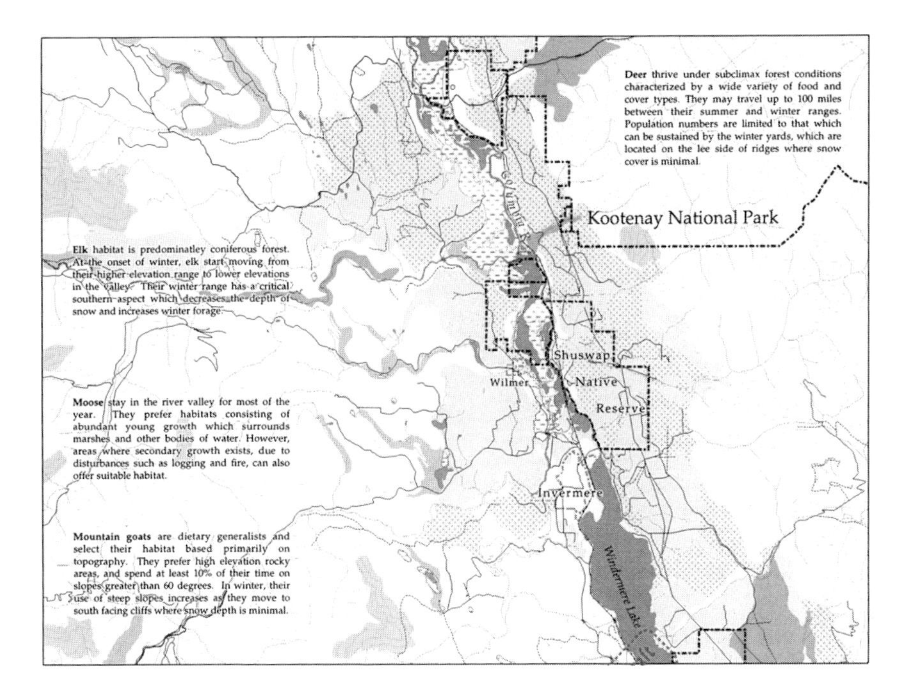

Figure 15.1 Ungulate habitat map from the environmental atlas. Ungulate habitat maps, along with seven other environmental theme maps, provide a clear distillation of complex data that are important to local environmental perception and decision-making. The exercise that led to the generation of these maps demonstrated, however, that there was considerable difficulty in getting *closure* on data sources. In other words, it seemed that we were always learning about a new potential data source, but we could not always get it, get it in a format that was usable, or get it with enough meta-data to verify its utility. This raised the fear that final maps may sometimes represent an arbitrary selection of data. (Atlas pages produced under the supervision of Richard Bachand.)

data meant canvassing agencies of three levels of government as well as crown corporations and private companies (logging and mining firms), and international agencies and NGOs (bird, wildlife, hiking, and hunting groups). Each new data source opened the door to other possibilities. The investigative effort (time, cost, and skill) is not within the grasp of most communities. This suggests that, perhaps inevitably, the data used in decision-making are not necessarily the best, but rather those most easily encountered! Addressing this barrier does not involve generating more information, but rather facilitating access to what already exists.

Legal barriers Forest inventories are expensive. It makes no sense for expensive data to be collected again each time a new user desires them. Yet,

Study Region:
The Windermere
Valley, B.C.

- 1991 Landsat TM image draped over elevation data

Figure 15.2 Satellite image draped over a DEM of the Upper Columbia Valley. Satellite imagery, with appropriate technical manipulation, provides dramatic new perspectives for local residents. With classification and expert interpretation, it can also provide them with valuable new information. However, the complexities involved in generating and interpreting images can still leave the public dependent on experts whose assumptions and technical limitations they may not fully understand. Complexity remains a barrier to full public participation. (Image produced by G. Yetman.)

there is no obvious basis for shared access to such data. The data a logging company may require for economic planning may also have considerable significance in environmental conflicts. Private survey data and opinion polls, likewise, may not be openly accessible despite their potential importance, and information about corporate activities (past, present, or planned) may be closely guarded. Census data cannot legally be disaggregated to the level that makes it meaningful at the local level. This mean that each stakeholder group works with a maximum data set that is only a subset of the total. Consequent disparities may be significant.

Financial barriers Financial limits to data access are inevitable. These limits may be at a very low level (e.g. it may be impossible to hire a project worker to conduct basic background library searches) or at a high level (research groups may not be able to buy expensive imagery or hire technical experts capable of using it). In this project, we were told that some map files could be made available to us only if we covered the wage of the technician required to retrieve them. Some data, including satellite imagery, are collected at public expense and then sold on a cost recovery basis. This barrier can be discriminatory: the real costs of collecting images are high, but the marginal costs of using them once collected are modest.

The question of unequal access to public data is an important one. These first three barriers suggest that, especially with PPGIS, there are real risks that data sets used may be severely constrained or even arbitrarily determined.

Technical barriers The rate of change in electronic data acquisition, storage, transmission, analysis and presentation is such that only trained specialists stay at the cutting edge of progress. Clearly, it is not possible for all potential users to acquire and maintain the requisite technical skills to use them. This technical barrier is inevitable. The question is not whether, but rather where, it exists and what its implications are. In the land-cover change study, we expected to conduct a demonstration exercise that could be replicated in the future within the community. In the process of classifying and comparing two satellite images, 1974 and 1991, we encountered problems of rectification, pixel size differences, band differences, image positioning, haze correction and aspect compensation (Yetman 1999). All of this meant that the community-based work we had originally proposed sank deeper and deeper into the technical space of our GIS lab and further from the understanding of community partners. This limited community control of the process as well as their capacity to verify results. The paradox of the desirability and simultaneous inaccessibility of advanced technology is further discussed below.

Paradigms of interpretation There are elements of local environmental change that may appear disparate and unconnected, but which are in fact consistent with existing theories or models. In this case, two such issues emerged: the relation between recreational road access and the viability of grizzly bear habitat, and the relationship between forest practices and stream hydrology. Different stakeholder groups interpreted connections in ways that permitted very different conclusions.

Non-conventional data The outcome of negotiations are often predetermined by the definition of the context, the terms of reference and the pivotal issues. The ability to set the agenda of a negotiation process may be the single most important part of the negotiation. Community groups have access to many forms of data and information that describe qualities of the community itself and are therefore not available from any source other than from the community. Traditional ecological knowledge, local spirituality, aesthetic and amenity values are relevant in the Invermere case. These values may be downplayed by stakeholders who have other value sets and priorities. If environmental negotiations are couched in the established frameworks of the legal profession or the scientific community, local community groups may be accepting, a priori, a handicap. Building effective PPGIS may mean learning how to codify and communicate non-conventional data.

Barriers of communication GIS and the closely related tools of automated cartography and digital communication can make dubious information appear compelling. Conversely, many sound positions have been lost because they were not communicated effectively. PPGIS will certainly help community groups make information look better. This will mean, of course, that community groups will become as vulnerable as any other groups to the GIGO (garbage in, garbage out) hazard. As noted with respect to technical barriers, the cutting edge of communication technology is always advancing, and only the specialist will be comfortable working at the vanguard. By definition, without direct access to specialists, the general public could be marginalized.

15.2.1 PPGIS and the specialist

This list of barriers represents reasons to hope that equitable access to information can become a reality; that is, each barrier can be addressed and potentially overcome. The exercise in Invermere showed the size of the 'information mountain' that needs to be climbed, but it also helped move the community part way up the slope. The technical barriers are perhaps the most interesting as these are structurally embedded in the way technology advances. Clearly, specialists will always be aware of cutting edge technologies that may significantly enhance the capacity of analysts to interpret situations and reach decisions. By definition, this 'moving front' will always be out of reach of lay users. Very user-friendly systems that the public is, or can become, comfortable with are necessarily some way back from the leading, exploratory edge of the evolving field. This reality requires that a mechanism be incorporated in PPGIS to address the gap.

15.3 GIS AND THE SOCIO-CYBERNETICS OF HUITZILAC, MEXICO

The site in Mexico was selected because of local concerns about changes in, and current pressures on, forested areas that lie immediately to the south of Mexico City. This area like others nearby was isolated by steep topography.

But population increases and the opening of access roads have exposed these areas to pressures of urban expansion (Ezcurra 1990). The urban footprint of Mexico City expanded from less than 30 km^2 in 1910 to almost 1,200 km^2 by 1990 (Ezcurra and Mazari-Hiriart 1996). The forests have been protected by inherent properties of the landscape: the mountains rise steeply to almost 3,500 m, many areas are quite inaccessible and soils are young, thin and easily eroded. The site illustrates several aspects of the second-order socio-cybernetic process that can be directly supported through GIS. For example, in early discussion with the local conservation group, we concluded that three issues of scale could be treated through a partnership employing GIS. These are briefly outlined below. The 'self-steering' sequence of events that followed is then discussed.

Temporal scale – the past Ecological changes that take place over a human life span may be considered insignificant because they are so slow. The mountains of the area are about 400,000 years old. Pre-Hispanic civilization may have had some impact on the area for about 1 per cent of that time, recorded history accounts for another 0.1 per cent, living memory about 0.01 per cent, and the planning horizon is perhaps 0.001 per cent. We were able to locate early maps of forest cover and superimpose areas of forest loss on colour composite satellite images of the region. GIS images that show the present state of forest cover and losses over several decades help highlight ecologically important transitions.

Temporal scale – the future Because ecological systems are complex, it is often difficult to predict the cumulative or mid-range effects of human action. For example, how can one predict the effects on ground water in Huitzilac of a 10 per cent loss of the forests, 10 per cent more domestic waste, or a 10 per cent reduction in rainfall? Modeling with GIS permits investigation of alternative scenarios and can make communication of concerns more effective. It is very simple to demonstrate what the region will look like if residential expansion continues at the same rate for the next 20 years, or if as much forest is lost in the next 20 years as was lost in the last 20.

Spatial scale GIS can help explore relationships between local, regional and national or international perception of resource issues. The forests of Huitzilac are a source of fire wood and medicinal plants for local people; they are an important regional source of building materials; they regulate water supply for people in the south of Morelos; they serve as recreational and residential sites for the population of Mexico City; and, nationally and internationally, they are recognized as both genetic resources and carbon sinks. Spatial data were used to demonstrate two outward links: one with water management in the south of Morelos (showing how local drainages are linked with major rivers that supply other regions) and the other with the possible consequences of the expansion of adjacent urban areas. This visualization of spatial 'nestedness' helps to demonstrate connections that affect local decision-making.

15.3.1 Local adaptation

What has been most interesting about this work is the extent to which initiating, focusing and participating in community-based discussions has influenced the community. Consider the difference in impact if exactly the same steps in data management had been taken by an outside agency. It would have looked at rates of forest conversion, population trends, land use trends, stated policy objectives and other data sets and made predictions about what was likely to happen and, depending on what the outsiders had been told about what was desirable, they would have recommended policy action to convert what is happening on the ground to what they think should be happening. This is first-order cybernetics. Instead what has happened (though of course it is in early stages) is that an evolving self-regulation system has emerged as people begin to think about factors that affect them directly. For example, the first major concern identified through local consultation was waste management. By the time structures were in place to collect reliable data about the nature, scale, causes and consequences of the waste issue, people in the community had become waste conscious and had begun to eliminate the very problem they

(a)

(b)

Figure 15.3 Community mural painting effort grew out of the organizing process of PPGIS. In the Mexican case study, we started out with the intention of using GIS to support local environmental decisions, but we wanted local people to define priorities independently of our prior agenda. This led to the creation of partnerships with local environmental groups, to widespread discussion on an array of environmental issues, and to workshops that solicited community concern. Map-making through a community mural painting event demonstrates how a PPGIS initiative can have surprising outcomes. These outcomes may achieve what the PPGIS initiative was intended to achieve long before the computer systems are up and running. This suggests that the real *success* of a PPGIS may be in the process as much as in any final product, and it may therefore be very difficult to measure. (Photos by G. Frias.)

were proposing to study. The *process* of the study became an educational – and therefore an adaptive – part of the system under study (Figure 15.3). It was self-reflective, and represented second-order cybernetics.

The lesson from this study is that first- and second-order cybernetic approaches to environmental data management are completely different in their impact. From the standpoint of simplicity of design, ease of execution and detached simulation, studies that do not include communities as partners may be preferred. But just as rational expectation theory attempted to reconnect economics with the real world – and in so doing made it messier and more susceptible to the vicissitudes of human will – so must community-based environmental research demonstrate that community participation in information management is essential.

15.4 CONCLUSION

The second-order cybernetic model implies that some form of PPGIS is essential for sustainable development. Three observations can be derived from these case experiences. First, there are real dangers that the effectiveness of local decision-making may be seriously curtailed by data limitations. Financial and legal barriers may prevent access to critical data, but the problems of gaining access even to widely diffused public domain data create the possibility that groups will use data not because they are the best, but because they are easy to get or are the first they encountered.

Second, in a rapidly evolving field like geographic information analysis, 'state-of-the-art' technology is exclusive, and by definition somewhat inaccessible to the general public. If the best available technology is desired, part of the design of an effective PPGIS has to be mechanisms for including effective liaison between specialist managers or resource personnel and the general public.

Finally, evoking Marshall McLuhan's famous dictum 'the medium is the message', at least in some cases of PPGIS, the 'process is the product'. The 'medium' of GIS requires new ways of thinking about environments and about spatial data. Programs that bring the public into the new medium change the public, and the 'message' of PPGIS is received *de facto*. Where this message is a new understanding of environmental decision-making, it may well be that the goal of empowerment through PPGIS will already have been achieved by the time a system is up and running. In fact it may not matter if the system never does get up and running. The process of building local capability and motivation – a process that is part of setting up a PPGIS – may be the very product we are hoping to produce.

REFERENCES

Adams, S. (1997) *Dilbert future: thriving on stupidity in the 21st century*, New York: Harper Collins.

Bennett, J. W. (1993) *Human Ecology as Human Behavior*, New Brunswick, USA: Transaction Publishers.

Ezcurra, E. (1990) 'The basin of Mexico', in B. L. Turner II (ed.) *The Earth as Transformed by Human Action*, Cambridge: Cambridge University Press, pp. 577–588.

Ezcurra, E. and Mazari-Hiriart, M. (1996) 'Are mega cities viable? A cautionary tale from Mexico City', *Environment* 38(1): 6–26.

Fisher, J. (1996) 'Grassroots organizations and grassroots support organizations', in Moran (ed.) q.v., pp. 57–101.

Geyer, F. (1994) 'The Challenge of Sociocybernetics', paper presented at 13th World Congress of Sociology, Bielefield, http://construct.haifa.ac.il/~dkalekin/cyber1.htm

Grumbine, E. (1997) 'Reflections on "What is Ecosystem Management?"' *Conservation Biology* 11(1): 41–47.

Johnson, K. (1993) 'Reconciling rural communities and resource conservation', *Environment* 35(8): 16–28.

Meredith, T. C. (1997a) 'Information limitations in participatory impast assessment', in John Sinclair (ed.) *Environmental Impact Assessment in Canada*, University of Waterloo Press, pp. 125–154.

Meredith, T. C. (1997b) 'Making knowledge powerful: Mexican village project uses environmental information technology to strengthen community voices in biodiversity conservation', *Alternatives* 23(4): 28–35.

Pepper, D. (1996) *Modern Environmentalism*, London: Routledge, 376pp.

Webler, T. (1995) '"Right" discourse in citizen participation', in O. Renn, T. Webler and P. Wiedermann, *Fairness and Competence in Citizen Participation. Evaluating Models for Environmental Discourse*, Boston: Kluwer Academic Publishers, pp. 35–88.

Yetman, G. (1999) 'Spatial information and environmental decision-making: the Windermere Valley, British Columbia', Masters Theses, McGill Department of Geography, McGill University, Montreal, 119pp.

Chapter 16

Promoting local community participation in forest management through a PPGIS application in Southern Ghana

Peter A. Kwaku Kyem

This chapter describes a PPGIS project at Kofiase in the Ashanti Region of Ghana that was implemented to help build collaborative forest management institutions in the community. The chapter begins with a brief introduction, followed by a description of the physical and socio-economic conditions of the study area. The GIS exercises that were implemented within the community are explained. The results of the study, including problems and challenges encountered in the implementation of the project are then discussed. The chapter concludes with a discussion of lessons drawn from the study.

16.1 INTRODUCTION

Ghana's forest reserves were established to protect the country's main export crop (cocoa) from the ravages of dry harmattan weather and to secure future supplies of timber and wood (Logan 1946). The state assumed responsibility for constituting and managing forest reserves on lands that were, and are still owned by ethnic groups and individual families (Belfield 1912; Asante 1975). Recently, the forest reserves have become important sources of capital for Ghana's economic development, and income for a majority of rural dwellers. At the same time, annual forest fires and an increasing reliance on the forests to meet domestic needs of natives who live close to the reserves have generated intense land-use conflicts (Ghana News Agency, 2 March 1993: 16). After coercive methods failed to prevent local inhabitants from destroying the forests, collaboration seemed the best option for protecting the country's rain forests. Consequently, in the late 1980s, Ghana adopted a new land-use policy and directed the country's Forest Department to integrate local communities into the management of the country's forest reserves (Norton 1989; CFMU 1993). The collaborative forest management project (CFMP) required a major reorientation of existing forest management practices to ensure responsiveness to the needs of local community groups, and to empower and enable communities to fully participate in the collaborative process.

By the time of the project, the Forest Department had acquired several GIS and remote sensing facilities. In 1993, the unit developed its own GIS called FROGGIE (Forest Reserves of Ghana: Geographical Information Exhibitor) for storing and displaying information on the biodiversity of plant species in the country's forests (Hawthorne 1993). The integration of GIS into the collaborative forest management project made it likely that issues of unequal access to data, technology, and expertise would reinforce the status quo and work against free and open discussions. On the other hand, active involvement of people in the communities was essential to the resolution of the conflicts that necessitated the establishment of the collaborative forest management project (CFMU 1993). While a graduate student at Clark University in Worcester, Massachusetts (USA), I attended an Africa GIS conference held at Tunis, Tunisia, in 1993, and heard about the forest management project. Motivated by the favourable changes in official forest management practice and the large-scale presence of GIS facilities within the Forest Department, I sought and obtained a grant from the Rockefeller Foundation in New York to implement a PPGIS project in the region to demonstrate an alternative, less elitist applications of the technology (Kyem 1997). A PPGIS methodology was designed around decision support procedures available in IDRISI for Windows GIS software. Assisted by a team of foresters from the Collaborative Forest Management Unit (CFMU) of the Forest Department, the methodology was implemented in four communities in the Ashanti region of Ghana, to facilitate the establishment of collaborative forest management institutions. This report covers PPGIS activities that took place at Kofiase.

16.2 STUDY AREA

Kofiase is located about 40 miles northeast of Kumasi, the capital of the Ashanti Region, and 250 miles north of Accra, the capital of Ghana (Figure 16.1). There are about 2,000 people in Kofiase and the surrounding villages. The majority of the inhabitants are Ashantis, who are part of the Akan people of West Africa. Also, the area has a large population of migrant farmers from other parts of the country. The majority of the people are subsistence farmers who currently produce staple crops such as plantain, maize, and assorted vegetables for sale in Kumasi and nearby towns. In the past, the area supported a booming cocoa industry. Logging used to be the second source of employment until 1983, when wildfires began destroying cocoa farms and the local forest reserve. The Aboma Forest is an important source of income, construction materials, protein supplements, drinking water, fuelwood, and many of the domestic needs of people in the community (Falconer *et al.* 1994).

The community is organized around matrilineal kinship groups (clans). Several clans usually live together in villages and towns. There were as

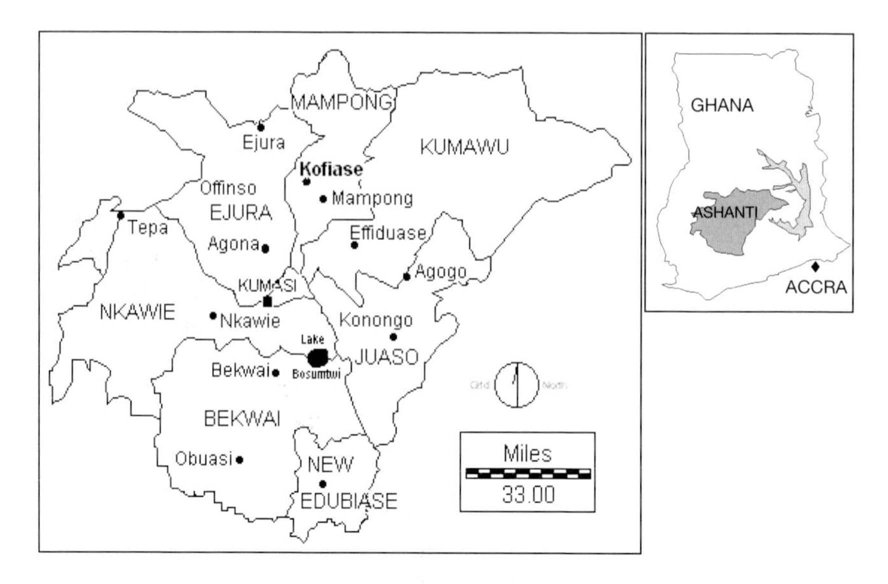

Figure 16.1 The study area: forest districts in the Ashanti Region of Ghana.

many as nine such Ashanti clans at Kofiase. Traditional authority in each village is represented by a male chief (Odikro) supported by a queen mother (Ohemaa) and a council of elders comprising heads of various clans and the migrant communities. Wiredu (1993) has noted that among traditional Akans, of which the Ashantis are a major subgroup, conflicts are resolved through dialogue. All parties are represented in councils of governance, and decisions are reached by consensus. He explains that consensus might not always mean total agreement, but it requires a dialogue in which all voices are heard and in which options proposed by minority constituents are seriously entertained. Thus, within the various traditional councils, both the majority and minority groups have substantial roles to play in the determination of policy. These traditional administrative arrangements prepared local representatives for the open discussions that occurred within the PPGIS project.

16.3 PREPARATORY PHASE OF THE PROJECT

Initially, we conducted a survey to identify institutional actors, including organizations, social groups, and individuals who had a direct and significant stake in the management of the local forest. An investigation was conducted into the socioeconomic life of the people, the main resource allocation problems, and power relations that existed among various forest user groups within the community. Stakeholder analysis, historical and participatory mapping, ranking, and brainstorming techniques were adopted to carry out the initial investigations.

16.3.1 Stakeholder analysis

Through interviews and direct involvement with people in the community, we identified several forest user groups (FUGs). Based on preliminary results of the survey, a Collaborative Forest Management Committee (CFMC) was established at Kofiase in early 1995 (Danso 1995). The committee comprised 21 members, 15 were natives of Kofiase and the remaining six were professional foresters from the local District Forest Office at Mampong. The local representatives (nine men and six women) included retired and active teachers, civil servants, traders, and representatives of various interest groups in the community. Many of the participants were literate, and several were well-informed about forestry issues in the community. In spite of their experience and the fact that they owned the land on which the Aboma Forest was located, the local people had no input into how the forest was managed (Baidoe 1972; England 1993). The collaborative forest management project therefore provided an opportunity for the community to have an impact on the management of the local forest. However, the outcome of the project was uncertain because of hostile relations between local inhabitants and the professional foresters who managed the local forest reserve.

16.3.2 Participatory mapping and ranking

With the local forest committee in place, we organized workshops to introduce members of the forest committee to participatory mapping and interpretation of maps produced with GIS. Each mapping session began with a discussion. The group then proceeded to draw maps showing the geographic location of the forest, its land-cover categories, and the location of specific forest resources traditionally used by the people (i.e. canes, building materials, game and wildlife). The maps were initially drawn on the floor and then transferred onto a flip chart using symbols that were understood by all members. We then input the maps into the GIS to make copies for distribution to participants to facilitate discussions. The mapping exercise provided an overview of the current state of the forest reserve, and served as the starting point for discussions of forestry problems in the area. Historical maps of the forest were compiled to provide evidence of changes that had occurred in the forest. Members of the forest committee were later asked to list and rank their preferences for uses of the forest. The ranking revealed strong agreement on the need to preserve the local Aboma Forest Reserve. However, the participants failed to agree on the allocation of some resources.

16.4 IMPLEMENTATION PHASE OF THE PROJECT

After the mapping exercise, the committee began discussions that focused on resources in the forest. The group was led to identify the community's main social and economic needs in regards to resources available in the forest. Foresters on the committee also explained the goal of official forest management to the group. Maps of the forest were then overlaid to help trace the linkages among the resources in the forest. A composite map was compiled from an overlay of several thematic maps of the forest, including the forest cover, streams, roads, towns, and other important features within the reserve. The group visited the sites of streams and many of the resources to gain firsthand knowledge of forest conditions represented in the resource maps. We used the resource mapping exercise to initiate discussions of ecological conditions within the forest. We drew the participants' attention to relationships that existed between such features as hills and local streams, slopes and erosion, and the sources of streams that supplied water to villages within the community. Other discussions focused on the impact that deforestation had on erosion and the flooding of local streams.

We then assisted the group to identify major threats to the Aboma Forest Reserve. The discussion generated multiple views about local forest problems and some solutions as well. Two of the most significant threats to the Aboma Forest Reserve identified by participants were: (1) the danger posed to the remaining forest by annual wild fires; and (2) a dispute over logging and forest preservation that threatened the peace enjoyed by people in the community. The group decided to use GIS to design a fire-monitoring plan, and to attempt to manage the conflict over simultaneous allocation of portions of the remaining forest to loggers and forest preservationists.

16.4.1 Designing a fire monitoring plan for the reserve

The Aboma Forest Reserve is located on the fringe of the dry savanna grassland of Ghana, and is subjected to damage from annual wildfires. As a result, a large portion has been destroyed by fire and converted to teak plantations. Prior to designing the fire hazard potential map with the GIS, the group visited the forest several times to assess and record previous damage caused by fire. A fire damage map of the forest was created that classified the reserve into four categories: (1) no fire damage; (2) little fire damage; (3) mild fire damage; and (4) severe fire damage (Figure 16.2). The map was digitized and stored in the GIS. We began the evaluation of the fire hazard potential of the forest with the GIS by helping the group identify evaluation criteria. Four factors (roads, villages, farms, and slopes) were identified and digitized from official maps of the forest. The factors were processed and standardized in the GIS, and were weighted to reflect each factor's significance in the evaluation. They were later processed in the GIS to produce a potential fire hazard map of the Aboma Forest Reserve. The map was classified in terms of fire risk using four categories: (1) very low risk; (2) low risk; (3) moderate risk; and (4) high risk (Figure 16.3).

Copies of the fire hazard potential and fire damage maps were distributed to participants to facilitate discussions. The group visited the forest to validate the fire hazard potential categories that resulted from the GIS analyses. After this, they adopted the map for the design of a fire-monitoring plan for the Aboma Forest Reserve.

16.4.2 Managing conflict of interests in forest resource exploitation

In the months leading to the Kofiase project, a logging company that held the concession to timber resources in the Aboma Forest Reserve decided to

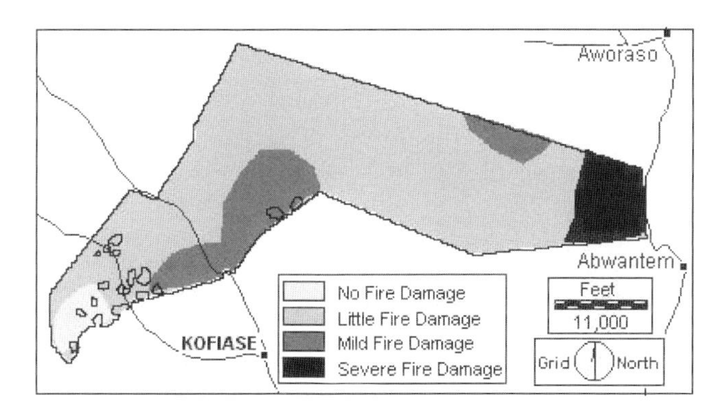

Figure 16.2 Aboma Forest Reserve, fire damage map.

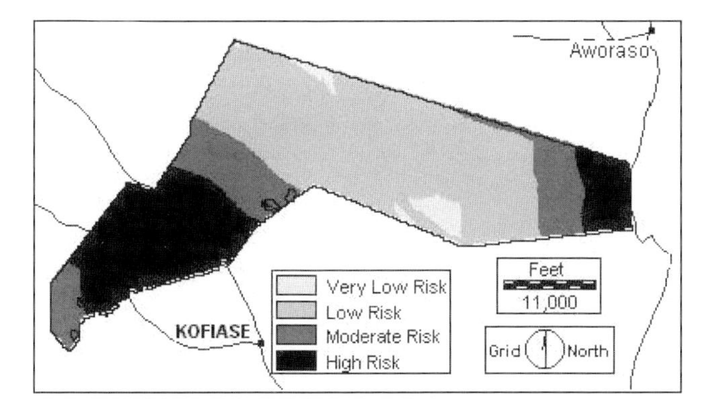

Figure 16.3 Aboma Forest Reserve, fire hazard potential map.

log the remaining trees. The decision divided the community between inhabitants who backed the loggers and those who wanted to preserve the remaining forest. Soon after the formation of the forest committee in the village, we invited supporters of both sides to a meeting to attempt to resolve the conflict using GIS. The demands of the two groups were in direct conflict, so the preferences of each party were treated as an independent multiple-criteria evaluation problem. The results of the single-objective solutions were then used to resolve conflicting claims to the forest resources (Eastman *et al.* 1993). Acting as focus group leaders, we held separate discussions with the parties to identify conditions that fulfilled their interests. The group that supported logging requested 350 hectares of forest that was stocked with some of the most valuable timber. We helped the group identify criteria that would be used to evaluate the suitability of the forest for logging. Four factors (roads, slopes, towns, and land-cover) were identified and processed into the GIS. Each factor was weighed, and after standardizing, the factors were multiplied by their associated weights. The GIS was then used to produce a logging suitability map for the forest reserve.

We held similar discussions with the group that objected to logging to identify conditions that satisfied the forest preservation objective. The group demanded preservation of 400 hectares of non-degraded forest to protect sources of local streams and other non-timber forest resources. Evaluation criteria (including streams, slopes, and roads located within or close to the reserve) were identified, mapped, and stored in the GIS. The factors were then processed as before to create a suitability map for forest preservation.

We then designed an exercise around the two suitability maps to explore the human values that sustained the conflict and to link such considerations

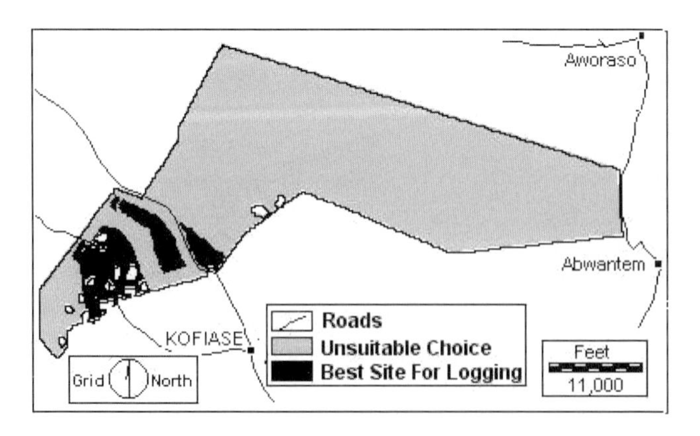

Figure 16.4 Best 350 hectares for logging.

to the choices the parties were making. First, we ranked cells in both suitability maps and extracted an adequate number of the highest ranked cells in each map to meet the area targets specified by the parties. The 350 hectares of forest demanded by loggers translated into 3,889 of the cells that constituted the map (Figure 16.4), while the 400 hectares demanded for preservation required 4,444 of the same cells in the data set (Figure 16.5).

The two maps were later cross-classified in the GIS to create a conflict map of preferences from both parties (Figure 16.6). In the conflict map,

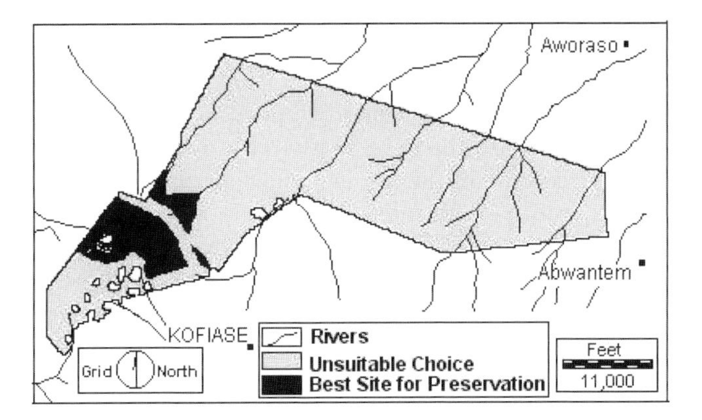

Figure 16.5 Best 400 hectares for preservation.

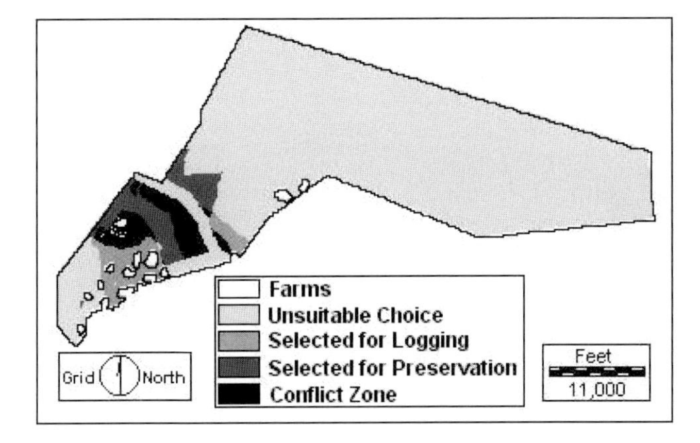

Figure 16.6 Conflict map.

areas in the forest that were not in dispute were separated from those areas that were in conflict, and hence relevant to the discussion. As expected when two interest-driven preferences are superimposed in a multi-dimensional decision space, four distinct categories resulted from the overlay of the most favored sites in the two suitability maps (Figure 16.6). These were:

- areas selected for logging only (non-conflicting),
- areas selected for preservation only (non-conflicting),
- a sizeable area not selected for logging or preservation (unsuitable choice), and
- an area selected for both logging and preservation (conflict zone).

It is clear from the conflict map shown in Figure 16.6 that the interests of the two parties converged in the southwestern portion of the forest reserve. This was the portion of the forest that had remained intact and was therefore suitable for logging. Unfortunately, it was also a zone where cocoa farms and sources of several resources including streams that supplied water to the villages were concentrated. The group visited the 'conflict zone' in the conflict map to study resources in the area. We used the conflict map to guide and refocus the discussion on areas in the forest where disagreement occurred between the two groups. This way, we were able to divert attention of the parties from broad philosophical positions onto specific and concrete evidence represented in the conflict map. Finally, we input the ranked suitability maps into the GIS, and a multi-objective land allocation procedure (MOLA) was used to allocate cells between the two conflicting objectives (Figure 16.7).

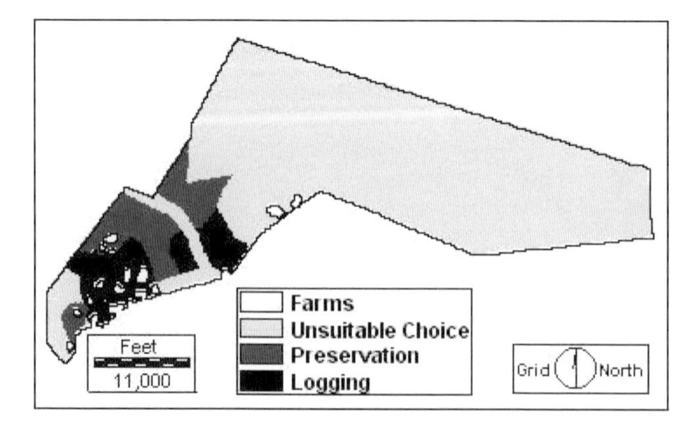

Figure 16.7 Final allocation map.

16.5 CONCLUDING PHASE OF THE PROJECT

The immediate reaction to the final resolution map was mixed. The group that supported logging agreed to the final allocations. Some of the representatives who favored preservation also agreed, but a handful insisted on preserving trees in the conflict zone. The conflict was finally resolved a few weeks after the PPGIS project during a meeting with the chief and elders of the town. The two parties agreed to a compromise solution. After several months of bickering, the conflict management exercise prepared the groups for compromise and laid the foundation for an amicable resolution of the conflict. This achievement was in part due to the opportunity the project provided for parties to form coalitions within which they jointly collected data, shared resources, and exchanged ideas around issues that affected the forest. The coalitions became very useful to the resolution of the conflict during the meeting with elders of the town.

Responses to a questionnaire administered before and immediately after the completion of the project revealed that members of the forest committee were satisfied with the role GIS played in the discussions (Kyem 1999). Members of the CFMC were particularly happy that the discussions helped to direct their focus onto the search for evidence contained in maps they had helped to prepare. The project laid a foundation for the establishment of future institutions for collaborative forest management in the community, but this did not happen without problems. In the remaining section of this chapter, we focus on some of the problems and challenges we encountered during the implementation of the Kofiase PPGIS project.

16.6 PROBLEMS AND CHALLENGES

As with many technology transfer projects that occur in communities other than those in which the technology evolved and is extensively applied, we encountered several problems during implementation. Many of the problems discussed below were peculiar to conditions within the locality but the foreign origin of the technology was also a contributory factor. Often if the environment-specific problems are identified early in the process, they can be handled better to ensure successful implementation of projects.

16.6.1 Cultural and institutional obstacles

The social and economic lives of people in the study area are shaped by tradition, customs, and local belief systems (Pogucki 1968). For example, there was a common belief among people in the community that rivers, commercial timber species, and some resources in the forest possessed spirits that must be appeased before the resources could be exploited. Some of the beliefs were attached to resources that became objects for mapping (i.e. streams, timber), and hence vital ingredients in the GIS analyses. We were therefore compelled to contend with the representation of values and beliefs using a technology that draws its strength from the analyses of empirical facts. Although our familiarity with local customs made it easy for us to understand and model many of these beliefs in the GIS, accurate interpretation of local customs remains a problem for PPGIS experts who undertake projects in unfamiliar communities.

There were occasions when local customary practices facilitated the implementation of the PPGIS project. For example, the Ashanti clans came in handy as the rallying point for assembling representatives during the formation of the Kofiase Forest Committee. However, working through ethnic groups in the community raised concerns about the resuscitation of past struggles over land ownership among clans (Benneh 1970; Ilegbume 1976; Agyeman 1993). Additionally, as clans in the community existed for a wide variety of economic, social, and political reasons, it is doubtful whether the Kofiase experiment can be used as a model for PPGIS adoption elsewhere in the country.

The absence of issue-oriented organizations in the community also made it difficult for us to organize people for the PPGIS project. Individual family units in the community were relatively self-sufficient, producing what they needed for themselves and some cash crops (mainly cocoa) for income. The major part of their subsistence was produced through farming rather than manifold social relations. As a result, the people were isolated, and could not actively participate in issues that affected the whole community. It is our observation that, as long as there are only local interconnections among the villagers and their common interests do not give rise to a regional bond that would overcome their isolation, it would be difficult to organize and significantly empower such communities through PPGIS applications.

Another formidable obstacle to the realization of the potential GIS offers for resource management in the community was the lack of effective administrative mechanisms for managing natural resources. The study revealed a disconnection between formal and traditional institutions of land resource administration in the country. The Forest Department was a relic of past colonial administration and, as such, it did not command the loyalty of the people. On the other hand, the people revered traditional institutions such as the local chief and his council of elders, but years of neglect by successive Ghanaian governments had caused the demise of many such local institutions. Thus, there were no effective mechanisms for administering resource policy decisions reached within the PPGIS project. We also realized that some basis of power must be established for groups who participate in a PPGIS project to level the playing field and ensure equal and full participation for all parties. The problem is that there can be no guarantees for equal participation in many PPGIS projects. For example, in the Kofiase project, the Forest

Department designed and was in charge of the implementation of the collaborative forest management programme. The foresters on the committee therefore had an advantage over their counterparts. They used their familiarity with data and information about the Aboma Forest Reserve to dominate discussions on different occasions during the implementation of the project.

16.6.2 Sustaining PPGIS projects in the community

The commitment of GIS experts and the inhabitants of Kofiase to the PPGIS project was essential to the successful completion of the study. For people in the study area, such a commitment depended upon the trust they had for us, or the interest they later developed for the project. Unfortunately, many people in the community were skeptical of innovation, and particularly distrustful of public officials and foreign experts. The natives felt reluctant to depart from routines and skills they had established over time. Some were afraid that the innovation would be enfeebling to them, and that there would be no immediate benefits to their participation in the project. The possibility of acquiring some portion of the degraded forest to cultivate food crops while attending to timber seedlings (for rehabilitating the Aboma Forest Reserve) convinced many of the villagers to enroll in the forest committee and participate in the PPGIS project. Based on our experience, we would recommend that scientists who implement PPGIS projects in the community should present tangible benefits to the people to create the incentives that would sustain the project.

We received little official support for the PPGIS applications even though the project was implemented within the official collaborative forest management project. We also realized that the rich and powerful people in the community objected to the open and participatory uses of GIS. Some were particularly resentful of the inclusion on the forest committee of representatives of local farmers. The resentment about participatory uses of GIS could hinder future adoptions of the technology for the empowerment of less privileged groups. The negative official reaction and lack of funds for the PPGIS project made it difficult for us to achieve some of our stated goals (i.e. providing GIS facilities for each forest committee to ensure continued practice).

16.7 LESSONS FROM THE STUDY

The implementation of PPGIS projects in local and indigenous communities can be difficult, and the impacts that such applications produce on people and organizations might not be easy to ascertain in the short term. The lack of infrastructure and skilled personnel to support PPGIS development, and opposition to PPGIS applications from state officials and influential people in the communities, present significant obstacles. These obstacles notwithstanding, the interdependence of nations, the powerful electronic medium within which GIS operates, and the information context within which the technology has emerged present a strong case for PPGIS adoption in local communities. Additionally, GIS application software are becoming more sophisticated, but also cheaper and more user-friendly. With the increasing availability of computers, advocates for underprivileged groups have access to tools for creating, storing, and analysing information in a way that makes it easier for local groups to participate effectively in public discourse.

The Kofiase project demonstrated that if GIS is utilized properly, the technology could bring benefits to local groups. The rapid transformation of data on local resources into easily readable maps in the GIS helped raise awareness about issues that affected the allocation and use of local resources. The project demonstrated that if GIS is embedded in a truly open and participatory planning process, it could empower local community representatives by actively involving them in public policy debates. In the project, the representatives participated fully in the discussions and moved from a position where they had no input into forest policy-making to one where they fully participated and influenced decisions about how the local forest should be managed.

It is important to state that creating a supportive climate for communities to actively participate in policy debates requires not only a commitment, funds, and some computer hardware enhancements, but also simple methodologies and procedures that could effectively involve stakeholders in open discussions. To make such procedures applicable for a wide range of local uses, they need to be iterative, simple, and intuitively appealing without sacrificing the mathematical rigour of the GIS analyses. It is also important in PPGIS applications that GIS is employed mainly as a mechanism for processing data and making information available to participants in easily understandable formats. The technology might not be used as a substitute to the participatory processes. In addition, sustained and efficient PPGIS adoption at Kofiase and similar communities in Ghana would require local capacity building in GIS expertise. A pool of local GIS experts could provide some safeguards to ensure appropriate and efficient uses of the technology. Local experts would be in a better position than their foreign counterparts to interpret and model local conditions and belief systems in GIS. They would also be able to initiate programmes and applications that could reform GIS development and applications from within their respective communities.

REFERENCES

Agyeman, V. K. (1993) 'Land, tree, and forest tenure systems: implications for forest development in Ghana, a summary report (May)' Kumasi, Forest Resource Institute Ghana.

Asante, S. K. B. (1975) *Property Law, and Social Goals in Ghana, 1844–1966*, Accra: Ghana University Press.

Baidoe, J. F. (1972) 'The management of the natural forests of Ghana', *Proceedings of the 7th World Forestry Congress*, vol. 2, Centro Cultural General, Sam Martin, Buenos Aires, Argentina, 4–18 October.

Belfield, C. H. (1912) 'Report on the legislation governing the alienation of native lands in the Gold Coast Colony and Ashanti, with some observations on the forest ordinance 1911', Cd. 6278, Accra: Government Printing Press.

Benneh, G. (1970) 'The impact of cocoa cultivation on the traditional land tenure system of the Akan of Ghana', *Ghana Journal of Sociology* 6(1): 43–61.

CFMU (1993) 'Collaborative forest management news', *Collaborative Forest Management's Newsletter* 2, Kumasi, Ghana, 3–5

Danso, E. Y. (1995) 'Support to collaborative forest management: progress quarterly report (January–June 1995)', Report to The Overseas Development Administration and the Planning Branch, Forest Department, Kumasi, Ghana.

Eastman, J. R., Kyem, P. A. K., Toledano, J. and Weigin J. (1993) *GIS and Decision Making*, UNITAR, Geneva.

England, P. (1993) 'Forest protection and the rights of cocoa farmers in Western Ghana', *Journal of African Law* 37(2): 164–176.

Falconer, J., Wilson, E. P., Asante, P., Jampo, E. L., Acquah, S. D., Blover, E., Beeko, C., Osson, K. and Lamptey, E. (1994) *Non-Timber Forest Products of Southern Ghana*, O.D.A. Forestry Series No. 2. Prepared by ODA on behalf of Forestry Department, Republic of Ghana.

Ghana News Agency: Peoples Daily Graphic, (1993) Staff Writer's Report, Issue of Tuesday, 2 March, Accra: Daily Graphic Press, p. 16.

Hawthorne (1993) *FROGGIE: Forest Reserves of Ghana, Geographic Information Exhibitor*, Planning Branch, Forest Department, Kumasi, Ghana.

Ilegbume, C. U. (1976) 'Concessions, scramble and land alienation in British Southern Ghana 1885–1915', *African Studies Review* 29(3): 24–36.

Kyem, P. A. K. (1997) 'A GIS-based strategy for improving local community participation in resource management, allocation, and planning: the case of institution building for collaborative forest management in Southern Ghana', Ph.D. Thesis, Graduate School of Geography, Clark University, Worcester, Massachusetts.

Kyem, P. A. K. (1999) 'Diagnosing users' perception of change in the transfer and adoption of geographic information system's technology', *Applied Geographic Studies* 3(2): 121–136.

Logan, W. E. M. (1946) 'The Gold Coast forestry department', a report on the State of Forest Reservation in the Gold Coast, Accra, Ghana.

Norton, A. (1989) *Participatory Forest Management in Ghana*, Overseas Development Association, UK.

Pogucki, R. J. H. (1968) *Ghana Land Tenure: a Handbook of main principles of Rural Land Tenure*, Accra, Ghana, Lands Department.

Wiredu, K. (1993) 'African philosophical tradition: a case study of the Akan', *The Philosophical Forum* 24(1–3): 33–62.

Chapter 17

GIS for community forestry user groups in Nepal: putting people before the technology

Gavin Jordan

17.1 INTRODUCTION

For more than a decade, Nepal has been consistently ranked in the ten poorest and least developed nations in the world. It is currently estimated that over 50 per cent of the population live below the absolute poverty line, and this percentage has barely changed in the last three decades (World Bank 1987; 1998). Agriculture is the key economic activity and over 80 per cent of economically active Nepalese are farmers, a significantly higher percentage than for most less-developed countries. This, coupled with a high population growth, puts great demands on the natural resources of Nepal for fuel, fodder, fertilizer, and building materials. These products (wood, leaves, and grasses) are obtained from forests, which are essential to Nepalese rural livelihoods. Forests in a Nepalese context are not an industrial resource, but are a critical source of inputs into farming systems (Figure 17.1). It is appropriate for these resources to be managed as far as possible by local people, a type of forestry that has become known as community forestry.

17.1.1 Community forestry

Nepal can be regarded as the 'home' of community forestry, with a legislative history dating back to the mid-1970s (Hobley and Malla 1996). Initially the move towards increasing community control over local forest resources was based on a realism by the government that it did not have the funds to support state controlled forestry throughout the whole country. Additionally, there was a perceived forest super-crisis situation in Nepal (Eckholm 1975). The World Bank went as far as publishing a report that predicted that there would be no accessible forests by the year 2000 (World Bank 1979). The combination of state-supported community forestry and perceived super-crisis led to massive external donor support for community forestry projects. With this donor support came an increased focus on community participation, representation and social development.

Figure 17.1 Farm–forest interactions. Farmers collecting animal fodder and bedding materials from a community forest.

Community Forestry is a form of 'social' forestry that has its roots in the change in development theory from industrial forestry, based on the Northern European macroeconomic model (Van Gelder and O'Keefe 1995), towards local-level forestry geared towards the subsistence needs of local communities. It has been said that community forestry has more to do with people than trees (Gilmour and Fisher 1991), and this has been reflected in an approach traditionally dominated by the social sciences. Participatory techniques have been the primary tool for obtaining community and resource information, and participation, empowerment and facilitation of the Forest User Group (FUG, a village-based forest management committee, which includes all forest users of a community forest) the main objectives.

At the same time there has been a need for obtaining more traditional quantitative information for forest management purposes. There are a number of reasons for this, principally:

- to assess responsible ('sustainable') forest management,
- to allow a sustainable yield of timber to be calculated,
- for local specific needs,
- to examine tenure rights and rights to resources,
- for conflict resolution purposes,
- for compensation claims,

- for monitoring biodiversity,
- to meet the requirements of International agreements, and
- for identifying potential economically viable Non-Timber Forest Products (NTFPs).

The normal developmental approach has been to keep qualitative and quantitative data collection and management separate. This may be due to the different disciplines they are associated with; social scientists have continued to conduct the participatory information gathering and analysis, whilst colleagues from the natural sciences and IT have managed the quantitative information.

District or national level studies often map socio-economic indicators, commonly called 'indicators of development', although the people targeted for the development process are entirely unaware of these indicators. Indicators are used for policy planning to identify both development priorities and geographic regions of activity. Therefore the 'developmental' role of GIS is often one of disempowerment of local people, involving a very low level of participation. It encourages the separation of the planning process from the people affected. There is little or no discussion with FUGs and other villagers regarding what information would be useful to them, and what information a GIS could provide. The GIS information is not *meant* for them. It is for the policy-makers, planners and researchers.

The most charitable way of looking at this lack of participation associated with the traditional use of GIS in development work is to view GIS as enabling decision-makers to correctly evaluate the required development input. But this is *putting the technology before the people*. Whilst it appears that GIS is being used for classic decision support purposes, the decision-making process itself is fundamentally flawed. There is little or no consultative process with communities. Their needs have not been identified, and the information gathered does not reflect their requirements. The old top-down development paradigm is being actively encouraged (Hobley 1996).

17.1.2 GIS in community forestry

Although it is technically and organizationally possible to integrate much participatory information into a GIS, this has seldom been attempted in development work, with a limited number of applications. The lack of use of GIS for local-level needs when compared to national or regional use has been commented on (Haase 1992; Simonett 1992; Carter 1996). This may be due to social scientists' mistrust of GIS technologists, who often have a simplistic understanding of the complexity of community forest resource management, coupled with their scepticism of a technology that is both centralizing and based on logical, deductive and empirical principles (Abbot *et al.* 1998; Hutchinson and Toledano 1993). Much other work that could be expected to have an element of participatory research relies on secondary data sources. This is true of most socio-economic research associated with natural resource management (Daplyn *et al.* 1994; ICIMOD 1996; Alspach 1999). An observation made nearly a decade ago for developmental work in sub-Saharan Africa still holds true today; most GIS applications are driven by a desire to demonstrate the technological capability rather than a desire for real-life problem solving (Falloux 1989).

There are a limited number of examples of GIS being used as a public participatory tool for community forest management. The Kayan Mentarang Nature Reserve Project in Indonesia combined oral histories, sketch maps, GPS and GIS for customary land-use mapping (Stockdale and Ambrose 1996; Sirait *et al.* 1994). It was noted that a constraint was the ability of social scientists and map-makers to accurately capture and portray the complex relationships of traditional resource management systems. Work in northwest Zambia by Jordan and DeWitt (SNV 1996) incorporated RRA (see next section) techniques to determine where villagers collected constructional timber, a participatory inventory to determine resource quality, and a GIS database for analysing this information and determining whether sustained yield management was being practised. Whilst this proved to be an effective management tool for examining village level forest resource utilization patterns by local communities, it is felt that the participatory element of this work could have been increased, as decision-making was largely the task of 'outsiders'.

17.2 PPGIS IN NEPAL

PPGIS in the field of community forest management is still in its infancy, and many issues still need to be identified and evaluated. This study was initiated in Nepal, with the aim of assessing the applicability and relevance of a PPGIS. The initial objectives were to:

- identify stakeholder information needs. This uses the classic Rapid Rural Appraisal (RRA) techniques of focus groups, semi-structured interviews, group walks and participatory mapping (McCracken *et al.* 1988; Chambers 1994),
- obtain the necessary information using general participatory techniques, geomatics techniques (participatory photo mapping, GPS), and participatory inventory techniques,
- analyse information and present it in a format and language that is appropriate for FUGs,
- feed it back to FUGs and determine the usefulness of the information to them, and
- examine the potential and problems of the PPGIS as an empowerment tool for FUGs.

However, as the study progressed, it became apparent that a more process-orientated approach was necessary. The focus shifted towards examining a systematic approach for participatory forest management combining the collection of quantitative, objective information with qualitative, subjective information in a way that was beneficial for the FUG.

17.2.1 The study area

The study was conducted in the Yarsha Khola watershed, Dolakha District of Nepal. It is an area of the high mountains of Nepal, and the watershed varies in altitude from *c*. 1000–3000 m. This is a predominantly rural economy, with some extra income earned from working in the tourist industry in Kathmandu, a day away by bus. There are a variety of ethnic groups, including Brahmins and Chettri in the lower altitudes and Sherpas at higher levels. Community forestry is an important component of an integrated farming system, with the majority of animals being stall-fed, fodder and bedding coming from forest products. Dung is used to fertilize terraced fields for intensive crop production. There is great interest in community forestry at a village level, and the FUG has an important role to play. It has a committee which liaises closely with the local forest ranger and the District Forest Officer (DFO), both from the Nepalese Department of Forests. The FUG has to demonstrate a capacity to conduct forestry operations in order for the DFO to authorize its forest management practices. A limiting factor for the FUG is the availability of management information about the forest, and spatial information on the extent of the resource. Hence the potential of PPGIS for empowering the FUG.

17.2.2 Methods

The methodological framework employed is outlined in Figure 17.2. It is interdisciplinary in its approach, combining the use of social science participatory techniques with geomatics technology and participatory assessment procedures. The methodology is on the interface between social approaches to community forestry and more traditional quantitative techniques to resource assessment. This is regarded as essential owing to the increasingly demanding and diverse information needs for community forestry in Nepal. It should be noted that a greater emphasis is placed on the means of collecting and disseminating information than the technical design of the GIS database, as it believed that a PPGIS is fundamentally dependent on obtaining community needs, perceptions and ideas. Indeed, it will be seen from Figure 17.2 that the role of GIS in its traditional capacity for data input, storage, retrieval, transformation and display (Burrough 1986; Grimshaw 1994) is limited, and the other aspects of an information system, namely

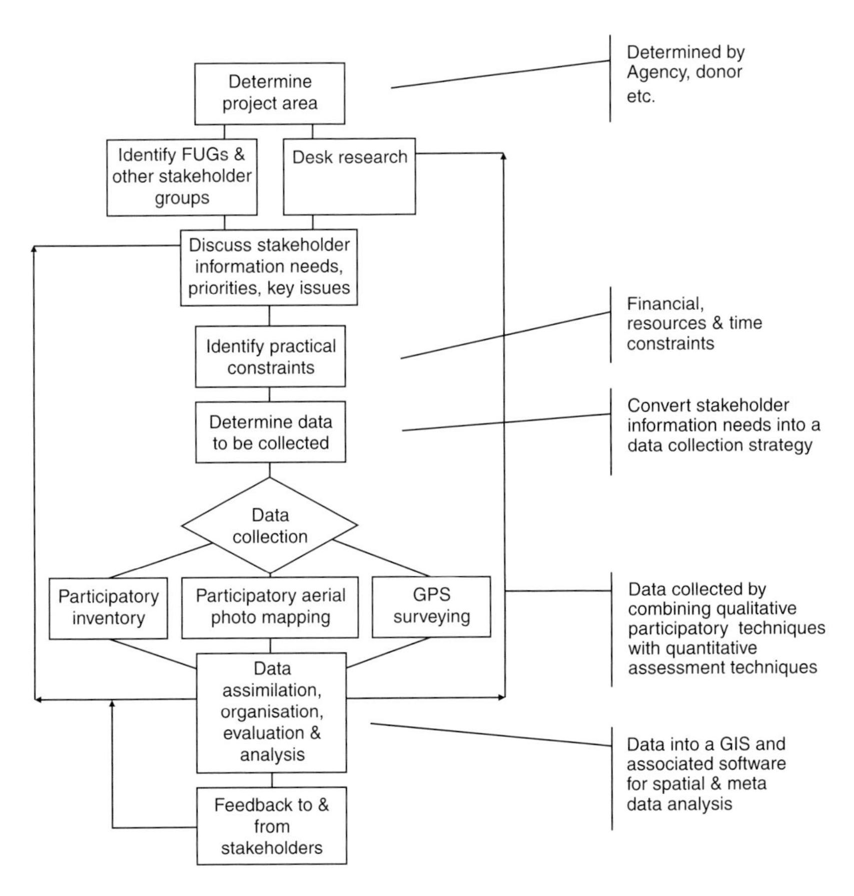

Figure 17.2 A systematic methodology for a community forestry PPGIS.

planning, user need identification, data collection and information feedback are of equal importance.

It should also be noted that as well as examining the information needs of the FUG, this work also looks at the information needs of other stakeholder groups, including the District Forest Office, national policy-makers, and international monitoring organizations. The PPGIS is designed to provide information to all these diverse stakeholders, at an appropriate level, and assist with decision-making. This is an additional attribute of GIS; it allows the information to be effectively stored, analysed and prepared for dissemination in a means appropriate for each stakeholder group.

The above methodological framework was tested with five FUGs from October 1997 to May 1998. Owing to the participatory nature of the work, the exact methodology varied across FUGs. The initial participatory session with the FUG examined their specific requirements. These ranged from maps of the community forest for boundary dispute issues to inventory information to assist them in planning sustained-yield harvesting for commercial purposes. Additionally, FUGs sometimes requested information on the sustained-yield of fodder (grass, leaves and shrubs for stall-fed livestock), when they could start removing fuelwood, and the general condition of their forest. This information was requested to enable the FUG to utilize their forests more intensively. Contrary to popular opinion, community forests in Nepal are usually managed in a very conservative manner. The information requirements were usually a combination of basic spatial information and management information; how best to manage their resource. This is where a combination of quantitative and qualitative information is essential. It is impossible to offer useful management advice without understanding the FUG's requirements and usage patterns.

Once the information needs of the FUG were established the data collection process was developed. This was based on a participatory forest resource assessment, designed as a multi-resource inventory to meet the information requirements of all stakeholders (Lund 1998). The resource assessment procedure contained one or more of the following elements: a participatory photo-mapping session, a participatory inventory (always conducted) and a GPS survey of internal and external boundaries. The specific procedure adopted with each FUG depended on information needs, availability of aerial photographs and terrain considerations. Of these methods, perhaps the least known is participatory photo-mapping. This is similar in philosophy to participatory sketch-mapping (Messerschmidt 1995), but uses a large-scale aerial photograph as a participatory tool (Fox 1986; Jordan and Shrestha 1998; Mather 1998). This has the participatory advantages of sketch-mapping, but greatly increases the spatial accuracy of information obtained.

The community interacted fairly well with the new technology and information. Villagers were generally able to interpret aerial photographs rapidly and effectively, with initial assistance from a facilitator. They were capable of faster and more accurate interpretation than many western graduate students. This may be because of the mountainous nature of the Nepalese terrain providing 'aerial' perspectives of villages when climbing ridges and spurs, or because villagers are used to dealing with non-literate media. Aerial photographs were an excellent facilitation tool, particularly due to their non-literate nature making them accessible to the whole community (see Figure 17.3). Villagers also found them exciting, and they greatly animated participatory sessions. Villagers also learned the basic techniques of forestry inventory very rapidly. They were able to lay out sample plots, measure trees, slope angles and take bearings within a day. By the end of a week, they were entirely capable of conducting the inventory work themselves

Figure 17.3 Women members of a Forest User Group conducting a participatory photo mapping exercise.

(including the record keeping, usually performed by a teacher or shop-keeper). GPS was much harder for the community to use, and required a professional to operate. However, villagers quickly grasped the general idea of GPS, and used it to map forest boundaries.

Once the information was gathered it was organized using a GIS and other basic software. Descriptive information obtained from the participatory research, such as indigenous management, FUG requirements and problems was recorded. Inventory information was entered into a database, and the spatial information was entered into a GIS (IDRISI™). This can be regarded as the information management component of the participatory information system.

17.3 DISCUSSION

The PPGIS is now functional as a basic pilot version. For a given FUG it has a georeferenced boundary of the community forest, with the area of the forest (something that is in itself often unavailable for community forests), internal community designated boundaries, and associated basic information, such as key species. Files can be called up for each internal compartment that have information on the sustained yield, recommended management practices, community uses and importance of spatial sectors of the resource for the community. Additionally, the raw inventory data is available for researchers and policy-makers who wish to examine biodiversity issues, slope angles or other issues. At present the PPGIS is clumsy, involving non-integrated software packages, and interfaces need to be developed. For the FUGs who have no access to IT, the appropriate images and management information can be used to form the basis of a visual report which the FUG committee can use for its forest management. Initial work indicates that FUGs appreciate the maps as a tool they perceive can help them in their negotiations with the Forestry Department. Inventory information is converted into basic management information to allow the FUG to participate in discussions with the local forest ranger and the DFO, regarding forest management, which the FUGs also felt was useful. The feedback is of critical importance: a PPGIS is there for its users, the participants. Some stakeholder groups have been very satisfied with their role, but the evaluation process is not yet complete. It should be noted that although the initial evaluation was based on the ability to produce and organise data for FUG use, this is only one benefit. The participatory work involved in community consultation, obtaining resource information, and the feedback meetings gave the FUG a sense of ownership and involvement with the process. This acted as an agent of empowerment, raising community expectations of what the FUG and individuals could achieve. These 'social' processes are felt to be of great importance, and should not be ignored by concentrating solely on the technical performance of the PPGIS. Evaluation issues for PPGIS are discussed in more detail below.

An important general discussion area is the need to consider the community agendas in boundary mapping. FUG members may not wish to survey the exact boundary of the community forest, but may survey what 'should' be the boundary. During GPS surveys, there is a tendency for FUG members to include 'doubtful' areas within the surveyed FUG area, in case this legitimizes their claim. This, of course, has important implications, as the 'true' boundary surveyed with GPS can be just as subjective as any other method. The GPS may be impartial, but the hand that holds it is not. There may be disputed areas, or disused farm or scrubland that FUG members feel should be part of the community forest. Therefore, it is important to realize that maps developed from participatory work are highly political, and they are far from being an objective portrayal of reality (Monmonier 1996; Wood and Fels 1992). A PPGIS can easily become part of a power struggle or village dispute. There is no easy way of assessing this. It is important to realize that hidden agendas in boundary surveying may exist, and to ensure that there is adequate participation to allow multiple views and realities to be demonstrated.

The initial objectives of this study have been satisfactorily met, and initial evaluation of the PPGIS indicates that it is an appropriate and beneficial tool for providing stakeholders with information regarding community forest management issues in Nepal. While this does support the validity of PPGIS in this context, a number of further additional issues have been raised that need both discussing and further evaluation.

17.3.1 PPGIS as a process

Whilst a PPGIS can produce information that is useful for the FUG, it can be viewed as extractive in nature, rather than achieving the Participatory Rural Appraisal (PRA) goal of utilizing local people's analytical capabilities as well as their knowledge base (Chambers 1994). This may seem academic, but it is important to note that any technology which requires data to be taken away for analysis rather than encouraging people to undertake their own investigations and analysis limits participation to some extent. This ties in with the consideration of whether GIS is appropriate technology for participatory development work where access to GIS is severely limited. Does the use of GIS encourage an alienation between participants and their information? Does it remove them from much of the decision-making process? If GIS is viewed as software and hardware, this could be a valid interpretation. But it is felt that a PPGIS should be a *process*; it starts with the public participation procedure and intrinsically involves feedback to, and from, the FUG. Decision-making should not be made centrally; the PPGIS should be a decision support tool for the FUG, providing information they can use for their management decisions. Although the software and decision analysis processes are outside of the sphere of access of the FUGs, with associated problems (Harris *et al.* 1995), it can be argued that the decision-making process can be brought back to the FUG. This is a central issue in making a PPGIS genuinely people-orientated.

17.3.2 Representing village level reality

There can be a loss of detail when entering descriptive information obtained by participatory methods into a GIS. Qualitative information is not easily entered into a GIS, and the rich social, economic, and environmental fabric of resource management at a village level is impossible to replicate. A people-orientated PPGIS must have a capability for storing some of this descriptive information. This may be not just as textual and diagrammatic information, multimedia offers a variety of interesting ways to represent this more realistically. But it is important to realize that all the information will still not be captured. What is necessary is to involve local people and incorporate their knowledge and decision-making into the PPGIS. The task is not to capture and replicate all the village information, but to organize and present pertinent information that was not previously available.

17.3.3 The need for participation

It is felt that a fundamental requirement of PPGIS is an emphasis on participation. GIS is a useful tool for enabling the participation and empowerment of FUGs, through providing them with increased information for decision-making, but only if it is focused on community needs. The technical performance of the GIS, spatial accuracy and quality of output are all secondary to the need for a participatory approach. This can easily be forgotten, particularly as this is a reversal of the traditional GIS priorities.

PPGIS is a systems-based process. The focus is on participation. Although the system will vary greatly from situation to situation, it should be based around identifying user information needs, and providing this information to support their decision-making. Figure 17.2 indicates a workable system, but it should be noted that as this work progressed, the emphasis switched from the technological considerations towards participatory issues.

17.3.4 Evaluating a PPGIS

The discussion above partly focuses on the evaluation of the PPGIS, and many of the issues discussed implicitly suggest a need for evaluation. The evaluation conducted during this research mainly involved feedback from stakeholder groups and technical issues relating to data quality. Whilst this represents an

Table 17.1 Evaluation areas for a PPGIS

Evaluation issues	*Means of evaluation*
PPGIS data issues	
Spatial accuracy	Spatial statistics
Relevance of data	Stakeholder feedback meetings
Quality issues	Data assessment
Error budgets & sources	Statistical analysis, data assessment
PPGIS process issues	
Level of participation (at each stage of process)	RRA, PRA & social science techniques
Stakeholder satisfaction	Stakeholder feedback meetings Examine usage of data provided by PPGIS
Ability to produce & organize data for stakeholder use	Examine usage of data provided by PPGIS
Assessment of long-term empowerment	RRA, PRA & social science techniques Examine outcomes of meetings/ discussions using provided data
Assess how stakeholder expectations have been raised	Stakeholder feedback meetings Examine outcomes of meetings/ discussions using provided data
Value of GIS to the process	Cost benefit analysis of the added value contributed by using GIS
Overall value of PPGIS	Social cost-benefit analysis

example of current best practice, it is felt that further work needs to be conducted in this area. PPGIS evaluation in general is not conducted with enough rigour. Without detailed systematic evaluation, PPGIS could easily fall into the trap of combining sloppy GIS practices with sloppy social science.

The thorough evaluation of a PPGIS is complex. The PPGIS has to be examined both as a systems-based process, and in terms of the information utilized and generated by it. Whilst the emphasis should be on participation and the process, data issues also need considering. Areas that require consideration during the evaluation are presented in Table 17.1.

17.4 CONCLUSION

In general, it appears that PPGIS is an appropriate and advantageous tool for community forestry in Nepal and should have much wider applications in participatory development work. It has a number of distinct advantages over more traditional approaches to this type of complex management issue:

- If it is viewed as a participatory process, it can empower the FUG by involving them in the decision-making process and raise their expectations of information available for them.
- It can be used to effectively combine quantitative and qualitative approaches to community forestry and rural development.
- Maps, resource management information and other spatial data can be given to an FUG to aid with their decision-making and negotiations without the need for them to have access to a GIS.
- Information can be easily collated, analysed and returned to stakeholders.
- The appropriate level of information can be returned to stakeholders.

However, the technology does have the potential to assist extractive collection of information, and GIS can disempower disadvantaged groups and further distance them from the decision-making process. It was found that the emphasis had to be firmly on participation rather than technical issues. A system-based approach that actively encouraged participation was found to be the key requirement for a useful PPGIS.

ACKNOWLEDGEMENTS

This chapter is based on work conducted whilst working with the People and Resource Dynamics Project (PARDYP), at the International Centre for Integrated Mountain Development (ICIMOD), Kathmandu, and the Nepal Australia Community Forestry Project (NACFP), whilst the author was employed by the Department of Environmental & Geographical Sciences,

Manchester Metropolitan University. The author would particularly like to thank Ian Heywood (MMU), Bhuban Shrestha, P. B. Shar and Richard Allen (PARDYP), Steve Hunt and Mr Prajapati (NACFP), and staff of the Nepal UK Community Forestry Project.

REFERENCES

Abbot, J., Chambers, R., Dunn, C., Harris, T., de Merode, E., Porter, G., Townsend, J. and Weiner, D. (1998) 'Participatory GIS: opportunity or oxymoron?' *PLA Notes 33*, International Institute for Environment and Development, October.

Alspach, A. J. (1999) 'Integration of geographic information systems (GIS) in regional participatory development projects', in *Geographic information and society: an international conference'*, Conference papers, 20–22 June, Department of Geography and the National Science Foundation, University of Minnesota, Minneapolis.

Burrough, P. A. (1986) 'Principles of GIS for land resource assessment', *Monographs on Soil and Resources Survey No. 12*, Oxford: Oxford Science Publications, Clarendon Press.

Carter, J. (1996) 'Defining the issues', in J. Carter, (ed.) *Recent approaches to participatory forest resource assessment*, Overseas Development Institute, London, pp. 1–32.

Chambers, R. (1994) 'The origins and practice of participatory rural appraisal', *World Development* 22(7): 953–969.

Daplyn, P., Cropley, J., Treagust, S. and Gordon, A. (1994) *The use of geographical information systems in socio-economic studies*, NRI Socio-economic series 4, Natural Resources Institute, Chatham Maritime.

Eckholm, E. (1975) 'The deterioration of mountain environments', *Science* 189: 764–770.

Falloux, F. (1989), 'Land information and remote sensing for renewable resource management in sub-Saharan Africa: a demand driven approach', World Bank Technical Paper No. 108, The World Bank, Washington.

Fox, J. (1986) 'Aerial photographs and thematic maps for social forestry', *Social Forestry Network Paper 2c*, Overseas Development Institute, London.

Gilmour, D. A. and Fisher, R. J. (1991) *Villagers, forests and foresters: the philosophy, process and practice of community forestry in Nepal*, Kathmandu: Sahayogi Press.

Grimshaw, D. J. (1994) *Bringing geographical information systems into business*, London: Longman.

Haase, S. (1992) 'Geographic information systems and community resource management for sub-Saharan Africa: opportunity lost?' A report submitted to the Department of Engineering and Policy of Washington University in partial fulfilment of the requirements for the Degree of Master of Science in Technology and Human Affairs, Washington University.

Harris, T., Weiner, D., Warner, T. and Levin, R. (1995) 'Pursuing social goals through participatory geographic information systems: redressing South Africa's historical political ecology', in J. Pickles (ed.) *Ground truth: the social implications of geographic information systems*, New York: The Guildford Press, pp. 196–222.

Hobley, M. (1996) 'Why participatory forestry?' in M. Hobley (ed.) *Participatory forestry: the process of change in India and Nepal*, Rural development forestry guide three, Overseas Development Institute, London, pp. 1–24.

Hobley, M. and Malla, Y. B. (1996) 'From forests to forestry – the three ages of forestry in Nepal: privatisation, nationalisation and populism', in M. Hobley, (ed.) *Participatory forestry: the process of change in India and Nepal*, Overseas Development Institute, London, pp. 65–92.

Hutchinson, C. F. and Toledano, J. (1993) 'Guidelines for demonstrating geographical information systems based on participatory development', *International Journal of Geographical Information Systems* 7(5): 453–461.

ICIMOD (1996) *GIS database of key indicators of sustainable mountain development in Nepal*, MENRIS/ICIMOD, Kathmandu.

Jordan, G. H. and Shrestha, B. (1998) *Integrating geomatics and participatory techniques for community forest management: case studies from the Yarsha Khola watershed, Dolakha district, Nepal*, ICIMOD, Kathmandu.

Lund, G. H. (ed.) (1998) *IUFRO guidelines for designing multipurpose resource inventories*, International Union of Forest Research Organisations, Vienna.

McCracken, J. A., Pretty, J. N. and Conway, G. R. (1988) *An introduction to rapid rural appraisal for agricultural development*, Sustainable Agriculture Programmes, International Institute for Environment and Development, London.

Mather, R. A. (1998) 'Evaluation of the potential for GIS based technologies to support the forest management information requirements of the FUG institution, Nepal', UK Community Forestry Project, Kathmandu.

Messerschmidt, D. A. (1995) *Rapid appraisal for community forestry*, International Institute for Environment and Development, London.

Monmonier, M. (1996) *How to lie with maps*, 2nd edition, Chicago: University of Chicago Press.

Simonett, O. (1992) 'Geographic information systems for environment and development', Summary paper, UNEP/GRID, United Nations Environment Programme, Carouge, Switzerland.

Sirait, M. T., Prasodjo, S., Podger, N., Flavelle, A. and Fox, J. (1994) 'Mapping customary land in east Kalimantan, Indonesia: a tool for forest management', *Ambio* 23(7): 411–417.

SNV (1996) 'The GIS component of MUZAMA Crafts Limited', Internal report, SNV (Netherlands Development Organisation), Zambia, Lusaka.

Stockdale, M. C. and Ambrose, B. (1996) 'Mapping and NTFP inventory: participatory assessment methods for forest-dwelling communities in east Kalimantan, Indonesia', in J. Carter (ed.) *Recent approaches to participatory forest resource assessment*, Overseas Development Institute, London, pp. 170–211.

Van Gelder, B. and O'Keefe, P. (1995) *The new forester*, London: Intermediate Technology Publications.

Wood, D. and Fels, J. (1992) *The power of maps*, London: The Guildford Press.

World Bank (1979) *Nepal: Development Performance and Prospects*, A World Bank country study, South Asia Regional Office, World Bank, Washington, DC.

World Bank (1987) *World Bank social indicator data sheet, Nepal*, World Bank, Washington, DC.

World Bank (1998) 'The World Bank Group Development Data 1998: Nepal at a glance', http://www.cdinet.com/DEC/wdi98/new/countrydata/aag/npl_aag.pdf.

Chapter 18

Implementing a community-integrated GIS: perspectives from South African fieldwork

Trevor M. Harris and Daniel Weiner

18.1 INTRODUCTION

A conceptual framework for PPGIS has been well developed during the last decade, and several crucial elements of both the public participation and GIS components have been identified (Abbot *et al.* 1998; Obermeyer 1998). Until quite recently, however, there were few examples to demonstrate how the implementation of PPGIS might actually proceed (Craig *et al.* 1999; Talen 1999). The purpose of this chapter is to identify one such approach based on fieldwork undertaken in Mpumalanga Province, South Africa. It was in addressing the complexities of PPGIS implementation that we coined the term community-integrated GIS (CiGIS), intended to represent a slightly different mode of PPGIS implementation than that previously envisaged (Harris and Weiner 1998). In this chapter, we briefly outline the basic concepts behind CiGIS, and present an application in support of land and agrarian reform in South Africa.

18.2 GIS, SOCIETY, AND PPGIS

Early thoughts about PPGIS implementation envisaged placing a GIS into the hands of communities almost as a counterpart to the systems operated by public and private agencies. In our fieldwork in South Africa, we quickly rejected this approach as infeasible and shifted to a CiGIS orientation. To provide some context for this conceptual and operational change, we now identify several key issues raised in the literature on *GIS and society*, and explain how they impacted our attempt to connect community participation with a GIS.

First, and perhaps foremost, was the issue of how to address or overcome differential access to hardware, software, data, and expertise. Much of the discussion on this topic in the literature was strikingly played out during our work in transitional South Africa (Harris *et al.* 1995; Weiner *et al.* 1995). Many communities in the case study area were struggling to acquire basic necessities of life such as water, shelter, food, and fuel. Most communities did not have access to electricity, and education had been deliberately withheld. Despite the tremendous enthusiasm and desire of local communities to participate, the chronic and endemic problems of community access to basic resources in Mpumalanga Province necessitated a GIS implementation strategy that drew upon GIS capability, support, and willingness from outside the communities themselves. In this case, a project team from West Virginia University (WVU), in collaboration with the South African Department of Land Affairs, filled this role.

Second, the issue of structural knowledge distortion in post-apartheid South Africa became of paramount concern in developing a community response to land and agrarian reform. The major central government and provincial agencies of the apartheid regime had fully embraced the new technologies of GIS and remote sensing, and under government mandates had operated them in support of an oppressive state regime. Digital spatial data were available from these agencies, but they were unreliable and expensive to purchase. Some of the data were also deemed confidential and were not readily available. Some data simply had not been collected. For example, land claims are a major component of the post-apartheid land reform process, yet no official documentation of forced removals exists. The overlapping tribal and community land claims that we encountered suggests several phases of forced removals occurred, none of which were officially recorded or documented. Thus, the data collected by the state, the available geo-spatial databases, and their content were representative of the goals of an apartheid state, and reflected the conceptions of space of an 'elite' (predominantly white) sector of society.

These factors highlighted a third area of concern frequently discussed in the literature: the desire to complement 'official' and 'expert' digital spatial information with local knowledge held by members of the community. Seeking to redress structural knowledge distortion through the inclusion of local knowledge held by members of black communities themselves thus became a primary challenge of the project. Much of a community's knowledge is heavily qualitative in nature and invariably based on oral history and the experience of having lived in a place for some time. Capturing this knowledge in a GIS that relies heavily on the spatial primitives of point, line, and polygon and the quantitative ordering of information is no easy task. These issues forced us to address both qualitative and quantitative aspects of local knowledge acquisition, and the integration of this knowledge into a GIS.

Fourth, it is erroneous to think of local knowledge as homogenous or uniform. In many respects, it would be better to use the term 'knowledges' as a way of recognizing that community information is varied and socially differentiated (Mahiri 1998). In seeking to include local knowledge within a GIS, the problems of identifying and incorporating the socially differentiated perspectives of community participants had to be confronted. Doing so explicitly acknowledges the very real and constraining problem of differential access to information, and underscores that in many instances the focus of PPGIS may well be to address and ensure that the varying perspectives of community members are incorporated into a GIS.

18.3 CONCEPTUAL FRAMEWORK FOR THE SOUTH AFRICA CiGIS

The South Africa CiGIS is guided by three broad conceptual principles: popular community participation; local, social and spatial differentiation; and regional political ecology. Community participation has become a mantra in development planning and field-based academic research. Unfortunately, most participation associated with development planning is essentially *participation as legitimization*. Community meetings are held, local input is gathered, reports are produced, and top-down planning is maintained. In this context, participation helps to legitimize decisions that are not necessarily 'popular' within impacted communities. In the academic world, participation has come to designate a configuration of qualitative methods designed to understand complex social processes better than conventional quantitative or qualitative methods. Efforts to hear the voices of 'ordinary' people and 'capture local knowledge' are well intentioned, but in many instances these are forms of *participation for publication*, in which academics undertake research to produce books and journal articles while leaving the subject communities with little (if any) tangible benefits.

Popular participation is an attempt to locate community participation in the context of particular local configurations of power within civil society. Participatory processes become part of the structures of everyday life, and ordinary people are able to express their opinions as openly as possible. The South African CiGIS has its roots in a participatory land reform project initiated in 1991 during a period of intense political struggle and violence (Levin and Weiner 1997). As a result of our participation in that project, we are known in the community and viewed as friends and advocates of popular local causes. The participatory process is thus central to our work, and the issues addressed in the CiGIS are community issues that have significant local importance. CiGIS implementation assumes, therefore, that tangible community needs are being addressed and that the project is political by its very nature.

Our conceptual and methodological framework for CiGIS development and implementation also assumes social and spatial differentiation. As suggested above, communities are not homogenous and GIS can inadvertently maintain unequal development. In the South Africa study, spatial differentiation is represented by the inclusion of diverse forms of participant social groups, including land reform organizations, peri-urban former homelands groups, farmworkers, large-scale (white) commercial farmers, and local chiefs. Race- and gender-based forms of social differentiation are also included in the South African CiGIS, and age, class, and other forms of difference will be added to the analysis in the future.

The South African CiGIS is also guided by an appreciation of regional political ecology. This conceptual framework helps researchers to analyse the social histories and landscape politics of the participant communities, and to reflect on their own academic interests in these areas. Our relationship with these South African communities began in 1991 when community elders explained how grand apartheid social engineering had dispossessed them of land, water and biomass resources in their former Lebowa homeland. The contemporary poverty of these groups was clearly linked to the historical geography of forced removals and to the production of local and regional apartheid geographies.

18.4 MPUMALANGA CASE STUDY

The Mpumalanga Province is a transitional area between the relatively cool and moist highveld plateau (over 1200 m in altitude) and the hot, dry lowveld (200–600 m in altitude). Mean annual rainfall ranges between 400 and 700 mm in the lowveld and between 1000 and 1500 mm on the escarpment and parts of the highveld. These environmental features, combined with the history of forced removals and forced urbanization under colonialism and apartheid, have produced a landscape of extreme social and ecological variation. The total population of the Province is over 3 million, of whom one-third live in urban areas and almost half reside in the former homelands. The case study area, the Central Lowveld subregion, is located mainly within the Lowveld Escarpment District of Mpumalanga Province, and includes a small portion of Bushbackridge to the north (Figure 18.1). The latter is disputed territory in Northern Province, and includes portions of the former Lebowa and Gazankulu homelands.

Intensive and exotic industrial forest plantations and large-scale commercial fruit and vegetable farms dominate the western third of the case study area. Some of these are located on highly arable land. Forestry companies control large tracts of state land, and this raises substantive issues regarding socially and ecologically appropriate land-uses. Forest plantations and large-scale commercial farms thrive because of a highly skewed system of water access. During the apartheid era, the social production of this watershed was centred on a complex system of dams and tributaries that capture valuable water for (mostly white) large-scale commercial farms (Woodhouse 1997).

The former homelands of KaNgwane, Gazankulu, and Lebowa are located east of the agriculture and forestry plantations. These bantustans are overcrowded and poorly serviced relics of grand apartheid. Land demand is

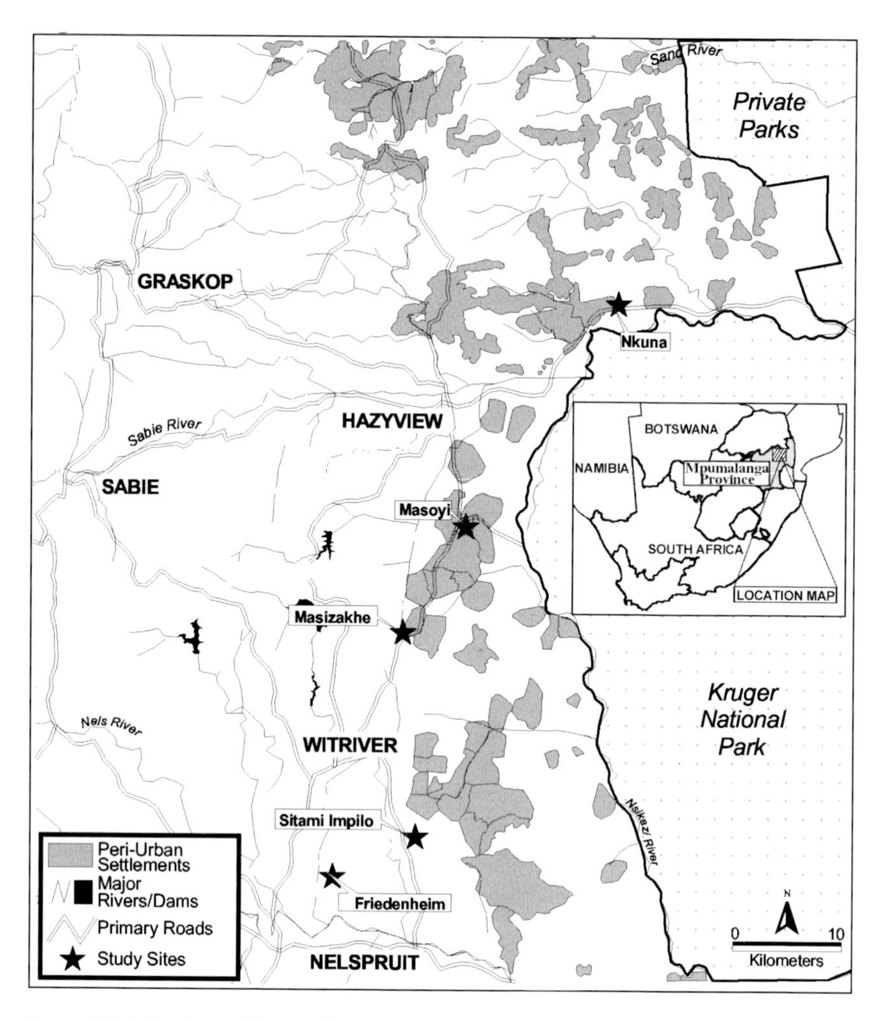

Figure 18.1 The Central Lowveld case study area, South Africa.

high, water is in short supply, and the history of forced removals remains fresh in peoples' memories and imaginations. Historically, political struggles have been connected to the decline in access to land, water, and biomass resources (Levin and Weiner 1997). The Kruger National Park and several private game parks occupy the eastern portions of the case study area. Since 1994, tourism has again become a growth industry, and visitors to the Mpumalanga and Northern Province Lowveld are growing. The use of land for game tourism has generated interesting discussions within the region regarding the potential for community-based range management models.

Many of the study participants, however, perceive limited personal benefit from the adjacent game parks.

18.5 CiGIS IMPLEMENTATION

Populating the CiGIS database was a central issue in the South African project. Acquiring spatial data is always a challenge for GIS practitioners; however, in CiGIS, this challenge is compounded by the need to draw heavily on local knowledge obtained from within communities. Developing the CiGIS database thus focused on obtaining the more traditional GIS coverages that detail the physical and cultural infrastructure of a region, and obtaining qualitative knowledge from members of the local community. Incorporating traditional data involved the familiar search for existing data of sufficient quality, attribution, relevance, and scale to meet the needs of the project. Digital spatial data were obtained from official government sources and private data providers, and by scanning and digitizing existing analogue maps. The digital data comprised both vector and image files. Digital raster graphics were generated from the Ordnance Survey 1:50,000 topographic map sheets and were based on mid-1980s source maps. Built-up and peri-urban settlement patterns were obtained in digital form from private vendors and were based on 1997, 1:10,000 orthophoto imagery. Data on land contour, hydrology and dams, roads, railroads, jurisdictional boundaries, and state-owned lands were obtained in digital form from government agencies at a scale of 1:50,000. Land-cover data were obtained at 1:250,000, and land-type data were captured at 1:50,000.

CiGIS requires that traditional top-down 'expert' information be complemented by information garnered from local community groups. In this case study, the latter consisted of local groups within the former homelands with various relationships to the government's land reform programme. These groups are characterized by a diversity of rural production systems and relations of production, and include Cork Village and Nkuna Tribal Authority; Friedenheim Farmworkers; Masoyi Tribal Authority; Masizakhe Land Redistribution Project; Sitama Impilo Land Redistribution Project (Figure 18.1) and six (white) large-scale commercial farmers in the area. Where appropriate, each of these groups were further subdivided into groups of men, women, and tribal leaders in order to capture the crucial socially differentiated local knowledge each group held.

Community workshops were held to compile information from each group based on five broad political ecology concerns:

1 the historical geography of forced removals,
2 identifying and comparing 'expert' and local understandings of land potential,

3 perceptions of socially appropriate and inappropriate land-use,
4 access to natural resources, and
5 community views about where land reform should take place.

Each community was contacted in advance to arrange the half-day workshops. Base topographic maps of the areas were created at a variety of scales and brought to the workshops. The maps were overlaid with tracing paper, and each group was asked to record their perspectives on the above questions using colour-coded markers. Each group, comprised of approximately eight to ten people, met separately, and included a facilitator.[1] The resulting 'mental maps' were digitized, attributed, and incorporated into the GIS database. In addition, photographs, video recordings, and voice recordings of the interviews were taken and georeferenced to the mental maps.

The core of the CiGIS process was to integrate this information into a GIS database. As indicated, this represents a significant challenge because of the qualitative nature of much of the information. Initial work focused on embedding objects within an ArcView GIS, but an Internet-based GIS replaced this system. The Internet-based system provided a more suitable GIS environment within which to link quantitative GIS coverages with qualitative voice, photograph, text, and video data. An Internet-based system also permits ready access to other resources on the web. Perhaps most importantly, an Internet-based GIS provides a means by which to overcome some of the disadvantages of differential access. Although access to the Internet is not widespread in Mpumalanga Province, it provides the potential for greater access to GIS resources in the future. With an Internet-based GIS, maps and information can be downloaded using simple point-and-click procedures. An understanding of GIS concepts or software is not essential, removing another significant obstacle to communities having access to their information. Although we do not wish to minimize the very real obstacles to South African communities gaining access to the Internet, it is remarkable how quickly the Internet is becoming accessible in these regions through state agencies, the private sector, and community telecentres. We believe multimedia Internet GIS is a central component in the development of CiGIS, and that state agencies will have to take responsibility for providing access to resources and technology in order for such a system to be developed and for communities to be incorporated into GIS-based decision-making.

18.6 FIELD RESULTS

In this section, we focus on two examples to illustrate the type of information generated from the fieldwork and some critical issues posed by CiGIS implementation for land reform in South Africa. A more detailed presentation of the data can be found in Weiner and Harris (1999).

Land types data obtained from the Agricultural Resource Council of the South African Institute of Soil, Climate, and Water are presented in Figure 18.2. Based on the Institute's soil classification and slope information, land potential categories were established. In the study area, 43 per cent is classified as land of 'higher' agricultural potential, 17 per cent as 'medium' potential, and 40 per cent as 'lower' potential. This database of 'expert' knowledge about land potential was focused on the fertile river valleys from which black South Africans tell us they were forcibly displaced. In the course of comparing local knowledge of land potential with the official coverage from the Institute, several anomalies were identified. In several instances the mental maps of local communities, which included both

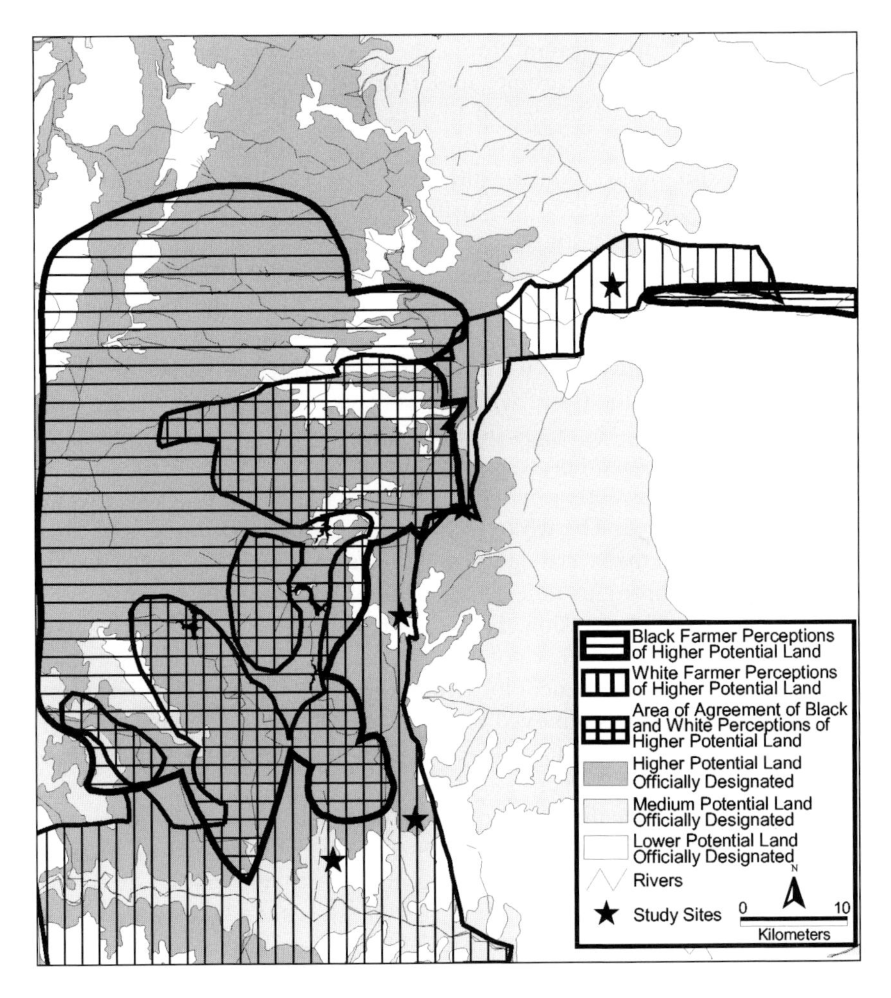

Figure 18.2 The multiple realities of land potential.

black and white participants, differed from the Institute's map. Specifically, local communities identified several areas of land as higher-quality land than was recorded by the Institute. The location of these areas is significant for their proximity to areas under scrutiny for potential reform. In essence, the differences occurred as a result of contrasting scales of data capture and differing perceptions of what constitutes high-quality land. Large portions of land that have a slope sufficient to make them inappropriate for mechanized agriculture were deemed of low quality by the Institute, but were considered very attractive by small-scale farmers who use animals and hoes. We would expect an operational CiGIS in the region to help locate viable high-slope areas with potential for small-scale agricultural production.

In the second case study example, there was no 'official' data on forced removals in the area. Braum Raubenheimer, former Minister of Water Affairs and a member of cabinet under Prime Minister Verwoerd, was a project participant, and he denied that black South Africans were forced to relocate as a result of white settlement or the actions of the apartheid government and police. A very different story emerged when this issue was discussed with black South African participants. One-quarter of the black population in the study area told us they experienced at least one forced removal in their lifetime (Levin and Weiner 1997). This is why CiGIS participants remain willing, even anxious, to talk about the historical geography of forced removals.

White farmers, however, were reluctant to discuss issues of forced removals. The mental maps of whites and blacks in the subregion are compared in Figure 18.3. The maps suggest very different perceptions of subregional landscape history. Forced removal mental maps for black participants delineated an extensive area of removals especially in the southwest region of the case study area. The white farmer mental maps acknowledged areas of forced removals that were significantly smaller in size. Unexpectedly, the white farmer responses also identified areas of white farmer removal, most likely for homeland expansion. Furthermore, because the peri-urban black settlement is located on higher potential arable land, black participants indicated that the area of settlement located to the immediate south of Hazyview is where the tribal chiefs removed blacks. This was done to enable members of the tribal authority and local black businessmen to gain better access to this high-quality land (Weiner *et al.* 1995).

Mental maps are qualitative representations that must be handled carefully. However, the mental maps were invaluable in representing the only known record of forced removals and for identifying phases of forced removals in which removed communities were subsequently relocated. These complementary interpretations of historical dispossession have provided the basis for understanding the existence of many overlapping land claims that have contributed to the slow pace of land restitution.

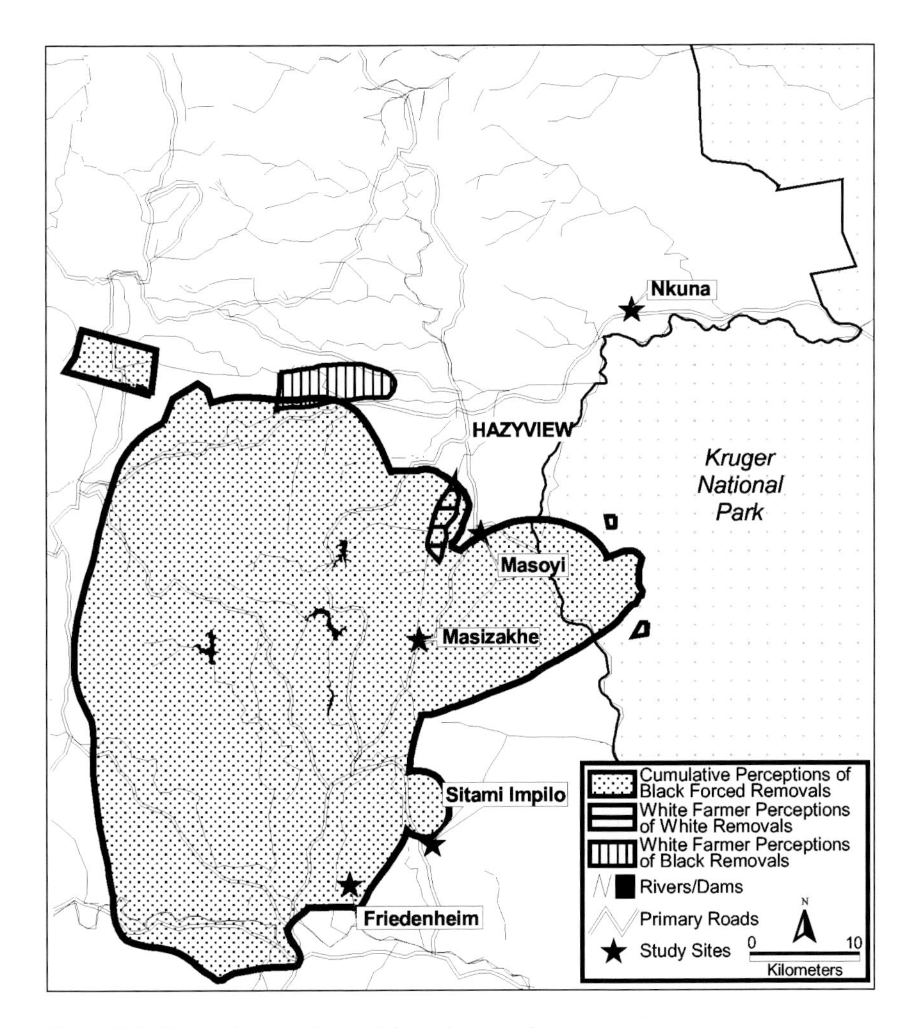

Figure 18.3 The multiple realities of forced removals.

18.7 CONCLUSION

The literature on *GIS and society* has generated rich conceptual and political questions for GIS developers, users, and practitioners about issues of database development and use, visualization and representation, and the power relations that affect system access. PPGIS is one outcome of such critiques and PPGIS efforts have, to date, been particularly concerned with issues of community empowerment, disempowerment, and the integration of quantitative and qualitative information.

The human, financial, and technical resources required for GIS development and operationalization suggest that in-house community GIS are unlikely to broaden access to spatial decision-making, especially in underdeveloped regions and poor communities (Hastings and Clark 1991). Rather, alternative mechanisms for data and GIS delivery that relieve communities of much of the hardware, software, data, expertise, and maintenance costs will most likely predominate. CiGIS seeks to address such concerns in GIS-based decision-making. In the future, Internet-based GIS will be a core component of such an access and delivery system, although the social context of Internet access will vary significantly from community to community.

It is the integration of community viewpoints that has dominated our work in South Africa. The reasons for this are perhaps more apparent in the socio-political context of South Africa than in other parts of the world. Under apartheid, official GIS data about communities was both selective and distorted, and served to justify agency mandates in support of grand apartheid social engineering. Addressing structural knowledge distortion in this context entails integrating alternative forms of local knowledge with the 'expert' data of government agencies. 'Capturing' this knowledge is one of the most challenging aspects for GIS practitioners, because local knowledge is invariably qualitative and spatially imprecise. Effectively collecting cognitive, visual, graphical, aural, and narrative forms of information and integrating them within a GIS entailed the use of what Schiffer (this volume) has called spatial multimedia. Furthermore, as demonstrated in this case study, community knowledge is socially differentiated and this raises important questions about how the final product might be used, and for whom.

The South Africa case study also demonstrates how a regional political ecology conceptual framework can be operationalized with a CiGIS. The mental maps depict local struggles for natural resource access, and an associated social reproduction crisis for some participants. An indication of this was the enthusiastic response of some participants, and their delight at simply being asked what they thought about their locale, historical geographies, and future aspirations. Black participants' consciousness about the landscape was formed though historical processes of dispossession, and many are anxious to discuss local natural resource politics and power relations. Large-scale commercial farmers, on the other hand, were much more reluctant to draw maps and discuss configurations of power within the local and regional landscape. It is important to note, however, that the Mpumalanga CiGIS did yield some interesting and important areas of agreement among the socially differentiated community participants.

A final point is that complementary knowledge acquired at differing scales creates a rich, valuable contextual resource for decision-making. In many instances the CiGIS incorporates information (such as the history of forced removals) that was previously unrecorded and that would almost certainly have been excluded from more traditional GIS databases. Spatial decision-making using CiGIS remains a significant challenge, particularly in the context of socially differentiated knowledge, perceptions about landscape and uneven access to GIS resources. Nevertheless, there is considerable interest in South Africa to link community participation with GIS (Hill and Strydom 2000; Mather 2000). To achieve this goal it is critical for local agencies to interact with communities on a continual basis and for GIS practitioners to seriously engage with the local structures of civil society. CiGIS contributes to greater access to, and sharing of, valuable spatial information, and provides multiple representations of past, present, and future landscapes. It also highlights the conflictual nature of spatial decision-making and acknowledges GIS practice as both technological and political (Lupton and Mather 1996).

ACKNOWLEDGEMENTS

This research was supported by NSF grant # SBR-951511: 'Integrating Regional Political Ecology and GIS for Rural Reconstruction in the South African Lowveld', NCGIA Initiative # 19, and the WVU Regional Research Institute. In South Africa, Indran Naidoo, Richard Levin and Rachel Masango within the Department of Land Affairs provided critical institutional and logistical support. Regina Dhlamini of Nelspruit was an effective interpreter and Edward Makhanya of the University of Zululand helped set up some of the field visits. Wendy Geary (former WVU graduate student), Ishmail Mahiri (lecturer at Kenyatta University, Kenya), Heidi van Deventer (lecturer at Rand Afrikaans University, South Africa) and Lloyd Mdakane (Department of Land Affairs, South Africa) are highly valued members of the research team. They have all contributed in important ways to this research. Thanks also to Tim Warner of WVU who helped interpret land types data for the land potential classification and Wilbert Karigomba (Geography Ph.D. student at WVU) for helping with GIS data analysis and map production.

NOTE

1. White farmers were interviewed individually or in small groups.

REFERENCES

Abbot, J., Chambers, R., Dunn, C., Harris, T., De Merode, E., Porter, G., Townsend J. and Weiner, D. (1998) 'Participatory GIS: opportunity or oxymoron', *PLA Notes* 33: 27–34.

Craig, W., Harris, T. and Weiner, D. (1999) 'Empowerment, marginalization and public participation GIS', Specialist Meeting Report compiled for Varenius:

NCGIA's Project to Advance Geographic Information Science, NCGIA, University of California, Santa Barbara, February.
Harris, T. and Weiner, D. (1998) 'Empowerment, marginalization and community-Integrated GIS', *Cartography and Geographic Information Systems* 25(2): 67–76.
Harris, T., Weiner, D., Warner, T. and Levin, R. (1995) 'Pursuing social goals through participatory geographical information systems: redressing South Africa's historical political ecology', in J. Pickles (ed.) *Ground Truth: The Social Implications of Geographic Information Systems,* New York: Guilford Publications, pp. 196–222.
Hastings and Clark (1991) 'GIS in Africa: problems, challenges and opportunities for co-operation', *International Journal of Geographical Information Systems* 5(1): 29–39.
Hill, T. R. and Strydom, J. M. (2000) 'A call for participatory GIS as an approach to rural development and spatial data dissemination, with reference to water management', *South African Journal of Surveying and Geo-Information* 1(4): 185–190.
Levin, R. and Weiner, D. (eds.) (1997) *'No More Tears': Struggles for Land in Mpumalanga, South Africa*, Trenton: Africa World Press.
Lupton, M. and Mather, C. (1996) 'The anti-politics machine: GIS and the reconstruction of the Johannesburg local state', *Political Geography* 16(7): 565–580.
Mahiri, I. (1998) 'The environmental knowledge frontier: transects with experts and villagers', *Journal of International Development* 10(4): 527–537
Mather, C. (2000) 'Geographical information systems and humanitarian demining', *South African Geographical Journal* 82: 56–63.
Obermeyer, N. (ed.) (1998) 'Special content: public participation GIS', *Cartography and Geographic Information Systems* 25(2): 65–66.
Talen, E. (1999) 'Constructing neighborhoods from the bottom up: the case for resident generated GIS', *Environment and Planning B* 26: 533–554.
Weiner, D., Warner, T., Harris, T. and Levin, R. (1995) 'Apartheid representations in a digital landscape: GIS, remote sensing and local knowledge in Kiepersol, South Africa', *Cartography and Geographic Information System* 22(1): 30–44.
Weiner, D. and Harris, T. (1999) 'Community-Integrated GIS for land reform in South Africa', WVU Regional Research Institute Research Paper # 9907, Morgantown W. V. (http://www.rri.wvu.edu/wpapers/1999wp.htm).
Woodhouse, P. (1997) 'Hydrology, soils and irrigation systems', in R. Levin and D. Weiner (eds), *'No More Tears:' Struggles for Land in Mpumalanga South Africa*. Trenton: Africa World Press.

Chapter 19

Information technologies, PPGIS, and advocacy: globalization of resistance to industrial shrimp farming

Susan C. Stonich

19.1 INTRODUCTION

This chapter investigates the potential of PPGIS to empower local communities, enhance global civil society, and contribute to public advocacy, especially in the Third World. It is based on lessons learned during an ongoing applied research project investigating the role of Information Technologies (ITs) in the globalization of resistance to industrial shrimp farming in tropical coastal zones in Asia, Latin America, and Africa. The project is multidisciplinary and highly collaborative – including the efforts of academics/scientists, NGOs, grassroots groups, and private/public donors – and is aimed at integrating research and practice. This chapter focuses on the challenges, issues, feasibility, and potential of scaling-up – that is, linking local/community-level PPGIS into a global PPGIS in order to advance advocacy, affect global environmental governance, and further alternative development.

To date, project activities have focused on conducting ethnographic and survey research among members of the global resistance coalition, with funding from the National Science Foundation (NSF), The University of California Pacific Rim Research Program, and the Rockefeller Brothers Fund. Preliminary research included assessment of existing access to and use of advanced ITs by individual coalition members, the global resistance coalition, and the worldwide network of industry supporters (the backlash movement). Project activities have also included a series of meetings and workshops for project collaborators. Preliminary results suggest the crucial role played by advanced information technologies such as e-mail, the Internet, and the World Wide Web (WWW) in the formation and maintenance of resistance and industry networks, in facilitating vital communication among members of each network, and in each network's strategy for achieving short- and long-term objectives. Preliminary work also reveals the virtually universal desire by the grassroots/non-governmental coalition members to increase access to, training in, and use of spatial ITs (maps, remotely sensed data, and GIS) and other ITs (e.g. e-mail, the Internet, the WWW) to achieve individual organizational and shared coalition objectives (Stonich 1998).

Discussion of the formation of the global resistance and counter industry coalitions, as well as the role of information and communication technologies in social change, can be found in Stonich and Bailey (2000) and Stonich (1998).

Current funding for the project includes an NSF planning grant. The scholarly aim of this phase of the research is to determine the social context and impacts of communications and spatial ITs on the formation, strategies, and effectiveness of the emerging global coalition of non-governmental and grassroots organizations that is resisting the expansion of the shrimp farming industry. Equally important are the applied objectives of the project: enhancing access to, and effective use of, these technologies by local individuals, communities, and organizations as well as the global network. These activities are a cooperative effort to collect, interpret, and communicate ecological information; to share information; to integrate scientific data with local knowledge; and to advance public/consumer campaigns. Simultaneously, field research is being conducted in a well-chosen sample of locales in Asia, Latin America, and Africa in order to identify information/data needs and assess how ITs, including PPGIS, might meet those needs.

By using an empirical approach that takes advantage of a dynamic, global phenomenon, this stage of the project will aim to enhance understanding and general explanations of information and spatial technologies. Such understanding is crucial for the design of more appropriate, accessible, and democratic ITs and systems.

19.2 GLOBALIZATION OF RESISTANCE TO INDUSTRIAL SHRIMP FARMING

Aquaculture often is promoted as 'The Blue Revolution', analogous to the Green Revolution in agriculture and essential to feed growing human populations in light of stagnating or declining yields of marine stocks (Figure 19.1). Recently, however, increased attention has been paid to aquaculture's social, economic, and environmental costs (e.g. Bailey *et al.* 1996). Particularly controversial is the explosive expansion of capital-intensive industrial shrimp farming in coastal brackish water ponds in Asia, Latin America, and Africa, which critics maintain has promoted social dislocations, ecological changes, and environmental destruction comparable to those caused by Green Revolution technologies (Bailey 1997) (see Figure 19.2).

Globalization of industrial shrimp farming has created new institutional linkages among international agencies, multinational corporations, governments, and national elites. Globalization also has provoked considerable violence and the emergence of grassroots resistance movements (principally NGOs) among the poor in coastal areas of Asia, Latin America, and Africa

Figure 19.1 Intensive shrimp farm in Thailand.

Figure 19.2 Constructing a shrimp farm along the coast of Honduras.

(Stonich 1996; Stonich and Bailey 2000). Aware of the powerful political and economic forces allied against them, resistance groups have sought regular contact with their counterparts in other countries, as well as support from organizations and individuals in industrial nations. Major environmental groups including Greenpeace, World Wildlife Fund, and the Natural Resources Defense Council, as well as private foundations such as the MacArthur Foundation and the Rockefeller Brothers Fund, have supported the network resisting industrial shrimp farming. These and other organizations and individuals have found common ground with several hundred community-based NGOs around the world. After a series of international meetings beginning from April 1996, these groups formed the Industrial Shrimp Action Network (ISA Net) on World Food Day in October 1997 during a week-long meeting in Santa Barbara, California. The aims of ISA Net include drawing international attention to the environmental and social costs of shrimp farming, and supporting the efforts of coastal communities to maintain viable communities, economies, and environments (Stonich and Bailey 2000).

Currently, ISA Net is made up of 25 member NGOs from 22 countries. Membership is divided almost equally between organizations in the North and South. ISA Net's operating structure includes its members, a secretariat (who works from ISA Net headquarters in the state of Washington), and a steering committee of nine individuals representing a subset of member organizations. Four working groups also exist: communities and communication, public education, science and industry, and international institutions and national governments. ISA Net maintains close ties with its parent organization, MAP[1] (Stonich 1998). Although ISA Net is focused on the human and environmental consequences of industrial shrimp farming, MAP has a much broader focus that includes the human impacts of destruction of mangrove and other coastal environments.

19.3 ASSESSMENT OF THE USE OF COMMUNICATIONS AND SPATIAL TECHNOLOGIES

An evaluation of access to and use of telecommunications and spatial technologies by members of the global shrimp farming resistance coalition was attempted in the following manner:

1 In-depth, semi-structured interviews with NGO leaders were conducted during various international meetings and at local sites between 1997 and 2000. These interviews focused on access to and use of these technologies, as well as on perceived obstacles to their use. Leaders were also asked about the usefulness of these technologies to advancing the goals/objectives of their organization.

2 MAP records, mailing lists, etc. were compiled, a master database was created using Microsoft Access, and a sampling frame of 811 MAP members was constructed.
3 Semi-structured questionnaires were sent by e-mail and postal mail to a subset of 64 key MAP members. This group was determined by attendance at one of the three international NGO strategy sessions, and included NGOs, academic advisors, and donors.
4 On the basis of various stakeholders identified earlier in the project (e.g. NGOs and grassroots organizations, the industry, academics, governments), an extensive search of WWW sites was done in order to evaluate and compare the use of the web by each group of stakeholders.
5 Two information technologies workshops conducted during the October 1997 meetings in Santa Barbara were used as a means to evaluate access to and use of associated technologies. One workshop focused on the use of e-mail, Internet, and WWW technologies, and the other on the use and potential of GIS and remotely sensed data.
6 A subsequent four-day meeting was convened in Santa Barbara in June 1998. This meeting was attended by the ISA Net secretariat and an international group of social scientists engaged in research on the human and environmental consequences of industrial shrimp farming. Many of the participants were using GIS in their work. Discussions at this meeting centred on the feasibility, potential content and alternative structures of an IT/PPGIS system that would meet the needs of ISA Net in a scientific, rigorous way.

19.3.1 Project results

Despite significant ideological, political, and strategic differences between members, members of the resistance coalition share a belief that ITs and spatial technologies are significant tools with which to achieve their divergent goals. They are becoming increasingly aware of the potential of ITs to facilitate communication among members and to advance political action. Although members rely somewhat upon direct networking in the field, the group is largely maintained electronically via e-mail and the Internet, and is loosely coordinated by MAP which maintains a member list-server and WWW site. Of the 64 organizations identified as key members of the coalition, 50 (80 per cent) reported that e-mail was their most frequent means of communication with MAP, ISA Net, and individual members.[2]

Although the majority (70 per cent) of NGOs surveyed used paper maps extensively, they reported little use of GIS and other advanced spatial technologies. However, their responses indicated a relatively sophisticated awareness of the potential of digital technologies to bolster their efforts. In addition, there is growing cognizance of how these technologies are used by governments and industry to justify the siting of shrimp ponds and other coastal development. In a letter to Dr Anjali Bahugauna at the India Space Applications Center on 9 September 1996, Indian NGO leader Dr Bittu Saghal wrote: 'I am on the Ministry of Environment's Coastal Task Force and am deeply distressed at the way in which technical experts are helping the government to interpret images to suit development projects. Mangrove and fragile coastal zones are easily being categorized as having "no ecological value" so as to facilitate their destruction by roads, jetties, and other kinds of development...I believe that our Coastal Regulation Zone Rules are vital to the survival of fishing grounds and therefore fisher folk.' (Saghal 1996).

According to Jorge Varela, Executive Director of the Committee for the Defense and Development of the Gulf of Fonseca (CODDEFFAGOLF) in Honduras:

> Our objects are the defense of coastal natural resources; development of local communities and environmental political activities...Clearly, we are interested in obtaining information and spatial technologies—and in improving our system of communication...Spatial technologies are critically important...With these technologies, we could determine, independent of the government and the shrimp industry, the grade of destruction or recuperation of various habitats which would make our studies, conclusions, and recommendations more objective
>
> personal communication 1998

19.3.2 Constraints on the use of ITs and spatial technologies

In spite of general recognition of the potential of ITs and spatial technologies by NGO leaders, several significant constraints regarding access and implementation were identified. These included: lack of economic resources to purchase equipment (reported by 75 per cent of respondents); inadequate training in wisely using and properly maintaining equipment (63 per cent of respondents); inadequate English language skills (56 per cent of respondents); and poor infrastructure in rural areas (56 per cent of respondents).

These constraints were even more significant in terms of spatial technologies, the use of which raised a number of additional questions: In the process of creating maps for local, non-science actors attempting to use GIS, will local users be forced to adopt the 'correct' scientific language to use these technologies? Will such requirements dissuade potential users because they believe they never win these kinds of science-based arguments? Does GIS involve more than simply using the tools, but also adopting a whole legal-science discourse? These questions involve accessibility and user-friendliness, both on the 'producing' and 'consuming' ends of the technology, and suggest that as much attention should be paid to public participation in production of PPGIS as in its use.

19.3.3 Current status of the use of ITs and spatial technologies

Since its founding in October 1997, ISA Net has made some significant strides in achieving its goal of advancing communication among members. These achievements are quite important in light of the extreme diversity of its membership. In 1999, with the help of external funding from private foundations, ISA Net created a WWW site, http://www.shrimp-action.org. This site includes information about ISA Net, action alerts and news, a list of ISA Net members and their individual websites, an archive of press releases, links to current research and scholars, and contact information. Almost all ISA Net members now maintain their own WWW sites, usually with external financial support. This is quite a change from 1997, when no Southern NGO member had its own WWW site and most did not have access to the Internet. In early 2000, ISA Net established a private e-mail list server through E-groups to expedite communication among members and associated academics. Although ISA Net concentrates on facilitating communication among its members, MAP has assumed the role of providing information to the public, primarily through a weekly electronic newsletter sent to several hundred individuals and organizations.

To date, ISA Net is not directly using GIS, although spatial technologies are central to the work of several of the academics who support ISA Net's efforts. A good example is the Shrimp Aquaculture Research Group at the University of Victoria in British Columbia (http://www.geog.uvic.ca/shrimp/). Under the direction of Dr Mark Flaherty, this multidisciplinary group is engaged in a number of projects focusing on the human and environmental dimensions of shrimp farming in Thailand. One of these projects uses GIS to integrate LANDSAT imagery and other digital data (hydrology, soils, political boundaries, irrigation infrastructure) to investigate inland shrimp farming. A related project investigates the potential of using RADARSAT imagery to identify shrimp ponds. These projects are part of a long-term effort to develop a training programme and build spatial analytical capacity in the Department of Aquatic Sciences at Burapha University.

19.4 CHALLENGES TO SCALING-UP

The many problems involved in public use of communications and spatial technologies are magnified when the community of users is a heterogeneous coalition of individuals and institutions throughout the world. In addition to the formidable financial, technical, and data constraints, significant social, cultural, and political obstacles also exist.

19.4.1 Lack of consensus among ISA Net members

Among the most prevalent and well-documented reasons for the failure of public participation efforts is failure to take into account and directly confront the diversity, contending perspectives, and unequal power relations among community members. Overcoming these obstacles is difficult at the local level; it becomes almost insurmountable when the community is constituted by a global coalition of diverse factors (Stonich and Bailey 2000). This is the primary obstacle to advancing a successful ICT system and PPGIS for ISA Net, and is more important than either financial or technological constraints. To some extent, ISA Net is an experiment in establishing a powerful global coalition of the poor in the local and global arenas of environmental governance.

The existing information and communication network among ISA Net members is the result of three years of concerted efforts to reach consensus among contending coalition members. During the Internet and GIS workshops and the subsequent focus groups conducted during the NGO planning meetings in October 1997, heated discussions occurred regarding potential alternative designs for a system. However, no consensus was reached at that time due in part to contending perspectives among participants about the structure and organization of the global network itself (Stonich and Bailey 2000). There has been much more agreement among members regarding the kinds of GIS or spatial technologies information and data that should be covered by such a system. These included identifying community management areas and use of local resources by artisanal fishers and farmers; integrating information about the distribution of shrimp farms, processing plants, and packing facilities, and about employment generation; undertaking a longitudinal study using historical aerial photos and satellite imagery to demonstrate the decline in mangrove forests and fish stocks; and creating a spatial database identifying the distribution and types of human rights violations and legal protections, and the areas of successful versus unsuccessful legal protections. Non-governmental representatives suggested that these types of information could be used to justify their community-based development and conservation programmes. Participants in the workshops also agreed that under the appropriate conditions, *access* to ITs, *training* in ITs, and *funding* for high-potential projects (e.g. projects focused on specific development, policy, or legal goals) could significantly increase the probability of successfully achieving their goals.

19.4.2 Heterogeneity and diversity among members, environments, and ecologies

Local communities affected by the expansion of the industrial shrimp industry inhabit locales that are somewhat similar environmentally (i.e.

coastal, tropical, etc.). However, they also are extremely diverse in terms of nationality, culture, language, technological capacity, wealth, and power. These differences (and related diversity of interests) among ISA Net members are apparent in ongoing debates about the network's goals and objectives, strategies for action, and appropriate ITs. Gender divisions also have emerged as an important factor. Thus the challenge of designing and implementing a successful global IT and PPGIS for ISA Net go beyond the considerable frustrations usually associated with attempts to link data from different scales and time periods to produce a more comprehensive understanding of how people regulate and manage resources. Despite the advantages of spatial ITs in linking different data sources, it remains difficult to identify for analysis particular social, economic, political, environmental, and ecological factors and relationships among diverse human, geographic, and environmental contexts. A critical requirement in this regard is to create a system that is at once sensitive to local diversity (among people, perceptions, knowledge, environments, ecologies, political systems, institutions, etc.) and capable of synthesizing information and demonstrating regional and global patterns and conclusions.

19.4.3 Technological capacity and training

Although confronting the constraints presented by diversity and lack of consensus among members is crucial, obstacles related to technology, financing, and training also are serious and must be addressed. Diversity in technological capacity is apparent among ISA Net members. Not surprisingly, Northern members such as World Wildlife Fund, Natural Resources Defense Fund, and the Environmental Defense Fund enjoy considerable technological advantage over their Southern colleagues. Representatives of these Northern NGOs also are fluent in English and have more resources at their disposal than do their Southern counterparts. Although it is tempting to work through these more powerful organizations, they do not necessarily represent the interests of the Southern NGOs, many of which are community-based organizations comprised largely of poor people from coastal zones. Working with the more powerful Northern NGOs may serve the North's interests at the expense of those of Southern NGOs. Making things more complex, Southern NGOs themselves differ significantly among themselves in their capacity to utilize advanced ITs. Thus, working with local groups in the South greatly reduces the speed of design and implementation, although doing so enhances representation and effective participation.

19.4.4 Cultural and social considerations

People from the South may be suspicious of technologies that are seen as Western or Northern, and with which Northern partners have a higher level of expertise, familiarity, and financial investment. Southern partners may not want to use such technologies because they feel that the playing field is not level. Even if they participate in e-mail discussions, some participants feel that the medium does as much harm as good because of the potential for misunderstanding inherent in e-mail messages. Research suggests that the Internet does not build trust as rapidly as face-to-face encounters. In the West, people have become very accustomed to phones, fax machines, and the Internet as communication media. But in international networks that involve people who do not use telephones or other technologies on a regular basis, it may take quite some time before the trust is there to use them. Nor is it simply a matter of others 'catching up' with technology. There are legitimate criticisms of the Internet and valid arguments in favour of face-to-face encounters. It will likely be impossible to sustain any Internet network without occasional face-to-face meetings and discussions. Although this may seem obvious, face-to-face meetings require large amounts of funding that simply may not be available to citizen action or international civil society networks. The Internet seems to work best in situations where people have already met in person one or more of the people they are communicating with, allowing Internet communications to develop from a personal basis, rather than the reverse.

19.4.5 Counter mapping, politics, and the ownership of information

As discussed elsewhere (Stonich 1998), the characterization of ITs, GIS, and other spatial ITs as capable of democratizing or reinforcing extant power relations is ambiguous. On which side of the line such efforts fall depends on a number of diverse factors. Mapping, especially counter-mapping, is frequently an extremely political endeavour, and must be viewed within the broader social, economic, and political framework in which it occurs. An IT/PPGIS as generally envisioned by members of ISA Net certainly enhances the potential danger of surveillance, conflict, and co-option of local knowledge and resources by power elites (including the shrimp industry). At the same time, it also has the potential to advance advocacy, make visible the claims of those most affected by the expansion of industrial shrimp farming, counter the claims of the industry, and promote new visions of development. The considerable agreement among resistance coalition members about the kinds of useful spatial knowledge and information that could be distributed through the Internet and at the local level remains a very significant (but unrealized) potential. The expansion of the shrimp farming industry already has provoked considerable conflict. The essential requirement here in terms of PPGIS is that the political consequences of such an effort must be thoroughly investigated and taken into account.

NOTES

1. MAP (Mangrove Action Project) is a non-governmental, environmental organization located in Washington State (outside Seattle). Its mission is to conserve mangrove and other coastal ecosystems. It is made up of more than 300 other NGOs, private foundations, and individuals from all over the world.
2. Other means of communication included: facsimile (9 per cent), postal mail (9 per cent), and telephone (2 per cent).

REFERENCES

Bailey, C., Jentoft, S. and Sinclair, P. (eds), (1996) *Aquacultural Development: Social Dimensions of an Emerging Industry*, Boulder: Westview Press.

Bailey, C. (1997) 'Aquaculture and basic human needs', *World Aquaculture* 28(3): 28–31.

Saghal, B. (1996) 'Letter to Dr Anjali Bahugauna at the India Space Application Center, on September 19, 1996' (cited in *Mangrove Action Project Quarterly Review*, Winter 4(3): 7).

Stonich, S. C. (1996) 'Reclaiming the commons: grassroots resistance and retaliation in Honduras', *Cultural Survival Quarterly* 20(1): 31–35.

Stonich, S. C. (1998) 'Information technologies, advocacy, and development: Resistance and backlash to industrial shrimp farming', *Cartography and Geographic Information Systems* 25(2): 113–122.

Stonich, S. C. and Bailey, C. (2000) 'Resisting the Blue Revolution: contending coalitions surrounding industrial shrimp farming', *Human Organization* 59(1): 23–36.

Chapter 20

Ensuring access to GIS for marginal societies

Melinda Laituri

20.1 INTRODUCTION

This chapter examines issues related to access and use of GIS by marginal societies. Marginal societies are defined as those groups that have been oppressed, exploited and denied access to the fundamental resources to enhance their everyday lives (Kozol 1991; Shiva 1997; Athanasiou 1996). Three different case studies are examined and evaluated to consider use of and access to GIS, as well as underlying issues related to data development, training and implementation. Common themes from the case studies are compared to identify larger conceptual issues related to GIS implementation. These case studies include the Maori communities of Panguru, Pawarenga and Whangape in Northland, New Zealand; the Arapaho-Shoshone Indian Nations of the Wind Rivers Reservation, Wyoming, United States; kindergarten through 12th grade (K–12) teachers in the Poudre School District, Ft Collins, Colorado, United States.[1]

Local knowledge is increasingly recognized as critical to resource management issues, but has not been adequately integrated into management strategies (Laituri and Harvey 1995). The two case studies involving indigenous peoples contribute to current work being conducted internationally to include indigenous biological knowledge within the Western framework of computerized knowledge systems used for resource management. Additionally, these projects explore the types of geographic information that is derived from different cultural groups for their explicit needs. An important thrust in recent geographic literature is environmental equity for disadvantaged and marginal populations (Ekins 1992). Increasingly, the use of GIS and information systems for resource management and development issues is a critical factor in allowing access to decision-making. Such databases must be constructed with equity in mind for all societal groups, and methods need to be developed that allow access and empower such groups through appropriate training and education. The case study involving K–12 GIS education provides a model for developing appropriate methods for specific user groups.

20.1.1 Politics of position: protocols of using alternative knowledge systems

Information is increasingly becoming a medium of exchange in technological society. Academics, scientific researchers and others have discovered that the knowledge indigenous people hold of Earth and its ecosystems, wildlife, fisheries, forests and integrated living systems is extensive and informed. Numerous efforts have been made to access such information (Duerden and Kuhn 1993; Denniston 1994). However, efforts to understand and utilize indigenous knowledge remain problematic due to cultural differences, lack of trust and controversies over who should collect such knowledge. As Katz (1992) has noted, representation of the 'other' is a serious play of power. Concern over social colonialism has developed into a debate over the 'crisis of representation' and questions of who should speak for whom. This crisis is synonymous with the struggle for indigenous land rights and ethnic identity, and its implications are far reaching in academic research (Jackson 1991).

The issue of representation is critical in devising research strategies that share indigenous knowledge. It is important to acknowledge and consider the implications of representing another group in technological-ethnographic research. All research is conducted from a position of observation. Defining the research position removes some of the power from the researcher, positioning them within the research as a visible part of the cultural representation which they construct. As an integral research component, partiality must be explicitly acknowledged in order to cross cultural boundaries and validate alternative perspectives. Indigenous perspectives often consider knowledge as sacred, while Western perspectives treat knowledge with skepticism and focus on evaluating and validating information (Patterson 1992). Given this difference in epistemology, it is important to respect different approaches to knowledge transmission. The challenge is to combine indigenous knowledge with Western technology in order to devise alternative natural resource management and conservation strategies that may be more efficient, and environmentally- and culturally sensitive.

However, strategies blending indigenous and scientific approaches need to be developed without privileging one culture over the other. The legitimization of local traditional knowledge is a promising avenue of empowerment in conservation decision-making. Supporting alternative knowledge systems of indigenous people may allow them to access foreign techniques as *they* choose. This is an essential caveat in the use of GIS by indigenous people: that the GIS is utilized by *them* for *their* needs. The need to assert self-determination in the research process itself is essential to the success of such efforts.

Several reviewers have identified specific problems with adopting indigenous knowledge in Western systems (Thrupp 1989; Watson and Chambers

1993). First, must indigenous knowledge be 'scientized' by Euro-American researchers to be legitimate? Euro-American scientific theories are commonly considered the dominant epistemology, and superior to alternative knowledge systems. Examining traditional knowledge through Euro-American methodologies may abstract such knowledge so that the complex subtleties (e.g. spiritual and mystical values and perceptions) are neither acknowledged nor recognized. Traditional knowledge could be further marginalized if it is considered 'unscientific'. However, romanticizing indigenous knowledge is equally problematic. A balance needs to be achieved that recognizes both the limitations and contributions of indigenous knowledge. Further confounding the relationship between Euro-American theories and traditional knowledge systems is language. Can cultural concepts transcend not only the barrier of cultural perspectives but of language differences as well, *and* be used within a computerized environment?

Second, is it appropriate for traditional knowledge to be extracted and used? A contradiction in the recording of traditional knowledge is that it can be both exploited for development purposes and used to protect culturally sensitive sites. There is a risk that institutions that sponsor conservation efforts may mine or exploit indigenous knowledge and develop projects inappropriate for local needs (e.g. the Green Revolution, or the use of biotechnology to create genetically different strains of crops that replace indigenous strains). Inappropriate use of sensitive traditional information (e.g. about sacred sites, traditional areas and hunting and gathering sites) may also pose problems. Restricted access to information may be critical not only to protect specific sites, but also to reinforce the integrity of knowledge systems dependent on ritualized processes of knowledge acquisition (Turnbull 1989).

Finally, what safeguards and assurances are built into the research or development process to ensure that the introduction of new technology does not represent another 'system of knowledge as a system of domination' or scientific colonialism (Cashman 1991: 49)? In addition, will safeguards ensure that new or existing local elites will not monopolize new technologies? Assurances are built on earned trust and goodwill, which obliges a researcher to a long-term commitment of time. Safeguards might include restricting access to products and outputs, creating monitoring committees made up of local representatives who oversee the introduction and use of new technology, and identifying returns that the community will receive from adopting such technology.

One purpose for blending indigenous and Western-based knowledge systems is to encourage participatory development and communication through 'knowledge-sharing' (Brendlinger 1992). This establishes an ongoing relationship with long-term goals rather than a single project goal. As information is jointly constructed through use of such tools as GIS, all participants gain a vested interest and knowledge acquisition is recognized as an evolving process.

20.2 CASE STUDIES

Three different case studies are now examined and evaluated to consider underlying issues of access related to data development, training and implementation. Common themes from the case studies are then compared to identify larger conceptual issues related to GIS implementation.

20.2.1 Maori communities of Panguru, Pawarenga and Whangape in Northland, New Zealand

The North Hokianga Maori Development Project is an ongoing joint research project at the University of Auckland with the Maori communities of Northland, New Zealand. The purpose of the project is to identify potential areas for economic development in the North Hokianga region of Northland, focusing on three communities, namely, Whangape, Panguru, and Pawarenga. A further component of the project is to develop culturally relevant data layers – specifically, the identification and assessment of Maori resources and identification of significant cultural and sacred sites through participatory mapping exercises – and then incorporate this information into a GIS. Maori input is important to the development of GIS in the New Zealand context due to the mandate of the Resource Management Act and the Treaty of Waitangi (Michaels and Laituri 1999). Future developments in technology transfer must occur with respect to transmitting GIS skills and the methods of capturing sensitive cultural information. GIS may facilitate the transmission, protection and maintenance of sensitive cultural information that is currently being lost. However, it is critical that local communities make their own decisions regarding the use, capture, and storage of such information.

The construction of a community-based GIS has provided the *iwi* (tribes) with the opportunity to identify meaningful applications of their particular areas from their own perspective rather than applications based only on Euro-American models of resource management and land-use. This has required close consultation with local Maori and especially the *komatua* (elders). The initial step toward creating a GIS demanded working through problems of methodology, accessibility, identification of the types of information to be included, and protecting sensitive cultural information. Tribal people were not only the GIS users but the GIS designers as well.

The database was designed to include two levels of access: (1) baseline data that includes all publicly accessible data (socio-economic, demographic, land resource inventory, valuation, cadaster and topographic); and (2) community information (traditional lands, hunting and fishing lands, subsistence land-use, historic and current agriculture fields, sensitive cultural information). Limited accessibility to the data was permitted dependent upon permission from community elders (Figure 20.1).

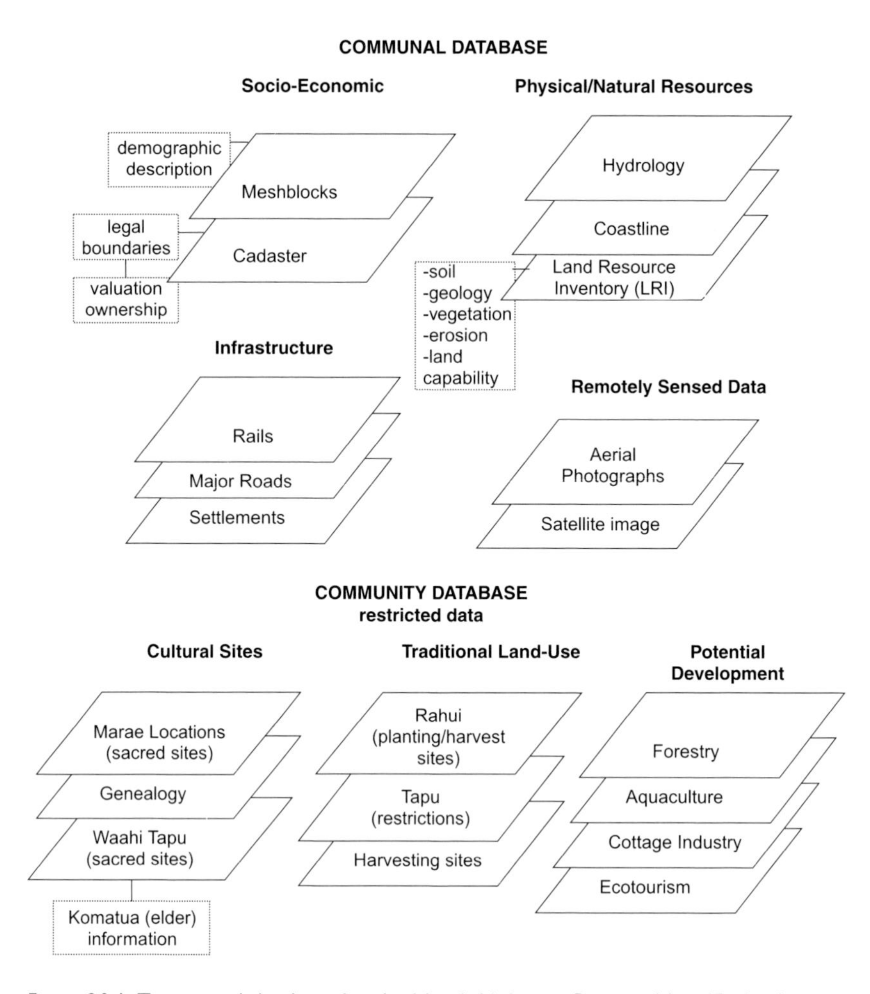

Figure 20.1 Two-tiered database for the North Hokianga Project, New Zealand.

The first tier of the database was completed and provided the community with a set of maps that evaluated the natural resources in each community. In addition, land tenure and parcel maps were created based upon existing digital data. These data were several years out of date (1979) and highlighted the need to update and integrate land ownership records. This is a daunting task as it demands the integration of data sets between several different governmental entities, including the Valuation Department, Lands and Deeds, the Department of Survey and Land Information, and the Maori Land Court. The second tier of the database design was agreed upon in principle; however, the actual inventorying and identification of culturally sensitive information proved problematic.

The community recognized that sensitive cultural information was being lost and that GIS technology had the potential to save, store and protect this information. Plans were initiated to serve both short-term and long-term needs for organizing and capturing such data. Short-term solutions focused on creating maps of buffered point locations. These point locations represented sensitive cultural sites but did not identify exact locations, hence the 'buffer'. Attribute information included the appropriate *komatua* who held the special knowledge of this site. Consequently, if a particular development project was to take place in a location assigned to an elder, the developer would know whom to contact to determine if the intended use was appropriate. Long-term solutions have focused on transferring technological GIS skills to these local groups.

During the course of this research it became evident that the local communities needed the technical ability to make their own decisions regarding the use, capture and storage of culturally sensitive information. Mechanisms to provide such technical ability can be developed through licensing arrangements with hardware and software vendors and data holders, and through grants to provide technical training. Access to technology and training in GIS skills is also necessary to assure implementation. More importantly, *iwi* recognize the link between GIS technology and self-determination. *Iwi* are mandated by the Resource Management Act to develop management plans for both development and conservation purposes. In New Zealand, GIS is a ubiquitous tool among resource managers and governmental agencies; Maori identified GIS as an enabling technology that would allow them to 'speak the same language' through culturally relevant data layers and mapped output.

20.2.2 Wind Rivers Indian Reservation

Models for incorporating culturally specific information into resource management decisions need to be developed to provide Native Americans and indigenous peoples around the world with equitable roles in the decision-making process. Mapping cultural areas in conjunction with ecological areas presents an opportunity to gather information that has generally not been included in databases for resource management or development purposes. Such databases are generally limited to cultural sites of historical interest. However, cultural areas are not only historic, they are dynamic, providing insights into and illustrating patterns of living of both the dominant culture and the indigenous population (Schoenhoff 1993). This case study focuses on the development of a culturally specific database of water and the river corridor in the Wind River Indian Reservation, Wyoming in the United States. Research efforts were aimed at identifying the spatial aspects of culturally specific resource management within the Shoshone and Arapaho tribes.

Collection of indigenous information was accomplished through participatory mapping, interviews and field visits with elders and community representatives. In addition, establishing community connections in the Wind Rivers Indian Reservation was of critical importance to earning trust and ensuring long-term commitment from all participants. Specifically, numerous meetings and interviews between elders and tribal councils were held to discuss water resource issues and methods to incorporate GIS.

This case study differs from the New Zealand project in that GIS technology is available and training has been conducted. The Wind Rivers Water Quality Council maintains a state-of-the-art GIS with technically trained personnel. However, GIS education has been limited largely to seminars offered by consultants, software vendors and the federal government. GIS applications tend to be based upon a Euro-American perspective of land-use; they do not incorporate information unique to Native American legal conditions, traditional tribal perceptions of land-use, or unique resource management goals such as the protection of sacred sites. Appropriate training for the specific needs, applications and analyses of tribal entities is needed.

Bringing Native American perspectives to GIS development and use should expand the range of GIS conceptions and research. As in the New Zealand case study, there was a reluctance to share sensitive cultural information. Both the Northern Arapaho and Eastern Shoshone expressed concern that 'outside' researchers would come to the reservation and 'take' their knowledge, giving nothing in return. The outcome was a negotiated research process that resulted in shared benefits such as data collection techniques, participatory mapping protocols, and identification of non-sensitive cultural information.

Through the negotiated research process, the Shoshone and Arapaho Tribes indicated that it was of paramount interest to them to understand the constraints placed on water resource management due to Western USA water laws. Our research focused on developing a database of Shoshone and Arapaho culturally specific information that could inform the Wind Rivers Tribal Water Code. Some spatial practices identified by the community are made explicit in the Tribal Water Code (Figure 20.2).

A comparative analysis was conducted to identify the relationship between the Tribal Water Code and Wyoming state water laws. The comparative analysis revealed that the beneficial use of water is narrowly defined by the state of Wyoming and has allowed for the allocation of water to non-Indian agricultural interests (Flanagan 2000).

This research has been critical in developing a cultural database from which to derive our spatially explicit information – such information is not digital and needs to be collected and organized into a database. The next step in this research is in progress and involves developing a GIS database that includes culturally specific spatial practices in order to determine ways of integrating cultural information into resource management and planning.

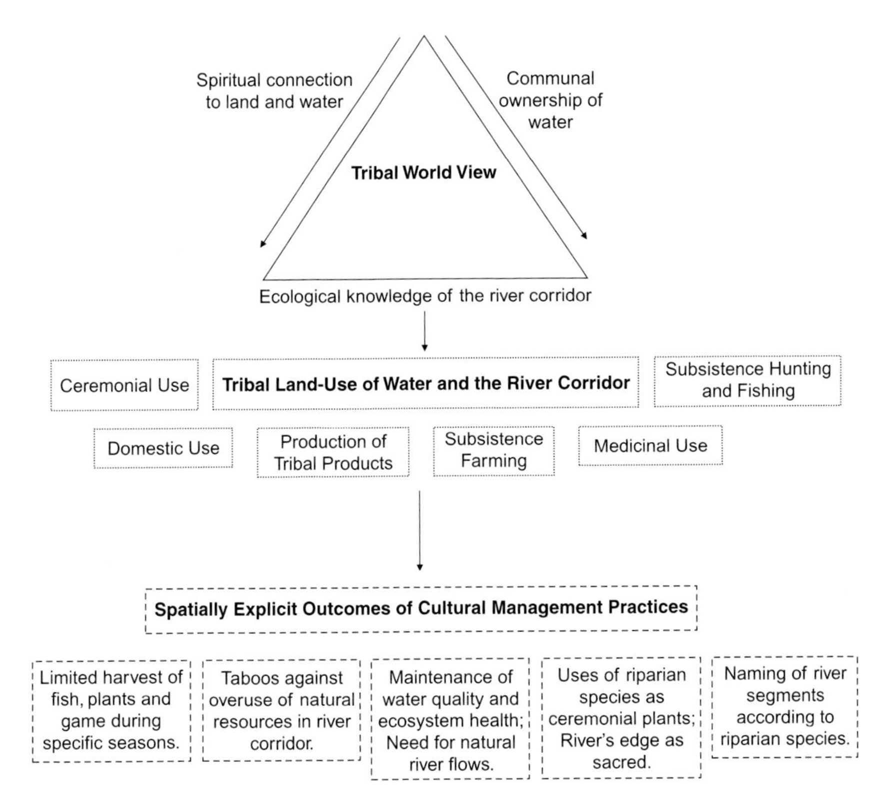

Figure 20.2 Culturally specific information based upon the World View informs Tribal water resource management.

A spatial analysis of competing cultural water management strategies will be invaluable in understanding the role of culture as well as the cultural conflict over water that continues in the Western United States.

20.2.3 Colorado State University–Poudre School District (CSU–PSD) Spatial Information Technologies Partnership (http://www.cnr.colostate.edu/avprojects/csu-psd)

This case study demonstrates how GIS can be integrated into a specific user community provided the necessary support is available. One of the challenges facing K–12 educators in the United States is the need to integrate technology into the classroom to educate an informed citizenry. One of the problems with the use of spatial information technologies in the classroom is the lack of teacher training and time to develop appropriate lessons and exercises for students. In addition, Colorado state education standards in geography demand a revamping of how geography is taught, particularly with regard to new technologies and methods. Following two assessment workshops that introduced spatial ITs, a training workshop for K–12 teachers was held in June 1998 and June 1999 using the Environmental Science Research Institute, Inc.'s ArcView 3.1 software package. Training was designed specifically to meet the needs of the teachers, taking into account their level of computer knowledge and expertise as well as the technological limitations of their school sites. This hands-on training has provided valuable insights into designing training to meet the needs of specific users. Through technology assessment workshops, discussion sessions for applications, and hands-on experience, teachers have begun to claim ownership of this complex technology.

Several elements of the CSU–PSD partnership have contributed to its success, and this partnership can serve as a model for how to facilitate the use of GIS within a community (Laituri and Linn 1999). These elements include: direct CSU support through partnering graduate students with K–12 teachers to create GIS-based lessons; content and technical specialists on site and on call to support teachers throughout the school term by providing advice on both software and geographic applications; a community advisory board made up of local experts who assist with internships, project development, class demonstration and data collection and analysis; and a webpage that describes the partnership, includes hands-on real world projects, and provides links to other Internet-based data and educational resources as well as a summer GIS Kids Camp for PSD students (Figure 20.3).

The project has been funded through numerous grants that have enabled software purchase, hardware upgrades, and teacher training and support. Teacher support has been in the form of content/technology specialists, but stipends have also enabled teachers to attend week-long summer training sessions. GIS has been integrated across disciplines (science, math, geography and business) as well as between grades (K–12) and schools within the district. For example, grade 6 teachers from three different schools are developing a project that compares the natural resources of the United States, Canada, and Mexico. The same database provides the basis for a project devised for high schools: a comparison of endangered species and ecosystem management of wilderness areas among these same countries.

The CSU–PSD case study indicates that exposure to GIS technology and appropriate training allows teachers to gain ownership of this tool, provided there is ongoing support, hands-on training and user-specific applications. The support personnel are a critical aspect of this project because they are representatives of the specific user community and can ensure the relevancy and appropriateness of the exercises developed. These findings are not so different from the responses elicited from both the Maori and Native American participants in the other two case studies. Community support for

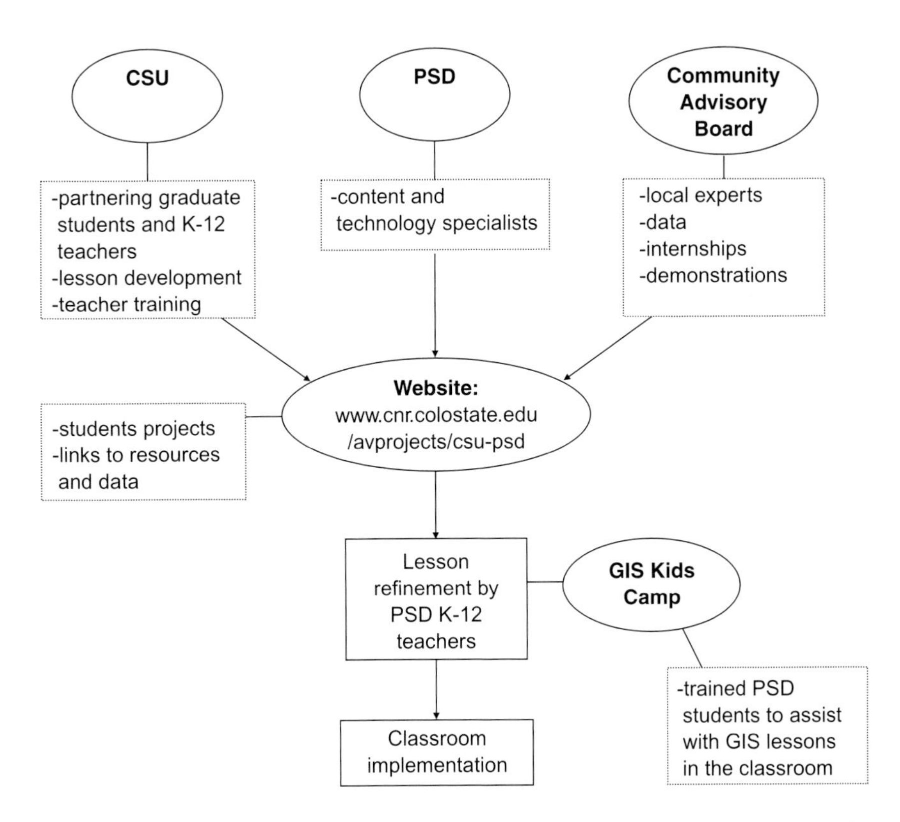

Figure 20.3 Elements of the CSU–PSD spatial information technologies and geographic education partnership.

the PSD project has been another important component. Local community experts from private business and governmental agencies provide guidance, expertise, data and 'real world' examples for GIS. However, community support for projects such as Maori economic development and Native American water resource management are not as evident. The politicized environment in which indigenous people and other marginalized groups operate will make access and equity with regard to the technology problematic.

20.3 ISSUES FOR CONSIDERATION

These case studies reveal important issues relating to the use of GIS and its impact on society in general and marginalized members of society in particular. The issues raised fall into three general areas: (1) limitations inherent in the technology; (2) access for different user capabilities; and (3) the homogenization of knowledge structures. One of the limitations of GIS is the emphasis on visual approaches and spatial reasoning inherent in a computerized environment. These systems are based on the premise that information is explicit, can be reduced to binary code and used in a relational database. In addition, GIS is taught within the confines of a specific platform and within the limitations of a particular software package. How do these characteristics influence the assumptions and compromises for implementation and analysis? How will GIS, with its visual emphasis and technological limitations, influence culture? Evidence from the case studies indicates that indigenous peoples are keen to utilize GIS, however there has been little discussion regarding long-term cultural repercussions of GIS use.

The second set of issues concerns access to GIS education for people with demographic characteristics that create barriers, such as race, age, gender, physical and learning impairments, or socio-economic status. What is the potential for understanding and incorporating different learning styles and cognitive abilities within this computerized environment? Millions of dollars are spent on creating datasets based upon remotely sensed data; will the same efforts be put forth to develop appropriate data sets of culturally specific information necessary for equitable management of resources?

Finally, if we agree that the world is increasingly driven by the flow of information, and that the interfaces and underlying codes that make information visible are becoming powerful societal forces, then it is imperative to educate an informed citizenry to understand the strengths and weaknesses of technology, computers and GIS. Standardized data are necessary for creating seamless databases that can interface; how will this data structure influence knowledge, and what implications will it have for diversity? Access to GIS may provide a venue to legitimize alternative knowledge systems within the dominant social paradigm. Will it also provide avenues for political empowerment?

Each of the preceding case studies raises serious questions that transcend GIS application and touch on issues of equity, access and appropriate technology. The case studies of indigenous peoples revealed specific questions with regard to the GIS research agenda in general and the application of GIS technology specifically:

1 How is research redefined within the context of technological-ethnographic efforts in a way that avoids scientific colonialism and creates knowledge-sharing cooperatives?
2 How will self-determination be operationalized in the research arena? What safeguards and assurances are built into the research or development process to ensure that the introduction of new technology does not represent another system of knowledge as a system of domination?
3 How will GIS projects be created that are of, by and for the people?

The third case study raises questions related to training and education that were also evident in the other case studies. The development of jointly constructed GIS applications for K–12 teachers and students raises issues of marginalization, empowerment and participation:

1 How do we begin to understand and assess the influence of technology on teaching and cognitive abilities? Who is actually empowered?
2 How do we ensure that marginalized members of society receive training and education within the current milieu of reduced funding for education?
3 How do we ensure that the use of GIS is context- and culture-dependent? What effect will spatial information technologies have on culturally dependent information?
4 Training sessions revealed that context-specific materials are necessary for the development of relevant and meaningful exercises and applications. How are context-materials developed? By whom? What biases are embedded in such materials?

20.4 OUTCOMES

Initial results from these case studies demonstrate that there is an intense interest in the use of GIS in marginalized societies, coupled with a healthy skepticism. Both the indigenous groups and K–12 teachers desire hands-on training and education for their specific needs. They recognize computer and spatial literacy as critical components of today's society and are determined to gain the necessary skills. However, they also recognize the practical, pedagogical, and philosophical stumbling blocks that accompany GIS technology.

NOTE

1. Within the context of this chapter K–12 teachers represent a group denied adequate resources to meet the demands of public education due to fiscal conservatism. K–12 teachers and administrators are forced to identify creative and alternative means to fund innovative projects to meet standards-based education goals (Bowers 1995; Kozol 1991).

REFERENCES

Athanasiou, T. (1996) *Divided Planet: The Ecology of Rich and Poor*, Boston: Little, Brown and Company.

Brendlinger, N. (1992) 'Ethics, communication models, and power in the agricultural community: thoughts about development communication', *Agriculture and Human Values* (Spring), pp. 86–94.

Cashman, K. (1991) 'Systems of knowledge as systems of domination: the limitations of established meaning', *Agriculture and Human Values* (Winter–Spring), pp. 49–58.

Denniston, D. (1994) 'Defending the land with maps', *Worldwatch* (January/February), pp. 27–31.

Duerden, F. and Kuhn, R. (1993) 'Introduction to the indigenous land-use information project', School of Applied Geography Occasional Paper, Ryerson Polytechnical Institute, Toronto.

Ekins, P. (1992) *A New World Order: Grassroots Movements for Global Change*, London: Routledge.

Flanagan, C. (2000) 'Culturally specific information in water and river corridor mangement: the Wind River Indian Reservation, Wyoming', unpublished Master's Thesis, Colorado State University.

Jackson, P. (1991) 'The crisis of representation and the politics of position', in Johnston (ed.) *The Challenge for Geography, A Changing World: A Changing Discipline*, Oxford: Blackwell, pp. 198–214.

Katz, C. (1992) 'All the world is staged: intellectuals and the projects of ethnography', *Environment and Planning D: Society and Space* 10: 495–510.

Kozol, J. (1991) *Savage Inequalities: Children in America's Schools*, New York: Harper Perennial.

Laituri, M. and Harvey, L. (1995) 'Bridging the space between indigenous ecological knowledge and New Zealand conservation management using geographical information systems', in D. A. Saunders, J. L. Craig, and E. M. Mattiske, (eds) *Nature Conservation 4: The Role of Networks*, Auckland, New Zealand: Surrey Beattie and Sons, pp. 122–131.

Laituri, M. and Linn, S. (1999) 'Graduate students + grade school (K–12) + geography standards + GIS = great success!?' *Proceedings of the Nineteenth Annual ESRI User Conference*, 26–30 July, San Diego, California.

Michaels, S. and Laituri, M. (1999) 'Indigenous and exogenous forces on sustainable development', *Sustainable Development* 7(2): 77–86.

Patterson, E. (1992) *Exploring Maori Values*, Palmerston North, New Zealand: Dunmore.

Schoenhoff, D. M. (1993) *The Barefoot Expert: The Interface of Computerized Knowledge Systems and Indigenous Knowledge Systems*, Westport, Conn: Greenwood Press.

Shiva, V. (1997) *Biopiracy: The Plunder of Nature and Knowledge*, Boston: South End Press.

Thrupp, L. (1989) 'Legitimizing local knowledge: from displacement to empowerment for third world people', *Agriculture and Human Values* (Summer), pp. 13–24.

Turnbull, D. (1989) *Maps Are Territories: Science Is an Atlas*, Geelong, Victoria: Deakin University Press.

Watson, H. and Chambers, D. (1993) *Singing the Land, Signing the Land*, Geelong, Victoria: Deakin University Press.

Chapter 21

The Cherokee Nation and tribal uses of GIS

Crystal Bond

21.1 INTRODUCTION

Geographic information technology was first introduced to American Indian Tribes by the Bureau of Indian Affairs (BIA). The Geographic Data Service Center (GDSC) was established by the BIA in Lakewood, Colorado in 1990. The mission of the GDSC was to bring geospatial technology to tribal people and teach them how to use it themselves.

The GDSC originally offered a variety of GIS implementation services to tribes. It provided cost-free training and a wide range of technical assistance to help tribes implement their own GIS. The highly trained BIA technical staff collected, compiled, enhanced, and standardized GPS, satellite and other digital geographic information relevant to tribal concerns. The resulting data sets became the foundation for the first tribal, geospatial database development projects in the United States.

The GDSC was a springboard for tribal people in the field of GIS. It is a credit to the foresight and mission planning of the BIA, and the skill and dedication of the technical staff employed by the GDSC, that their mission to implement tribal GIS programmes was accomplished. Due to the efforts of the GDSC staff, state-of-the-art GIS programmes have been successfully implemented for many American Indian tribes.

Over a 10-year period, other projects at the BIA have taken precedence over the GDSC. Services previously provided to tribes are extremely limited and steadily decreasing. As a result of the federally mandated downsizing of the GDSC and its services, there are now two distinct groups of American tribes with regard to geospatial technology; those who have it and those who don't. The first group used the GDSC to get started in GIS. This group now has a responsibility to assist the tribes not able to utilize the GDSC before it was downsized.

This sense of responsibility manifested itself in the form of the Intertribal GIS Council (IGC)[1]. The IGC was first established in 1993 with a seed grant from the First Nations Development Institute. It was created to educate tribal organizations and individuals about the various useful applications of spatial data technologies to the management of all types of resources. Over the course of several years, the IGC has adjusted its goals to provide those services no longer available to tribes through the BIA's GDSC.

21.2 GIS AT THE CHEROKEE NATION

One of the tribes lucky enough to have taken advantage of the GDSC in its prime was the Cherokee Nation of Oklahoma. The Cherokee Nation's headquarters are located in the foothills of the Ozark Mountains in a small town called Tahlequah (tal-eh-kwa). Indian population is about 117,000,[2] the Cherokee Nation has what is known as a 'checkerboard' land base; it is not a reservation with a single, perimeter-based boundary. The checkerboard of tribal land encompasses a 14-county area in Northeastern Oklahoma. This creates a series of complicated problems when dealing with tribal land and law enforcement jurisdiction questions (see Figure 21.1).

21.2.1 Cherokee Nation tribal land project

One of the most urgent questions the Cherokees are dealing with is the question of jurisdiction. Is it tribal land, individually-owned land, government land held in trust, restricted land or any number of other categories? The answer to this question determines which of the local law enforcement agencies has jurisdiction. The County Sheriff's Office cannot make an arrest on restricted tribal land. The Cherokee Marshal Service cannot enforce the law on individually owned property. The nature of this checkerboard land base is such that an Indian home on one side of the street may be within tribal jurisdiction, but the open field across the street where a crime is committed may be under the jurisdiction of the county. Law enforcement officials need to know exactly where the boundaries are. This is a serious problem for the Cherokee Nation. Inadequate tribal land information leads to a loss of convictions for known criminals. Those found guilty of crimes are often released because of jurisdictional questions and related legal technicalities. This perpetuates a crime rate that goes unchecked.

When the Cherokee Nation implemented their GIS, jurisdictional boundaries were the first data sets acquired. GIS staff used an original legal description provided to the Cherokee Realty Department by the BIA as the data source. ArcInfo's coordinate geometry module was used to input the boundary. The county and tribal voting district boundaries were taken from the 1990 US Census TIGER line files. In addition with other TIGER data and the locations of all the Indian health clinics, a map could be produced to show spatial relationships between the clinics and the county, voting district and jurisdictional boundaries. Within these

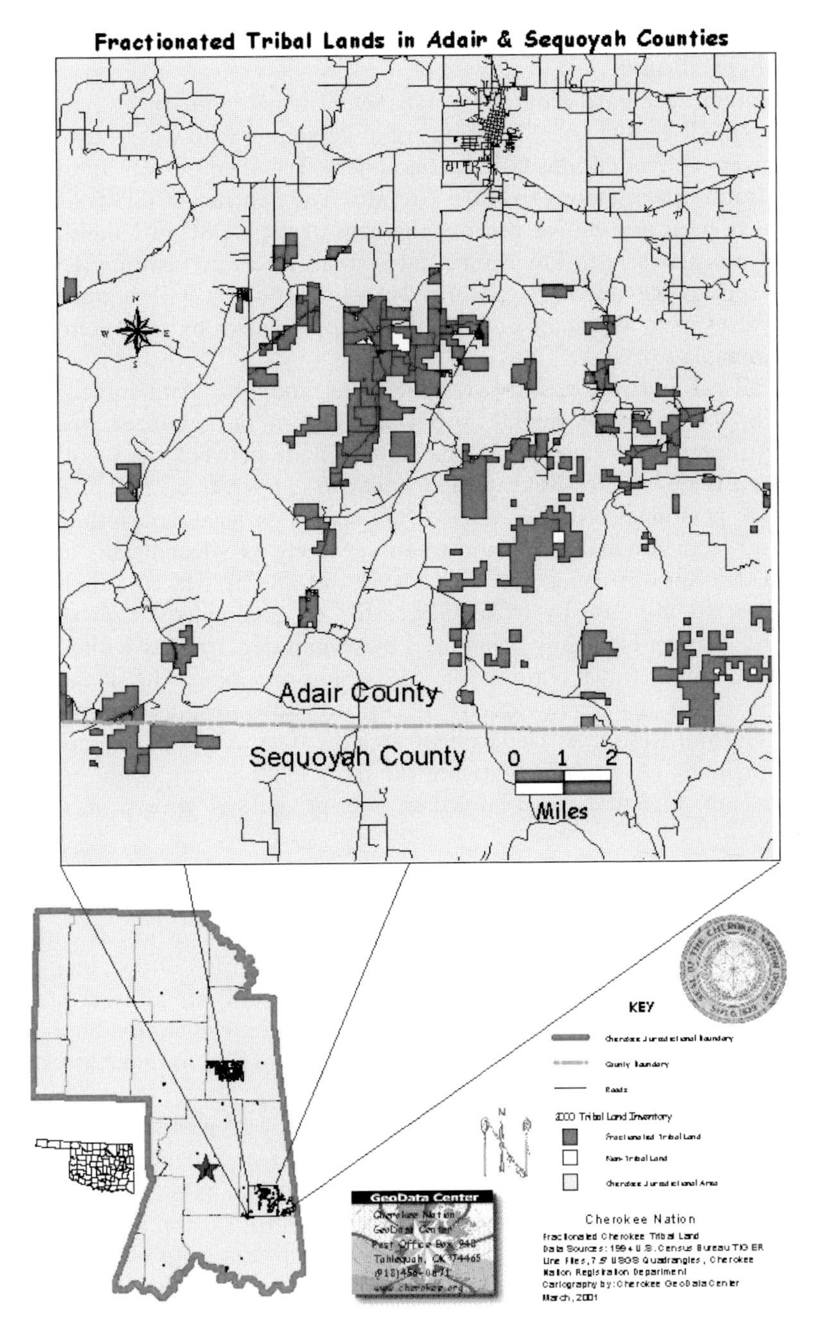

Figure 21.1 Fractionated tribal land in Adair and Sequoyah counties.

boundaries exist a checkerboard of tribal, non-tribal, restricted, non-restricted and many other categories of land ownership. This is where the real challenge begins.

To create an accurate geographic data set of tribal land, an investment had to be made by the tribe. Existing tribal personnel who were AutoCad drafters were converted into GIS cartographers. Tribal employees attended lengthy training workshops to learn ArcInfo. The expense of GIS training would have been prohibitive without the help of the GDSC in Lakewood. GIS staff was also sent to Environmental Systems Research Institute (ESRI) training centres for intensive, advanced ArcInfo training. GIS equipment and software were also major expenditures approved of by the Cherokee Tribal Council.

With all the hardware, software, personnel and base data intact, the Cherokee Nation's first digital, tribal lands data development project began. An inter-departmental agreement was made between the GeoData Center and the Cherokee Realty Department to cooperate on this project. Real estate personnel provide the GeoData staff with legal descriptions of tribal land from original deeds and treaties. Slowly, GIS technicians input the legal descriptions using ArcEdit COGO and an Arc Macro Language (AML) created in-house by tribal GIS staff.[3] An incomplete, in-progress tribal land data set is already being used by several departments within the Cherokee Nation for a variety of needs. For example, the Planning Department uses maps showing tribal land to illustrate grant proposals and the Natural Resources Department uses the data as a background for its GPS projects. (A disclaimer stating the incomplete nature of the tribal lands coverage and that tribal land data is in-progress is printed on each map.)

One of the most serious problems concerning tribal land boundary information is that of legal jurisdiction. Local law enforcement agencies need to know exactly where their legal jurisdiction begins and where it ends. The Cherokee Marshall Service, Cherokee County Sheriff's Office, city of Tahlequah, and the Federal Bureau of Investigation (FBI) are all interested in Cherokee tribal land ownership status. Undisputed jurisdictional boundaries are very important to these agencies and their ability to make arrests and follow up with convictions.

In order to map and perform analysis involving legal jurisdiction, each tract of land must be categorized and attributed with land ownership information. To help offset the cost of such labour-intensive work, the tribe is seeking federal grants and cooperative agreements with other agencies to get this done as soon as possible. Even with cooperation from other agencies and federal funding, this job is expected to take several years to complete. The nature of the Cherokee's checkerboard land base and the numerous tribal land ownership categories make it an exceptionally labour-intensive task.

When the information is in the GIS and ready to be used, the spatial jurisdiction information will be shared with law enforcement and other interested agencies outside the tribe. Although some tribal information is proprietary, most digital, geographic information can be shared. This will significantly reduce duplication of effort and contribute to successful partnerships with law enforcement in the future.

21.2.2 Transportation planning

Another issue being addressed by the Cherokee people is transportation planning. This is usually considered an urban problem found mostly in cities and more developed areas. The considerations for the tribe are more rural in nature. The solutions being sought are ways to accommodate the transportation of tribal elders from fairly isolated rural homes to Indian health clinics, community centres, churches, shopping areas, and other events located in the surrounding towns and cities. Good roads are also needed for school bus routes and access from Indian housing clusters and other areas of high Indian population.

The 1999 tribal roads project was a joint endeavour between the Cherokee Nation Roads Department and the Cherokee Nation GeoData Center. Through a series of informative meetings and GIS demonstrations, the Roads Department became educated regarding potential applications of GIS technology to their work. In 1998, the inter-departmental partnership was formed and the Cherokee Nation's first GIS-based tribal roads inventory project began.

A series of maps were needed to illustrate exactly where existing tribal roads were located, where proposed roads would be built and where maintenance and repairs were needed. Using ArcInfo software and enhanced TIGER line files,[4] a set of working maps was created. A list of attributes used by the BIA to score and prioritize specific road projects was incorporated into the descriptive, attribute data for the new tribal road coverage. Data layers representing the physical environment were used as a background for the maps. This allowed for quick spatial analysis at a glance.

A variety of factors are used in the Bureau of Indian Affairs' method of rating proposals for new roads. Included among these factors are things like tribal population, locations of schools, vicinities of churches and other cultural areas, medical facilities, mutual help housing clusters, and places of employment for tribal members.

The working maps were completed and used by management to plan and prioritize the work to be done. A detailed work schedule was developed and road crews began collecting attribute data from the field. This information was returned to the GeoData Center where it was input into an Access database and used to attribute the tribal road coverage. The in-depth,

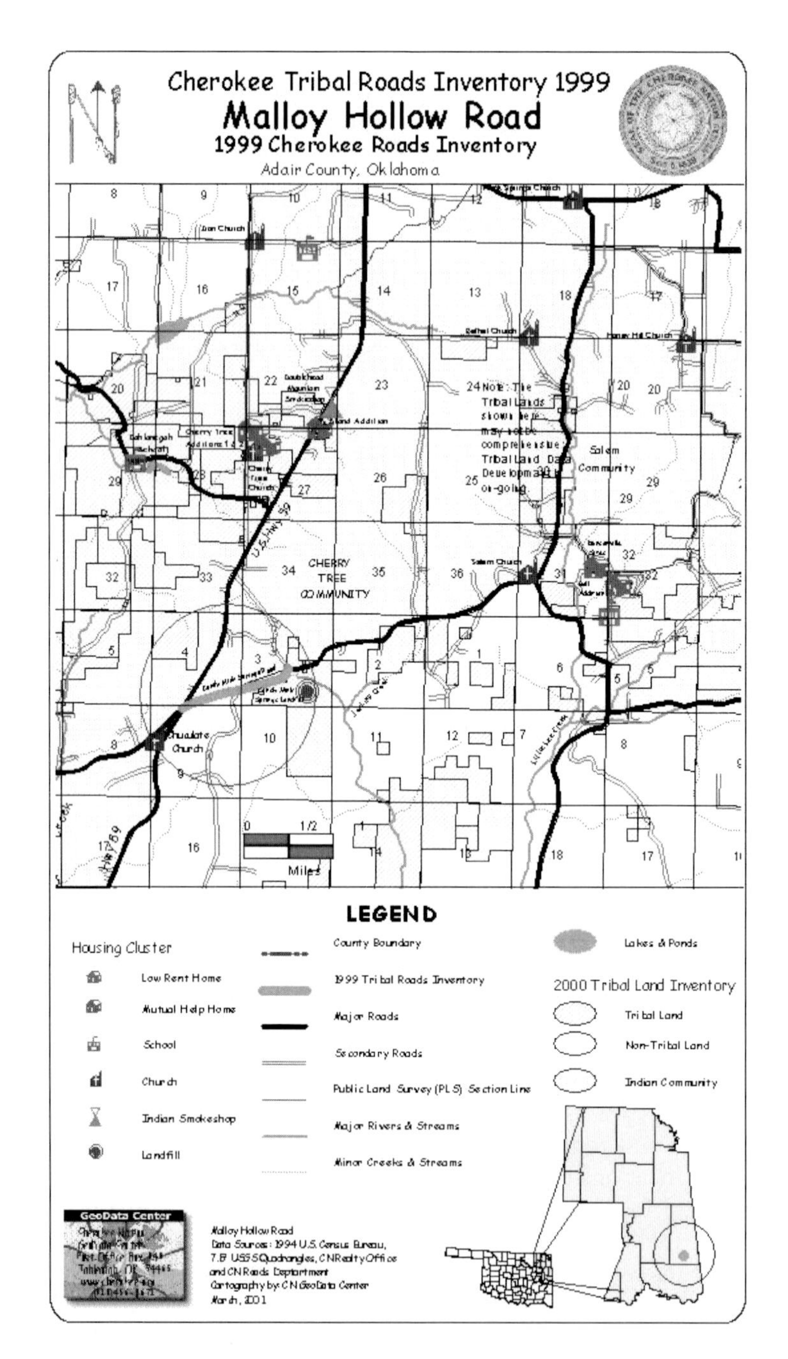

Figure 21.2 Malloy Hollow Road.

detailed attribute information linked to the road coverage was used to generate a variety of statistical reports.

With ArcInfo software, cartographers created buffers at specified distances surrounding key features, and the tribe utilized GIS technology to analyse their own transportation needs for the very first time. The particular roads and road segments were chosen for the 1999 inventory, and a new set of maps was needed. This group of maps showed a 'zoomed-in' view of each individual road with a background of all available data layers (see Figure 21.2). Each individual road map was paired up with its corresponding attribute information spreadsheet. By request, the maps were printed on the same page as the BIA's Road Needs Data Sheet. The spreadsheets were created using Microsoft Excel software. All the maps were generated in ArcInfo using the ArcTools module. The maps and their spreadsheets were incorporated into a book format for day-to-day use and to accommodate easy reproduction. Each county and all the tribal roads within it were individually mapped also and are included in the final product. The resulting 1999 Cherokee Nation Tribal Roads Inventory Book contains 149 maps depicting 139 individual roads and 10 Oklahoma counties. There are approximately 100 miles of tribal roads shown altogether.

21.2.3 Other applications in the Cherokee Nation

Within the Cherokee Nation, more departments are aiming for the capacity to use geospatial technology to make intelligent, highly informed decisions when planning their projects. The Roads Department has made the decision to begin learning and using ArcView routinely, as part of their regular work. The Natural Resources Department now has its own Global Positioning System and well-trained tribal employees who know how to use it. The Office of Environmental Services uses ArcView extensively to help perform site assessments and analyses for potential hazardous waste sites. The Center provides technical assistance and other mapping services for these departments and for those without GIS capability in their areas. In addition, the GeoData Center combines health data, crime data, social data, population data, and other demographic information with geospatial data to perform analyses and create maps to illustrate the results.

These projects have set examples for other sovereign tribes responsible for their own transportation planning and tribal land management.[5] GIS allows tribes an accurate, dependable method of inventorying and developing their own transportation networks, tribal land datasets, environmental baseline data, demographics, and other geographic information (see Figure 21.3).

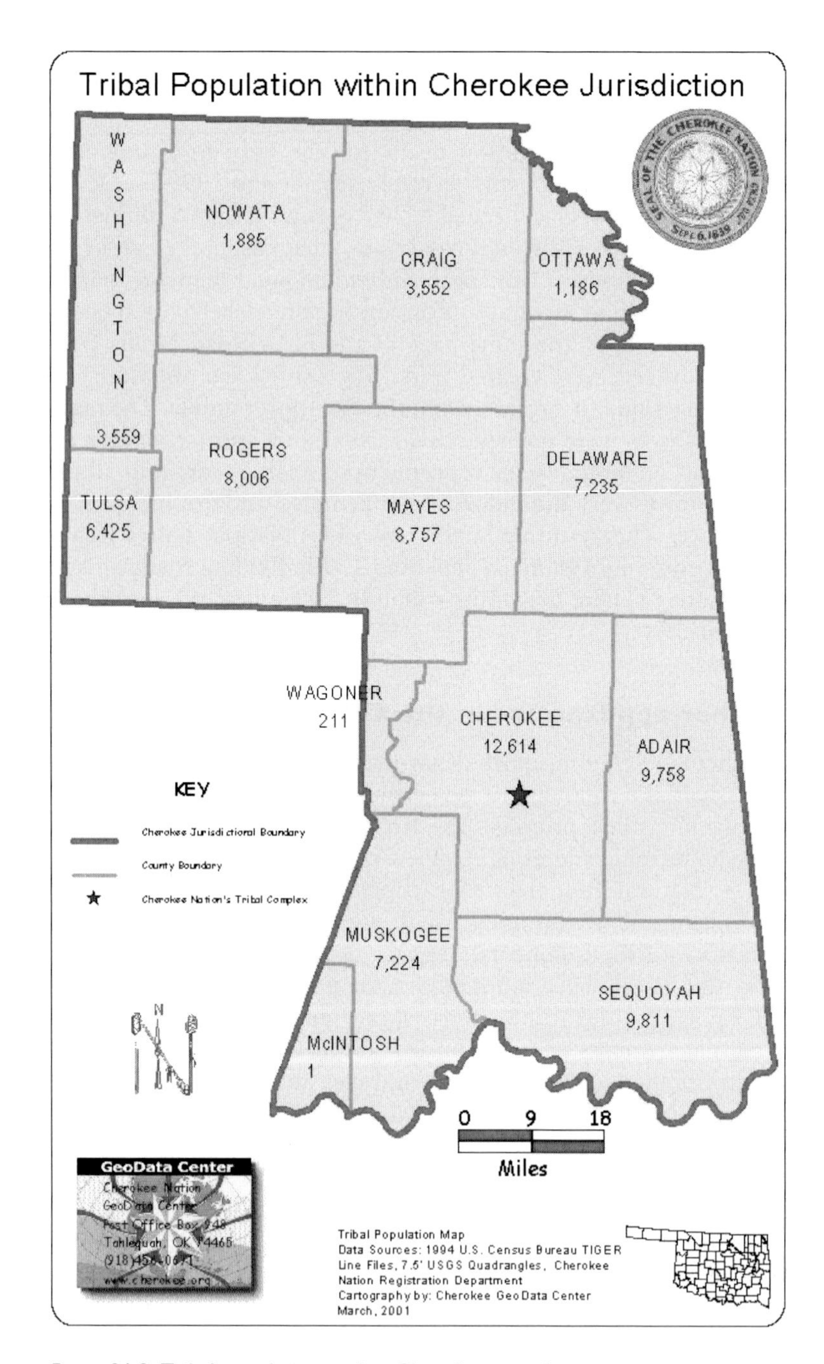

Figure 21.3 Tribal population within Cherokee jurisdiction.

21.3 GIS APPLICATIONS IN OTHER TRIBES

The previous situations are specific to the Cherokee Nation, but several other tribes have completed their tribal land data sets and have been performing advanced geospatial analysis and cartographic output for several years. The application of GIS within tribal governments can help empower tribes, especially with regard to natural resources management and land and water rights litigation. Some examples are listed here.

21.3.1 Change in land status

In 1987, in land litigation between Zuni Pueblo and the United States, a GIS was used to develop an automated spatial database identifying lands taken from the Zuni tribe. The GIS was used to develop an inventory of Zuni land taken from them between 1846 and 1939. The display and query abilities of the GIS allowed the attorneys to visualize and quantify how the Zuni sovereign area had changed over time. As a result, 255,266 acres were identified in the Zuni Aboriginal Area *over* what the Zuni had originally identified. This is significant when compensation for the land taken is a monetary value paid for each acre.[6]

21.3.2 Water rights litigation

In previous Indian water rights adjudication cases, aerial photos were used to identify and inventory lands where irrigation was practical and practiced. Often, hundreds of aerial photos have to be photointerpreted. This can become overwhelming. The *Wind River Reservation v. the State of Wyoming* water rights adjudication case lasted for 12 years.[7] Such projects will be significantly more manageable if the irrigated fields were mapped and entered into a GIS. The addition of a land ownership parcel layer is an important component of GIS water rights adjudication.[8]

21.3.3 Natural resources management

Many tribes are using geospatial technology to manage their natural and cultural resources. The Yakima Tribe in Washington State, the White Mountain Apache in Arizona and the Salish Kootenai in Montana utilize GIS technology to inventory, analyse, map and make decisions regarding tribal resources. Examples of such resources include timber production, grazing and farm land, water rights, wildlife, native plants, cultural sites, environmental data and hazardous site monitoring, historical preservation, health and human resources.

21.3.4 Planning & development

Some tribes utilize GIS technology for urban applications. The Agua Caliente Band of Cahuilla Indians in Palm Springs, California and the Cherokee Nation in Tahlequah, Oklahoma have developed urban projects specific to tribal needs. The decisions made using geospatial analysis can apply to a wide range of programmes and projects and include such things as: health clinic locations, school bus routes, housing, smoke shop and casino locations, tribal demographic analysis, transportation planning, tourism and others.

21.4 CONCLUSION

Now that the BIA has withdrawn from the tribal GIS implementation forefront, the IGC has stepped up to take its place. New developments in cooperative agreements with the ESRI, NASA, the US Dept of Commerce, the US Environmental Protection Agency (EPA) and other significant partners have made it possible for the IGC to offer quality assistance to tribes in need. In cooperation with the IGC, ESRI provides GIS software grants for tribes and registration fee waivers for tribal GIS professionals to attend the annual ESRI conference. NASA, in conjunction with the University of New Mexico and the EPA, works with the IGC to collect and develop baseline tribal environmental data for a long-term global warming project. Typical IGC services include: GIS education and training, tribal data acquisition and development, tribal economic impact/development studies and planning assistance, data management and maintenance of proprietary files and protection of sensitive data for tribal organizations and GIS work for tribes and organizations having no GIS capabilities. Experienced IGC personnel and board members act as liaisons to encourage technology transfer and cooperative agreements between tribal and non-tribal entities.

The annual IGC conference plays an integral part in the dissemination of GIS knowledge in Indian Country. Each year the IGC facilitates a week-long national conference geared toward the advancement of tribal GIS technology. Software training, GIS application seminars and workshops, project-oriented presentations, funding resources and networking with tribal GIS professionals are some of the benefits available to conference attendees. IGC membership and conference attendance is steadily increasing each year. Conference and IGC membership information is available on the website at www.itgisc.org. As tribes become increasingly aware of the capacities of GIS technology as a decision-making tool for tribal resource management, the IGC is expected to grow and expand its capacity to serve and assist all tribes.

With the passing of time, natural resources management and conservation will become more important to everyone. Tribes are the stewards of many natural resources that still remain in this country. It is to the advantage of our country's population as a whole that tribes are using state-of-the-art technology to manage, nurture, conserve and maintain their natural resources.

NOTES

1. Intertribal GIS Council, PO Box 1937, Pendleton, OR 97801, www.itgisc.org. igc@itgisc.org (541) 966–9097.
2. Census of Population, 2000, US Bureau of Census, Washington, DC.
3. Tony Glass, GIS Analyst, Cherokee Nation GeoData Center developed the Cherokee Nation Land Information System. AML. (CNLIS.AML) 1998.
4. US Bureau of Census, TIGER files 1992, enhanced by Wessex Company.
5. Public Law 102–240 provides for tribal determination of its own needs and priorities for the allocation of federal highway funds.
6. *American Indian Culture and Research Journal* 15(3): 81, 1991.
7. Teno Roncalio, Report Concerning Reserved Water Right Claims By and On Behalf of the Tribes of the Wind River Indian Reservation Wyoming, District Court of the Fifth Judicial District State of Wyoming, Civil No. 4993, 15 December 1982.
8. *American Indian Culture and Research Journal* 15(3): 84, 1991.

Part III

PPGIS futures

Chapter 22

Mutualism in strengthening GIS technologies and democratic principles: perspectives from a GIS software vendor

Jack Dangermond

22.1 INTRODUCTION

Educated and informed citizens are essential in a democracy where power is vested in the people and exercised by them. Policy and decision-making at all levels of government frequently involve geographically related issues such as the environment, transportation, natural resources, energy, agriculture, defense, trade, economics, and social welfare. GIS technology is the golden thread that is weaving its way through the fabric of democracy. Fundamental to many of the societal issues that are surfacing in the twenty-first century, the widespread use of GIS has value beyond simple efficiency, profitability, or even communication (Figure 22.1).

By combining a range of spatially referenced data, information media, and analytic tools, GIS technology enables citizens to prioritize issues, understand them, consider alternatives, and reach viable conclusions. When the public has access to timely, accurate information about the geographic aspects of the issues they seek to resolve, they and their representatives are better able to evaluate alternative courses of action, form opinions, and vote wisely.

22.2 PERCEIVING SPATIAL DATA

People seem to have a natural sense of place and for the significance of geography as it relates to human activities. This is perhaps best evidenced in the remarkable ability of the human eye-brain combination to recognize and understand the human environment. Humans are capable of quickly extracting great amounts of information from spatial images, but until recently, the tools for applying this sense of place in any detailed way to large-scale problems were either very difficult to use or entirely lacking.

Enter GIS technology, which helps make intricate and abstract problems real and concrete. While it does not simplify the problems, it does help to manage their complexity more effectively – far better than maps alone. GIS is making it possible for citizens to approach political problem solving with

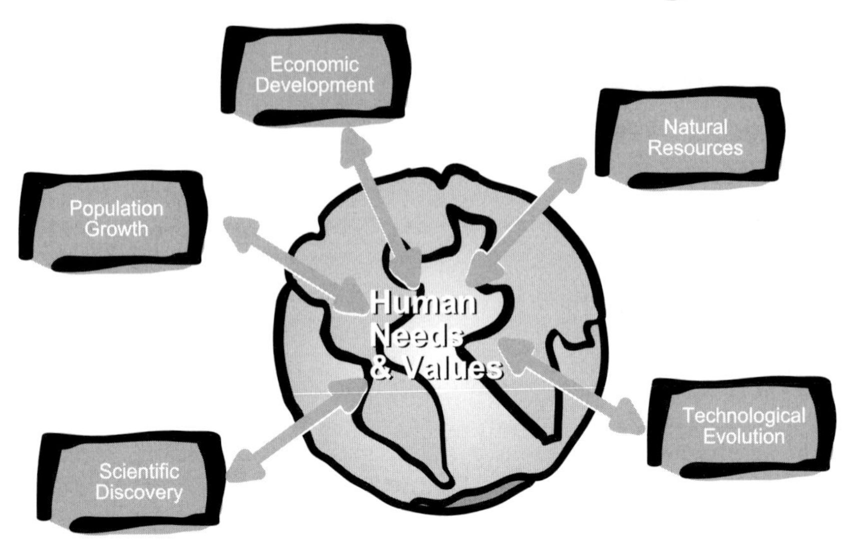

Figure 22.1 Geography is fundamentally affecting the major forces of the twenty-first century.

tools that even senior government officials lacked a decade ago. They can evaluate practical consequences of decisions, monitor the implementation of past decisions, and follow events as they unfold – all in real time.

Today, community residents are using GIS in a number of ways to evaluate their neighbourhoods with spatially referenced data, such as assessor and parcel data to compare their area's property values with those in other regions. Recognizing disparities helps to bring about changes in services and support such as infrastructure and crime prevention. The resources are now available for citizens to evaluate consequences of decisions, monitor implementations, and follow events as they unfold. Armed with this kind of relevant information, neighbourhoods are in better positions to lobby their elected officials.

22.3 NEW, BETTER APPLICATIONS FOSTER WIDESPREAD USE

With a history of less than 30 years, GIS software products have undergone an expansive transformation from highly customized one-of-a-kind products to less expensive out-of-the-box generic GIS products. Cheaper hardware along with more highly developed software programs provide an array of applications to users whose training need not be highly specialized. Developments in related technologies are also fueling the continual growth and expansion of GIS applications. These include wireless access to the Internet, higher data transfer rates, improved remote sensing, and the construction of global databases. All of these applications include geographic knowledge and the data and tools to leverage it.

ESRI has worked to introduce new, easier methodologies and tools, such as a richer data model that makes knowledge more accessible, and a strong and enriched data management technology. Software development efforts have focused on usability, software architecture, development environment, spatial analysis, modelling, cartography, data management objects, database models, metadata standards, interoperability, and dissemination of knowledge on the Internet.

ESRI's latest products, ArcInfo 8 and ArcIMS 3, promise to boost GIS into the mainstream of IT. Released in 1999, ArcInfo 8 marked a significant redesign in professional GIS software. It takes advantage of the modern concepts of software engineering and GIS theory, and is easy enough to be accessible to anyone familiar with desktop computing. User interfaces and

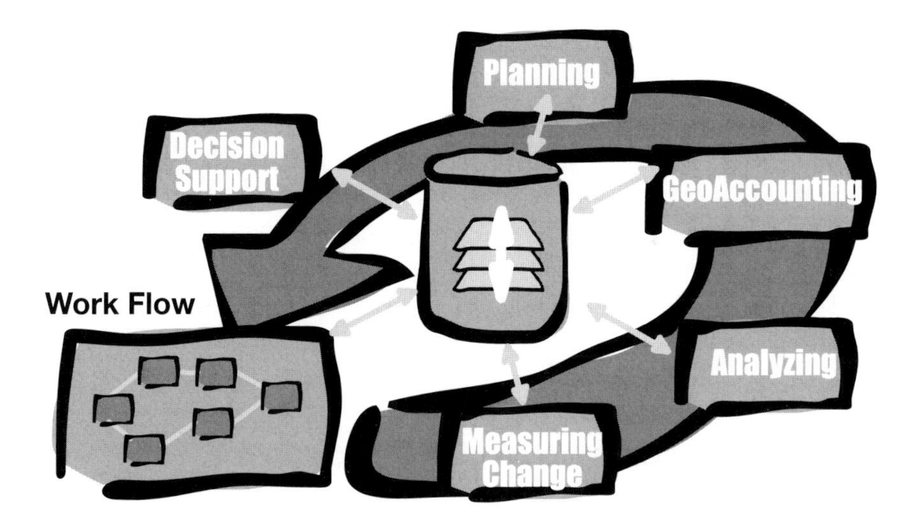

Figure 22.2 As GIS technology evolves, geographic data will be imbedded into most information applications and services.

wizards are key features of ArcInfo 8, which are accessed through three applications representing the fundamental methods of how people interact with a GIS – maps, data, and tools.

An important element affecting the growth of public involvement in GIS is the Internet. Immensely popular, the World Wide Web has stimulated development of GIS products that exploit its special capabilities. And the future is bringing improved access to the Internet. With wireless access to the Internet at very high data transfer rates, geographic information is becoming increasingly accessible to everyone, everywhere, at any time (Figure 22.2).

22.4 STRENGTHENING INTERNET MAPPING

For several years, ESRI has been growing its Internet mapping and GIS technology. During this time, the primary focus was developing server-based mapping and geoprocessing solutions by offering Internet extension solutions for ArcView GIS and MapObjects. Although this has been very successful with thousands of user deployments, ESRI is now launching the next phase of Internet Map Server (IMS) technology, ArcIMS 3, which enhances the server-based architecture with software that enables users to take advantage of clientside processing in addition to server processing.

A key feature of this new technology is that data is optionally streamed directly from servers to clients, and it can be combined with local data. ArcIMS 3 acts as an integration tool for reading local and network-based data in the same browser. In addition and equally important, data can be streamed simultaneously from multiple IMS sites. Leveraging GIS data to many users, the ArcIMS software represents a major step forward in creating a distributed GIS architecture. Agencies throughout the world can publish data and services for users to access directly via a simple browser interface. New ways of cooperation are opening as users integrate distributed data with their local data. By increasing the accessibility to GIS-based information, organizations and society in general are maximizing the use of their existing spatial data investments.

22.5 OPEN ACCESS TO GEOGRAPHIC DATA

Technological development is leading us toward a future in which all geographic change will be measured by various kinds of instruments. These measurements will flow into information networks where they will be accessible to everyone. This flow of information is destined to transform society just as it is profoundly changing how organizations operate. GIS provides the fundamental elements of any information system – geographic measurement, analysis, integrated decision-making, and support for coordinating work flow. It is also a remarkable visual spatial language with rapidly evolving capabilities and gives us a framework for systematic measurement of geography. One of the missions at ESRI is to build technology that facilitates open sharing of geographic knowledge freely and easily so that the power of thinking geographically can be brought to bear on many of the world's problems. For this vision to become a reality, it is essential that geographic data, geographic processing capability, and user expertise be easily available (Figure 22.3).

Widespread use of Web-based GIS is facilitating broader public participation and citizen empowerment as data producers begin to collect and manage geographic information more effectively and enable open access to it. Adopting this approach to open access to information enables interactive analysis and decision-making on the part of the public, agencies, and private organizations. And, it is moving GIS from being a group of small projects to becoming an integral part of organizations' information systems providing the means and structure for measuring change on any scale, even at the global level.

Collecting, storing, and sharing more of our information in digital forms are vital for decision-making, accountability, and success. As we share common knowledge, we become more effective. GIS technology enables us to

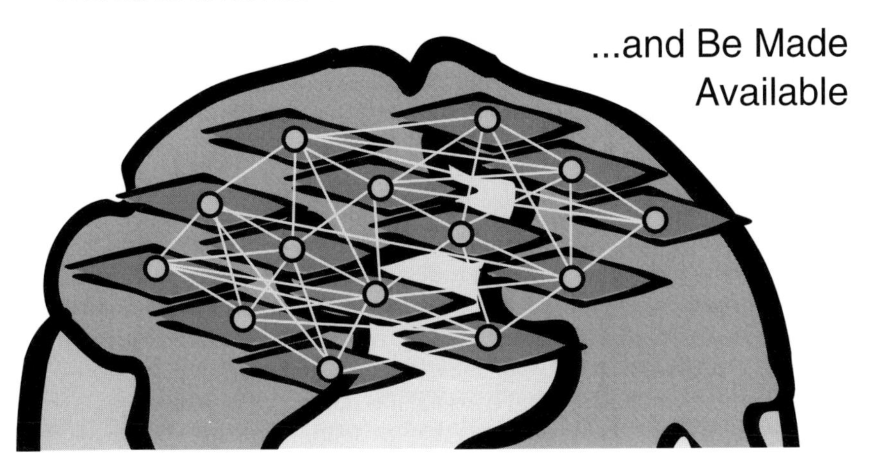

Figure 22.3 GIS provides the framework for the systematic measurement of geography.

integrate what we know into the flow of our work so that the whole is automatically considered in whatever we are doing. Providing a new way of being accountable, GIS is not just for the economic accounts of an organization, but also for the economy of a country, for its economic development, and for a country's biodiversity, its environmental protection efforts, its culture, and its national health. It is also accountability by community, by state, by region, or just by neighbourhood.

The ability to see the information – to see what is happening on a local, state, or national level – is making citizens more responsible, government more responsive, and all more responsive to one another. With larger databases and concurrent users, there has been a natural transition to database management system (DBMS) technology for storing geographic data. In the early 1980s, ESRI built the first commercially viable product that accessed data stored in a DBMS. Initially this was attribute data only but now encompasses geographic data. Today, the combined geodatabase and ArcSDE (spatial database engine) technology in ArcInfo 8 is an excellent data management solution capable of managing data stored in several different database management systems on multiple hardware platforms.

Open access to data in databases enables users to take advantage of DBMS technology to store and manage data, to support multiple users and applications concurrently on the same database, and to integrate heterogeneous data at the desktop. Using DBMS to store and manage data provides a superior solution for backup/recovery, replication, failover remote synchronization, and multiuser access. As more and more information is linked to these large, integrated, shared databases, people are exploring the data, analysing it, and finding new meaning in the patterns they observe. Instead of narrowly focused research, they are mining data from the vast spatial data resources, which leads to discovering new patterns and relationships and ultimately to new knowledge.

22.6 SHARING GEOGRAPHIC KNOWLEDGE FOR LIVABLE COMMUNITIES

U.S. federal policies have taken on a 'smart growth' theme, which is part of a livability agenda intended to help communities flourish in a strong, sustainable manner. The livability agenda is designed to strengthen the federal government's partnership with local governments as they strive to build livable communities by providing new tools and resources to preserve open space, ease traffic congestion, and implement regional smart growth strategies.

Information partnerships and consortiums composed of public and private agencies at all levels are developing complex spatial databases for larger geographic areas, which are eliminating database duplication and at the same time serving multipurposes within each organization. Sharing essential spatial data in this manner is enabling communities to make informed, collaborative decisions about their futures. As these databases are built from the 'bottom up', the role of local agencies increases as the grassroots level feeds information to regional, state, and national arenas that have the wherewithal to fund and administer the database (Figure 22.4). The National Spatial Data Infrastructure (NSDI) is based on this architecture.

In 1990, the Federal Geographic Data Committee (FGDC) was established by the U.S. Office of Management and Budget (OMB) to promote the national coordinated development, use, sharing, and dissemination of geospatial data. The OMB assigned responsibilities to specific federal agencies to coordinate the various themes of geospatial data that contribute to the development of NSDI. The NSDI seeks to link the technology, policies, standards, and resources that are necessary to improve the way geospatial data is acquired, stored, processed, disseminated, and used.

Designed to advance the NSDI by providing communities with the ability to create and use geospatial data, the Community/Federal Information Partnership (C/FIP) is making GIS technology available at the local level. ESRI supports the NSDI and the activities of the C/FIP, which demonstrate how cross-government, cross-functional geospatial data, maps, and

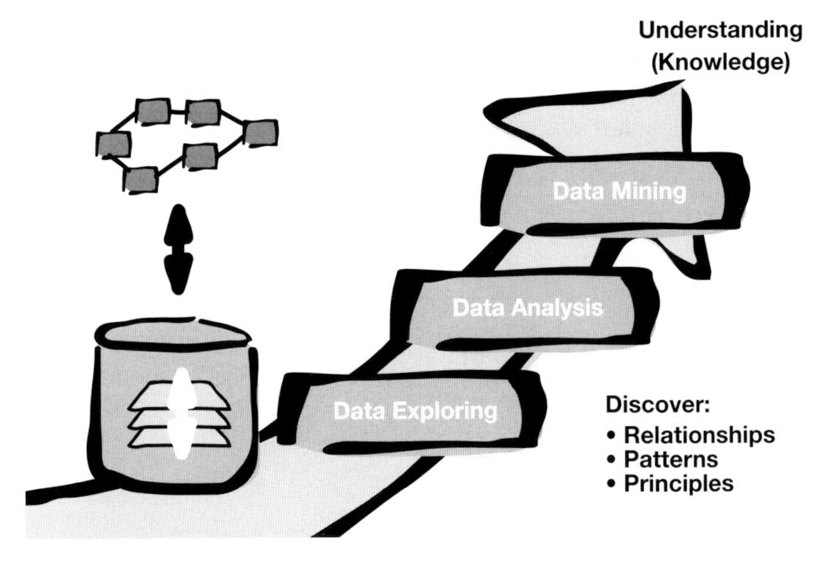

Figure 22.4 Building vast spatial data resources from the bottom-up fosters new scientific knowledge.

applications help solve community problems. As governments and private sources of information adopt policies of open access to geographic information, a range of geographic data becomes accessible at reasonable costs. And, as standards for metadata (data about GIS data) are adopted, it is easier to find data.

As part of its mission toward improving the quality of life and providing support for analytical decision-making, ESRI has earmarked millions of dollars in grants to local governments and agencies. These efforts foster the development of spatial databases and help communities implement programmes that champion increased public access to information and decision-making power. The assistance includes donations of software, training, ongoing technical support, and support services.

22.7 COMMUNITY DEMONSTRATION PROJECTS PROVE THE POWER OF GIS

Key to the success of the C/FIP are six NSDI Community Demonstration Projects, all of which ESRI is supporting at both the local and national levels. The demonstration projects each have an ongoing GIS programme with specific cross-regional challenges and are focusing on issues from water quality to crime analysis to land-use planning. Dane County, Wisconsin; Gallatin County, Montana; Tillamook County, Oregon; the Tijuana River Watershed in California; the Upper Susquehanna–Lackawanna River area in Pennsylvania; and the city of Baltimore have demonstration status. Each serves as an example of the benefits that can be realized through expanded cross-sharing of geographic information between federal and local agencies.

The Dane County, Wisconsin, project is creating a citizen-based, on-line, smart growth planning process to protect farmland and open space and address environmental concerns while sustaining continued growth. Gallatin County, Montana, just north of Yellowstone National Park, contains extensive areas of public lands and is experiencing rapid population growth. This community's project is developing tools for the county government to access integrated federal, state, and local information, consider population impacts, and understand alternatives for growth and the effects of their decisions on the community.

Tillamook County, Oregon, supports a public–private partnership by creating on-line Web-based tools for reporting and accountability. Citizens and local, state, and federal government agencies can monitor and report progress toward common goals for water quality, flood mitigation, and fish habitat restoration. The Susquehanna–Lackawanna River partnership in central and northeastern Pennsylvania provides an integrated regional GIS to help local communities support an environmental master plan, flood mitigation, and performance monitoring for one of the American Heritage Rivers.

The Tijuana River Watershed is one of the most populous and environmentally stressed areas along the U.S. and Mexico border. With new tools and integrated data, this local, state, federal, and international partnership is demonstrating an on-line decision-support capability to improve water quality and availability and to promote better health. The Baltimore, Maryland, City Police Department applies GIS tools and integrated data to support the development of CrimeStac, a comprehensive digital mapping centre to track crime and related trends (e.g. housing and public health), creating a world-class model for crime reduction information.

22.8 THE COMMUNITY 2020 GIS STANDARD

ESRI recently joined forces with the U.S. Department of Housing and Urban Development (HUD) to develop, install, and maintain an agency-wide standard GIS platform. Part of HUD's Community 2020 programme, the software is intended to enable communities to visually analyse, understand, and respond to opportunities and constraints by bringing to life demographic, economic, and HUD programme data via smart maps.

Community 2020 seeks to improve the ability of communities throughout the country to make strategic decisions, open the local planning process to community input, and increase the effectiveness of HUD programmes. The software package designed and implemented by ESRI will improve access to a range of information and expand the dialogue between citizens and their government. Representing a huge step forward for place-based planning and GIS technology as a data dissemination tool via the Internet, this project will help HUD leverage its investments in a shared federal, state, and local government-wide geospatial database.

ESRI is using two of its highly scalable software products for the HUD solution—ArcIMS 3 and ArcSDE to address HUD's requirements for database management, analysis, and dissemination. SDE is client–server software for storing, managing, and quickly retrieving spatial data from a single database management system. ArcIMS will establish a common Web-enabled platform for the exchange of HUD data and services.

22.9 GEOGRAPHIC THINKING – THE GEOGRAPHY NETWORK

As more and more people understand the value of using geographic thinking for structuring organizations and solving problems, the need for easy access to geographic information continues to rise, and the Internet has emerged as the best way to meet accelerating demand for spatial data and information. ESRI has worked toward providing easy access to a large and distributed collection of geographic data, data resources, and services. With its launch of the Geography Network in June 2000, this goal is being realized. Powered by the ArcIMS mapping technology, the Geography Network is the first application service provider (ASP) system focused on delivering GIS content and capabilities to users anywhere in the world, via the Internet.

Driven by interrelated forces, including the significant increase in computer speed, the rapid implementation of Internet technology, and the burgeoning growth in the collection of geographic data, the Geography Network is a global network of geographic information users and providers. It uses the infrastructure of the Internet to deliver geographic content to user browsers and desktops.

This new, network-based architecture is multiparticipant, collaborative, and enables organizations to openly share and directly use GIS information from many distributed sources at the same time. We call this the Geography Network architecture, or g.net, because it was first implemented for the Geography Network. It works with any scale of implementation in any size organization and promises to leverage the work of GIS professionals while radically enlarging the use of GIS in the world. The g.net architecture easily supports distributed data management, metadata searching, dynamic data integration, and departments and divisions using each other's information via the loosely coupled protocol of XML.

While a number of websites currently offer geographic data and mapping tools, the Geography Network has been designed to integrate a distributed set of spatial content to offer mapping and related geoservices (e.g. address matching, network routing, and spatial analysis) for use in websites, GIS software, and custom applications. The Geography Network site (www.geographynetwork.com) serves as a hub, providing access to a global network of complementary mapping systems – an on-line library of distributed GIS information, available to everyone designed to adhere to open standards for the dissemination and sharing of data and services.

Content may be provided in the form of raw data, maps, or more advanced services such as lifestyle mapping, flood risk mapping, address geocoding, and network routing. The Geography Network channel guide is a searchable index of the geographic information and services available to the network users. Information can be located on any server on the Internet and accessed from any Internet browser or Geography Network-enabled desktop GIS (e.g. ArcInfo, ArcView GIS, and ArcExplorer). Much of the content on the Geography Network is accessible free of charge, but commercial content is also provided and maintained by its owners.

The network is an open system. Anyone with an Internet browser can use the system. An open protocol is used for communication that is compatible with emerging Internet standards for geographic information sharing. The Geography Network technology has been used in the Open GIS

Consortium Web-mapping test bed, and offers citizens who lack direct GIS experience and who seek answers to geographic questions, high-level spatial solutions in the form of a suite of on-line application services relating to business, governmental, environmental, and educational concerns. On-line tools are available to define areas of interest and search for specific geographic content, and searches can extend to data held in the NSDI clearinghouse nodes. Other menus guide users to mapping services and geographic data for a selected study area. The Geography Network not only eliminates the need for protracted Web searches to find project data, but will also make the content available immediately in standard browser and desktop GIS applications.

Hundreds of data layers are currently available through the Geography Network. International offerings include jurisdictions, elevation, vegetation, land-use, socioeconomic statistics, and satellite imagery. U.S. data include administrative boundaries, detailed streets, topographic maps, natural hazards, environmental hazards, demographic statistics, crime statistics, and aerial photography. Many government agencies – national, state, and local – use the Geography Network to build GIS systems for their communities and constituents to access and use their information and services. These include applications on land ownership, land-use, and planning initiatives. Many of the world's NGOs (non-governmental organizations) also use the Geography Network for sharing and publishing their information. An example is the World Wildlife Fund's new project called Forest Watch, which provides the world with up-to-date information about the status of forests anywhere in the world.

Using standard product components and the g.net architecture concept, the United Nations Environmental Programme has built the next generation of its Global Environment Monitoring System, enabling members from different countries to share and provide information about their natural resources. Although stand-alone, this system can be integrated into the larger Geography Network community for broader information sharing. These initial efforts demonstrate that organizations and communities with distributed GIS data, such as states and local governments or national organizations, can practically implement their own server-based GIS networks.

Organizations, professionals, and citizens will be able to freely access, browse, and overlay this information for hundreds of practical applications, including education. Perhaps the most interesting and important implication of the Geography Network is that citizens from around the world will be able to share in the rich treasures of information currently maintained and accessed by only a few. The result will be that over time, everyone will learn and have a better understanding of how the world works. This will lead to better personal decisions and facilitate more participation and collaboration in the decisions that effect how the world evolves. Ultimately, people will become more conscious of how closely related and interconnected they are to the earth – like a bee to a flower.

22.10 PROVIDING A COMMON LANGUAGE

The evolution of 'geographic knowledge everywhere' is fostering the adoption of new methodologies and accelerating change for the better. Cutting across nearly all disciplines, GIS provides a common language for discussion and acts as a means to bring people together in the decision-making process. GIS is successful not only because it integrates data but because it enables us to share data in different societal segments. It helps us integrate these specializations, bringing information together – not just data but our organizations and people to help put the world's pieces back together again. Dynamic and inter-connected, the world is a living system and is constantly impacted by fast-paced technological advancements and an increasing population. As individual sciences and information systems become more specialized – fragmented, focused, and single-purposed in their conception and content – coordinating whole organizations becomes more difficult.

ESRI is committed to promoting the global benefits of GIS. By developing new GIS platforms, providing education and technical support, making spatial data accessible, and promoting GIS on the Web, ESRI is helping to make a difference by giving us the tools to organize our future. GIS helps to create geographically conscious societies that are able to consider problems in a holistic way. The technology is bringing people closer to their worlds and empowering them to define a future that reflects their values, hopes, and dreams.

Chapter 23

Spatial multimedia representations to support community participation

Michael J. Shiffer

23.1 INTRODUCTION

One of the criticisms levelled against GIS and similar spatial information systems (as they relate to public participation), is their relative inaccessibility and the lack of capacity to incorporate informal mental models, such as personal anecdotes and observations. This chapter explores some potential roles that complementary technologies to GIS can play in facilitating public participation in planning contexts. The complementary technologies, which afford the capacity to link images, text and sound to maps will be referred to in a general sense as spatial multimedia.

Three perceived impediments to participation comprise: (1) the inability to physically attend meetings; (2) being unable to understand others; and (3) struggling to have competing views understood by others. For each of these, various implementations of spatial multimedia are described that begin to overcome each impediment, thus achieving an enhanced degree of public participation in a specific context. Finally, several areas where additional research is needed will be identified.

23.2 IMPEDIMENTS TO PUBLIC PARTICIPATION

There are many political, organizational and institutional impediments to public participation in planning contexts. These have been explored extensively throughout the planning literature (cf. Forester 1989; Innes 1996; Day 1997; Tett and Wolfe 1991). This chapter focuses on the tangible impediments to public participation, and how IT can address those. This chapter does not suggest that IT can be an easy cure to participation challenges. Rather, it is proposed as a potential catalyst towards a more inclusive process.

23.3 IMPEDIMENT: JUST GETTING THERE...

The results of GIS analyses are often conveyed to the public in the context of meetings. Yet one of the more significant impediments to public participation in planning is simply being physically able to attend meetings. This can be particularly challenging for the elderly and physically challenged, especially in climates with diverse weather patterns. Furthermore, being at meetings can be a significant challenge to 'two-income' families or families with small children, where a willingness to attend such a meeting can often be superseded by simple practicality. Finally, there may be a degree of ambiguity surrounding how relevant the agenda of a particular meeting may be to a specific individual. All of this tends to lead to lessened public participation on routine matters, and perhaps even an unintentional lack of participation on matters of particular relevance.

23.3.1 How IT can help: virtual presence

In recent years, IT in the United States has matured to a point where it can be employed to bridge the physical gap between planning meetings and those who wish to attend them. A category of technologies can support a degree of *virtual presence* at a meeting – thereby complementing the data handling and mapping capability of GIS. The most relevant technologies include cable TV access, video teleconferencing, and WWW access. GIS can play an active role in all of these forms of virtual presence due to the capacity of the technologies to effectively transmit live displays from GIS applications. Hence, the user can often have the option of transmitting a realtime image of a human, a place, or a map.

23.3.1.1 Public access cable television channels

Cable TV has been popular in the United States since the 1970s. A key element involved in the granting of many local cable franchises is the required provision of a public affairs channel which enables broadcast of city or town council meeting and planning. Often these meetings involve a one-way interaction between the meeting and the viewer. This level of participation is relatively easy to attain, for many US households have cable TV access, and it's easy to use, as it simply involves switching on a TV and watching. Although most immediate interaction is one-way (from the meeting participants to the viewer), and traditional feedback mechanisms (such as letter writing and telephone calls) may be employed, some of the more forward-thinking municipalities may actually provide a mechanism for immediate viewer feedback by taking phone calls during the meeting. These situations, however, are rare.

A less-formalized mechanism for public participation that also involves cable television access and viewer participation, can be found within the public access model of local cable TV franchises in the United States. For example, several years ago, Cambridge (Massachusetts) Community Television developed an 'electronic soapbox' known as 'BeLive'. BeLive consists of a desk with a phone and two chairs. Any citizen wishing to broadcast on the local cable television channel simply sits at the desk, flicks a switch to turn on the camera, and instantly he or she is broadcasting to the local community. The citizen may announce the number of the phone at the desk and broadcast conversations with viewers. Although BeLive is a relatively 'low-tech' approach to public involvement, it demonstrates how an innovative use can be made of existing technologies.

Both the BeLive and the traditional model of cable TV access have the capacity to convey limited spatial information. This information can be made available through simple mapping tools or software that provides rapid access to a set of maps (such as Adobe Acrobat or Microsoft Powerpoint). Nevertheless, when using this mode of transmission, special attention must be paid to the fact that TV is an exceptionally low-resolution medium. This is further complicated by the fact that the NTSC video broadcast standard in North America scans every other line of a video signal in an alternating manner that leads to a 'flicker' of thin horizontal lines. Finally, the use of bright colours can be problematic for broadcast signals. All of this adds up to the fact that it can be useful to learn the following simple tactics from television news organizations about the broadcast of maps: (1) display maps at a scale suitable for simple and clear reading; (2) use muted colours whenever possible (i.e. maroon rather than red, mustard rather than yellow); (3) Draw lines thickly to avoid flicker; (4) remember that people viewing these maps will have very limited time to see them, so keep them simple. For example, if necessary use a slow succession of simple maps rather than a single map to convey multiple attributes.

23.3.1.2 Video conferencing

Another relevant form of IT with the capacity to enable 'virtual presence' is video conferencing. This technology has matured and is extensively used by private industry, where collaboration most often takes place over great distances. Video conferencing has only rarely been used to facilitate public involvement in local affairs, however, due to its relatively high cost (for both infrastructure and connection). It is more typically used by the private sector where elimination of travel over great distances is economically feasible. Access to the technology in the United States has changed significantly in recent years, due to the fact that many service bureaus (such as Kinko's Copy Shop) have installed video conferencing stations and charge a small fee for access to these facilities. However, use of this technology is dominated by the private sector, where activities such as remote interviews of job candidates might make economic sense.

However, video conferencing can be relevant to public participation in local government where it may be necessary to draw upon remote expertise to solve a local problem. In this context, cities have the capacity to meet remotely with consultants for progress reports on projects ranging from real estate development to transit improvements. When using video conferencing to convey spatial information, one needs to be cognizant of the same technical limitations as described above for cable TV access (regarding thickness of lines, use of colour, etc.). One exception to this is that since video conferencing is a two-way conversation, maps can be redisplayed for longer duration at the request of the other party, and it may be possible to use more complex maps in these situations. Cable TV and video conferencing may indeed bridge the gap of space; however, participation using these technologies still requires that all participants be involved at the same time.

23.3.1.3 'Getting There' through the Internet

The development of the Internet and the WWW is really a centerpiece addressing the physical inaccessibility of spatial information systems. Data and information are no longer tied to a discrete set of machines in a single location such as a planning office. Instead, Internet-based GIS makes it possible to access this data from virtually any suitable machine in the world provided that it has an effective network connection. 'Suitable machines' can range from desktop units to hand-held devices used in the field. By virtue of the WWW's global system of associative document links, as well as the Internet's capacity to support remote conferencing (or less formal 'threaded' conversations via discussion groups and mailing lists), this technology has profound implications for enhancing public participation by bridging the barriers of space and time. The Internet enables interactive conversations among multiple stakeholders. Furthermore, the WWW affords multiple channels of access to a variety of media in the form of linked maps, images, and documents. In this context, however, the concern is the capacity of the Internet to facilitate virtual presence.

Access to the Internet has been growing rapidly in the last few years. This is attributable to the many thousands of households that gain access to the Internet each day through their personal computer and either a telephone line or a cable TV connection. Furthermore, wireless access to the Internet, though still in its infancy, is rapidly growing. However, for those concerned about equitable access to public participation in local affairs, the Internet can be viewed as a barrier between those who are affluent enough to afford access from the home computers and those who can't. Fortunately, there has been movement in both the private and public sectors to address this issue. On the private side, several Internet providers have initiated programmes where they will give away personal computers to people who subscribe. This, however, is probably small consolation to those who are concerned with equity issues, since this model still requires a sustained outlay of cash to support Internet access. Some programmes require no cash and are instead fully supported through funds from online advertisements; however, these can be difficult to become involved with, due to high demand for these services. Another private sector model is 'pay-as-you-go' Internet access. This was initially popularized by so-called 'cyber-cafés', but has recently found its way to such routine places as the local McDonald's. Here one finds a kiosk-like machine that accepts cash (in this case, $1 for 20 minutes) for Internet browser time. Nevertheless, the use of a McDonald's Internet kiosk to access local-government-related information has yet to be observed!

In the public sector, access to the Internet, and by implication to local government information, has been made available through libraries and other public buildings where clusters of computer terminals can often be found. These are supplemented by a significant number of community computer centres that have arisen in various low-income neighbourhoods.

Finally, physical inaccessibility to spatial information systems can be addressed in a very 'low-tech' way by simply bringing a laptop computer, projector and an information specialist to a planning meeting.

23.4 IMPEDIMENT: UNDERSTANDING OTHERS

For many years, planning professionals have been challenged to describe technical information to non-technical audiences. Where abstractions have been used to convey concepts such as noise and traffic levels, more descriptive indicators have been somewhat elusive. Furthermore, it is often challenging for meeting participants to effectively describe a former, current or proposed physical environment to those who may be unfamiliar with the area or the time frame in question. Representational aids provide an implementation of IT that can support the gap of understanding that often exists between the speaker and the audience.

Representational aids are designed to make the abstract more concrete by employing a richer set of descriptions. They have evolved from gestural and verbal tactics such as waving of hands and copious use of adjectives, to artistic conceptualizations and the employment of linked media. The intent has been to close 'the gap of understanding' between technical specialists and key stakeholders. This has most recently been accomplished through the augmentation of typically abstract environmental representations with direct manipulation interfaces and multimedia representational aids, which have been made available in planning settings through increases in computing power over the last decade (cf. Câmara *et al.* 1991; Shiffer 1995 and many other works on this subject).

For instance, as one observes public participation contexts, a gap often becomes evident between what is being conveyed by a specialist at the head of a room and what the public understands. This gap can lead to misunderstandings, arguments and ultimately mistrust. A noise specialist may display a map of noise contours and describe to the public how each contour line represents a noise level. Furthermore, the specialist may describe this noise level as an aggregation of peak noise events, perhaps even going so far as to provide a quantitative representation of Ldn (frequently characterized as noise level averaged over time with night time events weighted more than day). Although talented and well-meaning professionals have the capacity to effectively convey these concepts to an attentive public, often times this may not be the case. Thus, it is important to employ some sort of representational aid to bridge the gap between what the specialist intends to convey and what the public can effectively understand. Such an aid might be as simple as a spreadsheet calculation that conveys concepts such as annoyance (Schultz 1978) or relative impacts on property values (Frankel 1981).

More complex representational aids might actually augment traditional noise contour representations with digitally sampled recordings of actual discrete noise events (Shiffer 1995). The combination of such representational aids is illustrated in Figure 23.1. Other representational aids might animate the peak noise level of a vehicle (such as a motorcycle) as it moves through an environment (Figure 23.2) (Ferrand 1999).

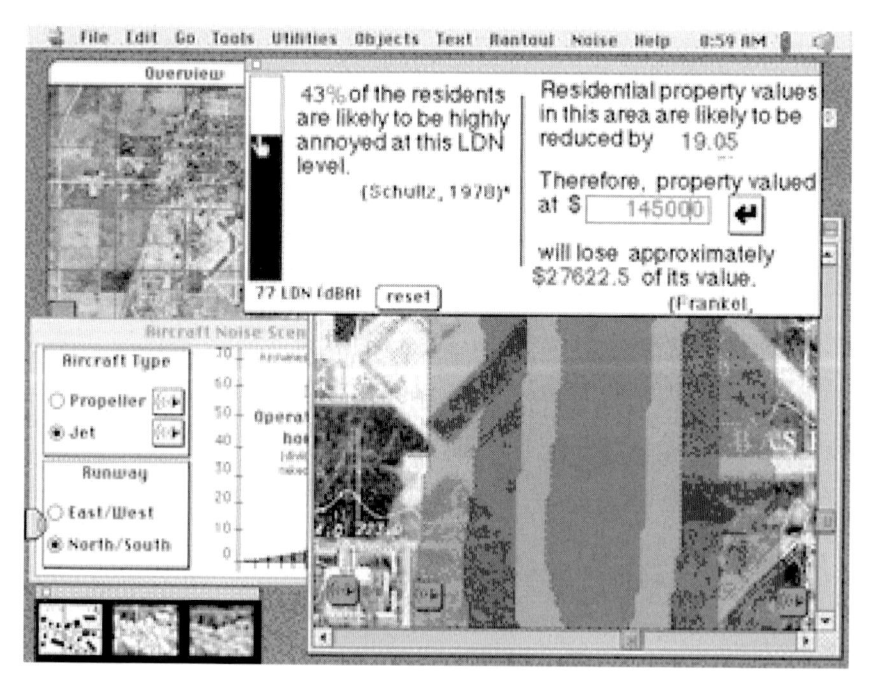

Figure 23.1 Aircraft noise representations for Rantoul, IL c. 1991.

Figure 23.2 A sequence of vehicular noise representation with a peak of approximately 85 dbA. Taken from an animation of a motorcycle on Newport Ave., Quincy, MA.

The intended result of representational aids is to make analytic tools and their outputs more manipulable, understandable, and appealing, so that information that would normally be inaccessible to the layperson can be comprehended more effectively. Nevertheless, just as this technology has the capacity to deliver compelling and descriptive representations, it can deliver compelling and descriptive misrepresentations. While this issue is not new to spatial analysis, it can be exacerbated through the use of multiple representations in collaborative contexts; therefore it is important to understand the potential pitfalls of unintentional misrepresentation so that measures can be taken to minimize it. A good place to start is with the works of Monmonier (1991), and Tufte (1983; 1990; 1997).

23.5 IMPEDIMENT: BEING HEARD

Spatial annotation tools allow users of an information system to relate their comments to a geographic (or spatial) area. These have essentially been with us since ancient times when humans would draw in sand to illustrate spatial relationships while telling stories. Annotation tools can be as simple as pens, pins or other devices that might be used to mark up a shared map or diagram. In fact, 3M's Post-It Note® is probably one of the most significant and accessible spatial annotation technologies to be developed in recent times.

Electronic annotation is made possible by the interoperable characteristics of contemporary software that allows the linking of various computer files across applications either on a single machine or through a network. For instance, one now has the ability to draw a polygon on a map and then link that polygon to a web page or other type of multimedia file with a broad variety of information pertaining to that location. This can be accomplished using either a GIS, a WWW-based map, or using a 'portable' document that supports multimedia linking.[1]

These tools give us the capacity to link ideas and comments to simple marks that we make on shared electronic maps either before, during, or after a meeting. This technology goes further by allowing us to link external resources, such as text, sound or imagery that may be either centrally located or distributed across a network to maps. Efforts at crime prevention can provide an example of this kind of information capture where, for instance, neighbourhood crime statistics can be overlayed on top of individual perceptions of the relative level of safety of a given area.

Planners can take advantage of various types of digital annotation. For instance, at the simplest level, such annotations might be basic graphical marks (such as lines, circles, dots, etc.) that are intended to convey spatial flow, physical alterations or a multitude of related concepts. Such graphical marks would likely be tied to a variety of more descriptive annotations. The remainder of this section will describe three types of annotation (text, audio, and video) along with some of the benefits and drawbacks associated with each.

Spatially linked textual annotation typically takes the form of an 'Internet-like' discussion thread that is linked to a location on a map shared over a network. In essence, a user could click on a map location to access a discussion thread that relates to that spatial area. As opposed to the other forms of annotation described below, textual annotation offers the benefits of exceptionally low storage overhead (and subsequently low demands on network resources). Furthermore, its levels of descriptiveness are limited only by the prosaic talents of the contributors.

Audio annotation allows one to link verbal comments to a location on a map. This is accomplished by speaking into a microphone that is linked to a computer-based digitizer. This can theoretically work more rapidly than other forms of annotation. However, early experiences have demonstrated a reluctance to annotate a map with one's voice due to the awkwardness of stopping a meeting and concern about how one's comments might be viewed out of context.

We also have the capacity to link video images of contributors to maps in a manner similar to audio annotation. Such a system might be employed as part of a kiosk installation in a public place where voluntary comments could be solicited from the public using an embedded camera. This has the effect of enabling one to view 'the face behind the name'.

On the positive side, this can lead, convey, expressive and compelling opinions about various proposals. On the negative side, such images can lead to eliciting unintended bias on the part of the viewer (who can, for instance, make judgements based on appearance). Furthermore, video annotation requires significant storage overhead and can be exceptionally difficult to convey through as network using existing technologies. This last concern is being addressed with continual advances in compression technologies and network bandwidth.

The archive that results from multiple annotations can assist with recollection during subsequent meetings. Access to this archive can be based on geographic relevance, chronological relevance and associative relevance. Geographic relevance allows users of an information system to search for annotations that are related to a specific region or subregion using typical GIS spatial selection operations. Chronological relevance allows a user to add the capacity to search for annotations made before, after or between two dates. Finally, associative relevance allows searching by keywords or related concepts that could be linked together in a WWW-like associative structure.

While it is certainly conceivable that GIS-based archival mechanisms can be set up to aid future recollective efforts, this requires that a substantial spatial data infrastructure be already in place. As we are only beginning to realize the development of substantial spatial data infrastructures around the world, we will need to continue to rely on the (frequently chapter-based) libraries of local historical societies for more specific spatial descriptions that can effectively convey the character of a local area. Even in this case, the issue becomes a question of what material is worth maintaining, which has profound implications for the scalability of such a system. For instance, is it reasonable to expect a planning council to archive a spatial representation of every proposal made along with the corresponding minutes of every planning meeting? If so, what is a reasonable time frame for keeping the record in the archive? Five years? Fifty Years? Forever? If not every proposal is archived, then how is the choice of 'what is relevant' made? These questions aside, such annotations have the capacity to significantly enhance recollection by providing a means of encoding informal memories of a location.

23.6 CONCLUSION

Through the use of the complementary technologies and implementation approaches described above, PPGIS can empower groups and individuals who have traditionally been informationally disadvantaged due to a lack of both cognitive and physical access to traditional spatial analysis tools. Exactly who benefits from such empowerment will depend on the situations in which the PPGIS is implemented.

Representational aids will not completely replace quantitative measures of environmental phenomena. Rather they will serve to supplement such measures through multiple representations. While it is possible that easier access to tools and information can offer easier access to their misuse, initial observations have demonstrated that ease of access can also promote experimentation and exploration. Such experimentation and exploration may result in the identification of related issues to a particular problem, or the generation of new alternative approaches to a problem or issue.

While one might argue that these technologies can lead to better-informed conversations, their use can also make it difficult (or impossible) for a group to reach consensus. Furthermore, while the use of annotations and representational aids have the capacity to minimize arguments based on misunderstanding, they can also confuse or mislead. Therefore, careful attention needs to be paid to strategies for the implementation of these tools in an institutional context.

NOTE

1. This last form of map delivery is made possible using software packages such as Acrobat from Adobe Systems. Acrobat is built on the capacity to deliver documents in a 'Portable Document Format' (PDF). This has tremendous implications for the delivery of maps due to a number of factors. First, this file format has a capacity to effectively scale vector and raster graphics. Second, the free distribution of a software plug-in that opens these files in WWW browsers and extensive use of this file format for private sector document distribution make this software relatively ubiqitious. Finally, the capacity to link multiple media (such as WWW pages, images, and sound) to locations on maps displayed in this format make it an exceptionally useful file format for capturing annotations to maps. Furthermore, the conversion of GIS maps to PDF files is a simple matter of selecting a 'virtual printer' that is bundled with Acrobat software.

REFERENCES

Câmara, A., Gomes, A. L., Fonseca, A. and Lucena e Vale, M. J. (1991) 'Hypersnige – a navigation system for geographic information', *Proceedings of the European GIS Conference*, Brussels, Belgium, April, pp. 175–179.

Day (1997) 'Citizen participation in the planning process: an essentially contested process?' *Journal of Planning Literature* 11(3): 424–434.

Ferrand, N. (1999) 'Emerging visualization technologies to support public participation in an urban mass transit planning context' (unpublished MIT master's of city planning thesis).

Forester, J. (1989) *Planning in the face of power*, Berkeley: University of California Press.

Frankel, M. (1998) 'The impact of aircraft noise on residential property markets', *Illinois Business Review* 45(5): 8–13.

Innes, J. (1996) 'Information in Communicative Planning', Number 679 in Working Chapters, University of California Berkeley, Institute of Urban and Regional Development, HT390.C153.

Monmonier, M. (1991) *How to Lie with Maps*, Chicago, Illinois: University of Chicago Press.

Schultz, T. J. (1978) 'Synthesis of social surveys on noise annoyance', *Journal of the Acoustical Society of America* 64: 377–405.

Shiffer, M. J. (1995) 'Environmental review with hypermedia systems', *Environment and Planning B: Planning and Design* 22: 359–372.

Tett, A. and Wolfe, J. (1991) 'Discourse analysis and city plans', *Journal of Planning Education and Research* 10: 195–199.
Tufte, E. R. (1983) *The Visual Display of Quantitative Information*, Cheshire, Conn.: Graphics Press.
Tufte, E. R. (1990) *Envisioning Information*, Cheshire, Conn.: Graphics Press.
Tufte, E. R. (1997) *Visual Explanations: Images and Quantities, Evidence and Narrative*, Cheshire, Conn.: Graphics Press.

Chapter 24

GIS and the artist: shaping the image of a neighbourhood through participatory environmental design

Kheir Al-Kodmany

24.1 INTRODUCTION

Participatory planning is fundamental to finding appropriate and effective solutions to community design and planning problems. The benefits of broad-based community involvement in planning and design are widely documented; they include enhancing the capacity of citizens to cultivate a stronger sense of commitment, increasing user satisfaction, creating realistic expectations of outcomes, and building trust (Altschuler 1970; McClure *et al.* 1997). However, these benefits do not come easily; a truly participatory planning process requires a serious commitment of time, energy, and resources on the part of both the technical expert and the community expert, as well as a mutual respect for the assets that the others bring. Planners and designers contribute technical skills and knowledge; citizens provide community history, local knowledge, cultural values and understanding. These types of expertise complement each other and result in richer, more comprehensive planning and design solutions.

The first significant task of the planner is to create a framework and a language for the planning process that motivates the public to participate and allows them to comfortably share their knowledge, ideas and vision for the community. Lynn McDowell (1987) argues that 'the public needs a language that can give its creativity a focus and help individuals turn their intuition and knowledge into a workable idea'. That language must also be able to bridge the gap between the vision of the community resident and the technical thinking and jargon of the architects' (p. 20). Stanley King (1989) contends that visualization provides just this – it is the only common language to which all participants, technical and non-technical, can relate. Visualization provides a focus for a community's discussion of design ideas; it guides community members through the design process, it raises their design awareness and facilitates better communication. Consequently, exploring alternative visualization techniques could be a key to promoting participatory planning and utilizing community expertise and local knowledge (Sanoff 1990; 1991; Nelessen 1994).

This chapter describes a collaborative community planning process involving faculty and staff at the University of Illinois at Chicago (UIC) and residents and community leaders from the Pilsen neighbourhood in Chicago. In the initial stages of the project, it became clear that the UIC team lacked appropriate visualization tools to engage the residents. It was also clear that creating these tools would be a major component of having a truly participatory planning process that built trust between the two parties. The UIC team then developed a process that involved a combination of high-tech and low-tech tools: a high-tech GIS and real-time sketching done by a skilled artist. A significant finding during the subsequent planning workshops was that the role of 'expert' constantly shifted as the professional planners and the residents shared their knowledge on a variety of issues.

24.2 THE PROJECT

The shared history of UIC and its neighbours includes not only the displacement of homes and businesses to accommodate the University's need for expansion, but also large, well-publicized, and eventually discontinued community service programmes. These issues have created a nearly universal distrust of the University in Pilsen, a neighbourhood just south of the UIC campus. In recent years, the UIC has been working to rebuild trust with its neighbours through collaborative community planning and design.

Pilsen is a largely Mexican-American and Mexican immigrant community of nearly 50,000 people. Leaders in the Pilsen community expressed an interest in a participatory, collaborative approach to the planning and design of their neighbourhood. A particular focus was 18th Street, the neighbourhood's main commercial district. Leaders were very interested in promoting commercial tourism along this business corridor and in addressing such problems as urban blight and decay, vacancies and crime. Community leaders were anxious to harness the creative energies of residents as a way to foster the enthusiasm required to take serious actions and improve the neighbourhood. Leaders felt that the meaningful involvement of all stakeholders – including the technical experts at the University of Illinois – would strengthen the sense of community and that a cooperative effort would help present a 'unified front' when funding opportunities arose. A planning team was formed that included 25 community residents, including representatives of the 18th Street Development Commission, two architects, two planners, and one artist.

The University team's objectives went beyond the actual neighbourhood planning and design process. UIC's objectives included creating a mutually respectful partnership with neighbourhood residents, preserving neighbourhood history, gaining a broader understanding of urban issues, and exploring effective visual communication methods. Building trust was the highest priority in the planning process. Trust arises from consistently meeting expectations and creating outcomes that all partners perceive as beneficial. One of the first lessons that the University team would learn was that effective visualization was a key to engaging residents and building trust.

24.3 THE PROBLEM

In his book *Designing with Community Participation*, Henry Sanoff (1978) writes that currently employed methods of user participation actually disenfranchise the user because the methods of communication do not accommodate a non-design-oriented population. This was true in the case of UIC and the Pilsen community. After a short period of involvement, University design professionals realized that the presentation and visualization techniques at their disposal were not promoting trust and meaningful public participation.

At the first working session, dozens of slide images of the neighbourhood were presented to display current site conditions. Slides were set and presented in a fixed sequence. As the discussion moved from the project introduction to the design development stage, there was no interaction between the images of the present conditions of the neighbourhood and the images of potential future design images. Because a slide projector lacks navigational capabilities, the images of site conditions were not readily available during the second half of the workshop, the design discussion. When participants requested to see a specific image, it was impractical to search for it in the slide tray. The process lacked a means of visualizing what was being proposed within the context of what currently existed. Residents and community representatives experienced context disorientation. Long-time community residents became overwhelmed trying to remember small details of specific sites, rather than applying their community knowledge and expertise to develop overall strategies and solutions. Planners, architects and artists also grew frustrated with the limitations of the design process.

It became clear that the role of the technical experts had to be expanded. It was insufficient merely to lend planning and design expertise to the process; specific tools had to be developed to enable community members to fully participate in the process. In order for community residents to participate as co-planners and co-designers, they needed access to the same tools as planners and designers, and these tools had to be developed for use in a public setting. The UIC team began searching for a visualization environment that could effectively connect the two traditional stages of a project: (1) orientation and (2) design development. An intimate relationship exists between proposed design alternatives and their physical context. A visual connection between the two had to be established to enable citizens to participate in evaluating these design alternatives to the fullest extent possible.

24.4 THE GIS

A system was needed that would illustrate geographically the neighbourhood's context – its geography, cultural and architectural history, as well as present conditions, including neighbourhood strengths, weaknesses, opportunities, and threats. The system also needed to provide some design prototypes to foster discussion about how the neighbourhood might look in 5, 10 and 20 years.

An interactive GIS image database was developed. First, a historic database was compiled, consisting of maps, images, tabular data, and textual information about the Pilsen neighbourhood and its surroundings. Thematic layers were created for plat maps, land use maps, zoning maps, base maps, historic maps, and current aerial photographs. Historic images showing the neighbourhood characteristics in various time periods were collected and hot-linked respectively to historic maps of various periods. Second, the database had to show existing conditions of the neighbourhood, particularly the 18th Street Corridor. A digital camera systematically documented the present condition of the neighbourhood, and images were hot linked to their geographic locations. Finally, a GIS library of environmental design prototypes was incorporated in the database. It consisted of photographs of key developments in Chicago's neighbourhoods, particularly those adjacent to the campus. The artist on the University design team annotated these photographs regarding quality, historical significance, architectural style, and building materials, and these photographs were hot linked to their geographic locations. This arrangement was intended to visually represent types, architectural styles, and locations of buildings and designs that could be incorporated into the neighbourhood plan (for viewing the GIS database, images and some of the artists' sketches, refer to http://www.evl.uic.edu/sopark/new/RA/).

24.5 THE ARTIST

The GIS provided critical contextual information, such as maps, demographic information, and neighbourhood images. But this technology was no substitute for human drawing capability that could quickly transform ideas into realistic drawings. To facilitate the design process, a graphic designer was recruited to the UIC design team. The artist was trained specifically to draw urban scenes including streets, parks, plazas, and retail areas, as well as detail elements such as shrubs, street signs, benches, and chairs.

She also depicted human activities in her sketches, to bring a human scale to the drawings. With a few lines, this highly skilled artist could capture the salient features of an image. The artist's role was to listen to participants and sketch their ideas as they articulated their preferences and desires for their community.

24.6 IMPLEMENTATION

Workshops were planned for four consecutive Saturdays, from 9 a.m. to 5 p.m. at a church in the Pilsen neighbourhood. The president of the Pilsen community organization served as the host for the event, welcoming and introducing community and university participants, detailing the goals and objectives of the workshops, and describing the contribution of participants in the long-term planning process. Inspired by the work of Stanley King (1989), researchers set ground rules for the workshop discussions, including: (1) speak only for yourself; let others speak for themselves; (2) don't criticize an idea; instead, suggest alternatives; and (3) don't focus on solutions; rather, brainstorm for alternatives.

Equipment for the planning workshops included a computer, an electronic sketchboard, two projectors, and two large screens. The electronic sketchboard is a drawing board with a forgiving surface – it is easy to erase. Sketches on this board can be saved as an electronic file in a graphic format, such as a TIFF or JPEG, to a zip drive. The sketches were projected onto a screen using a multimedia projector while, beside it, another screen was used to display the GIS images. The positioning of the screens allowed for cross-referencing for both the artist and the participants. The images on the large computer screen showed the existing condition of the street or building under consideration, or the 'before' scenario. The other large screen showed the artist's sketches as she changed the scene according to participant input, or the 'after' scenario. The positioning helped keep the artist and the residents in check with reality, to ensure that the emerging drawings were practical, applicable, and relevant.

This design process was in many ways counter to conventional practice. It is common for architects and planners to prepare a master plan and then proceed to address details. Often, the first communication with the public is the presentation of the final plan. In this example, however, residents and other key stakeholders were actively involved in the development of the design plans. In many cases, participants would become so involved in the discussion that they would proceed to the electronic sketchboard and draw their own ideas. The artist was then able to take their ideas and build upon them.

The GIS image database greatly assisted all members of the planning team in visualizing past and present conditions of the neighbourhood. It engaged community members in developing alternative design solutions and also helped in visualizing current urban development examples in the city. The constant reference to the image database – including maps, existing buildings and lots – made the discussion contextual and more realistic for everyone involved. The GIS and the artist working in tandem had the effect of 'leveling the playing field' between university planners and designers and the community. Residents truly became co-planners and co-designers in the process. A few examples illustrate how technical expertise and local knowledge were critical to the planning process.

One issue that arose was the lack of sidewalks in Pilsen. Some residents expressed a strong need for sidewalks; others said sidewalks were not a priority. A lengthy and heated debate ensued. The UIC team used the GIS to display streets in Pilsen with and without sidewalks. The data indicated that approximately half of the streets did not have functional sidewalks, highlighting them in bright yellow. Interestingly, the cluster of yellow matched the location of pedestrian/automobile accidents that appeared on a separate GIS layer. As mentioned in Section 24.4, electronic layers contained maps and information about the site, particularly the 18th street.

To further examine the issue, the UIC team browsed images of streets. One picture showed school children entering and exiting their school and walking on the street alongside cars. Another picture showed how some sidewalks were too small in the busy retail areas. These sidewalks were jammed with people, and pedestrians were encroaching on the right-of-way. Other pictures illustrated the deteriorated condition of the existing sidewalks, and as a result, pedestrians did not use those sidewalks. Instead, they used the right-of-way. The images showed that the elderly and disabled had a difficult time getting around the neighbourhood; one picture actually showed a blind person attempting to walk along the cars.

As a result of the maps and images, business owners became more supportive of sidewalks as they learned that they would facilitate better access to their businesses. Parents became more supportive as they learned that sidewalks would protect their children from traffic accidents. The community became more sensitive to the needs of their disabled population. These processes led to the collective agreement on the necessity for sidewalks as a top priority. The method helped identify important issues and build consensus. In this example, technical expertise of the planners and the visualization tools they developed helped residents reach an informed decision about an important safety issue.

Once consensus was reached that sidewalks were needed, the participants moved on to design. The artist incorporated sidewalks in sketches of future neighbourhood streetscapes. In sketching the sidewalks of a major thoroughfare, 18th Street, she added tall trees (Figure 24.1). Some members of the audience objected. The architect and planners in the UIC team supported the concept, explaining how trees would enhance the neighbourhood. One of the residents explained that it is impossible to plant trees

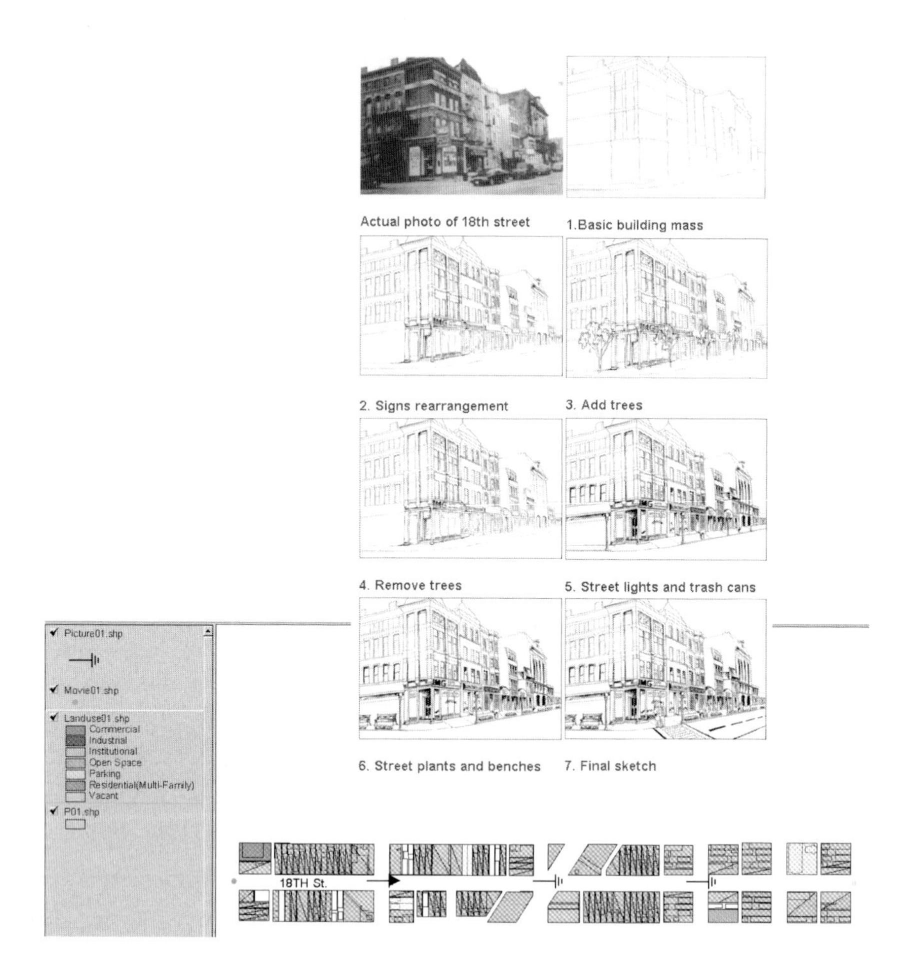

Figure 24.1 Integrating artists' sketches, street images, and maps in ArcView GIS.

because of the hollow vaults under these streets. Due to an elevation problem, the sewer system was built on the ground, and streets were built on top of the sewer lines in a vaulted structure. Instead, the community participants suggested plants. The artist drew beautiful plant beds with heavy vegetation and greenery. Some participants objected to the heaviness of the vegetation due to visibility and safety concerns. The artist adjusted the scale and intensity of the vegetation in the sidewalk plant beds. In this instance, community residents' knowledge of the area's history and safety issues led to effective urban design solutions. Planners and designers who did not live in the neighbourhood would not have necessarily known about these issues and could have made uninformed decisions.

These examples illustrate how community residents relied on the university professionals for planning and design expertise and how the planners relied on the residents for community expertise, such as cultural values, history, and context. Our experience reinforced the notion that designers and users frequently do not consider the same factors in planning and using a space. For example, designers often give careful attention to such things as the cost of construction, the definition of space, and construction methods (Hester, 1975). While users rarely consider these things, they do see other details that designers miss, particularly in how the environment will be used. In this project, designers found that the community residents suggested many more useful design ideas than they had expected. In this project, community residents suggested many useful design ideas that might have been overlooked by designers.

These examples also show how the GIS and the artist enhanced this process, providing the visualization tools necessary to engage all participants in the process. The positioning of screens allowed participants to observe constantly how the artist modified images and incorporated them into the design. Everyone was potentially able to voice an opinion or concern. Such a setting reinforced the reliance on visual cues and minimized reliance on jargon.

24.7 DISCUSSION

The GIS and the artist working in tandem helped community residents articulate their ideas in relation to the neighbourhood context. Together, the artist and the GIS image database reinforced each other in creating a common visual language. People who are not trained in the design professions sometimes have difficulty communicating ideas about architecture and urban design, but most people do have definite design preferences. The GIS and artist helped to draw out these preferences. The GIS contained images of buildings and new developments in other areas of the city and these were used to assist participants and the artist in discussing design alternatives. As participants suggested solutions, the planner would display images on the large screen that most closely matched the participants' ideas. Design examples were used to probe and support residents' ideas.

Second, the GIS helped highlight the importance of cultural values and history in the future design of the neighbourhood. One of the major concerns of the Pilsen community is to preserve cultural heritage as represented in the physical form. The GIS images reminded the artist, the planners, and the residents of the cultural artifacts and environmental el in Pilsen. These images supported discussion of cultural issues in the neighbourhood. Images helped the artist to incorporate some of the cultural symbolic features and artifacts of the community in the new designs. Also, the GIS showed the geographic distribution and clusters of buildings of significant cultural and historic value.

Finally, and most importantly, the workshops and the visualization tools helped build a relationship of trust between the university and the community. The GIS and the artist helped empower residents to plan and design for the future of their community. As one of the residents stated, 'As we saw ideas begin to take shape before our eyes we could feel excitement rise. The pulse begins to beat a bit faster!' The designs that were created by the planners and designers reflected the community's wishes and input and respected their cultural heritage. At the end of the process, several community members said that they felt that the University's purpose was not to destroy their lifestyle but rather to revitalize their community. This helped overcome some of the distrust experienced in the past. At the end of the workshops, several participants expressed an interest in attending classes at UIC in the Urban Planning and Art programmes. Several neighbourhood participants have been admitted to these programmes.

24.8 CONCLUSION

Sanoff (1990; 1991) writes that many environmental problems requiring technical guidance can best be solved through the active participation of those affected by the design decision. Interest in user needs or user participation is not rooted in romanticism about human involvement but rather in the recognition that users have a particular expertise different than, but equally important to, that of the designer. This expertise needs to be integrated into a process that concerns itself with environmental change. This project suggested that technical expertise alone is inadequate in solving community design problems and that involving community members ensures that effective, relevant strategies are created. However, without the proper communication tools, public participation exercises frustrate both designers and users. Technical experts must devise tools that 'level the playing field' so that residents can truly function as co-planners and co-designers.

This project found that a combination of basic sketching and visual techniques along with advanced GIS visualization power can create a visualization environment that enhances public participation. Visualization is crucial in a public participation process because it is the only common language to which all members of the community – young people and adults, poor and rich, powerless and powerful – can relate. Public participation is meaningless if people cannot understand what is being proposed (Towers 1995). With the advent of digital technology, new ways of eliciting public participation in planning and design have become possible. GIS, with its sophisticated mapping and visual display capabilities, is a powerful example of this. Its capacity to integrate many different layers of data, its user-friendly windows interface and its speed make public information accessible to people at the neighbourhood level and provide them with a way to integrate their own knowledge with additional data about the community. Clearly, the development of methods and skills in community design are still at the exploratory and discovery stage. This chapter is a step in the development of such skills – progressing towards the art of designing with people.

ACKNOWLEDGEMENT

The author wishes to thank several individuals from the University of Illinois for their help in this project including: Roberta Feldman (City Design Center and Architecture), George Hemmens (City Design Center and Planning), Robert Bruegmann and Peter Hales (Art History), Charles Hoch, Wim Wiewel and Tinwei Zhang (Urban Planning), James Hudson (Arc/Info Technology Lab), Yequao Wang (Geography Department), Kate Pravera (Great Cities Institute). The author also wishes to deeply thank Angie Marks for editing the text. Finally, the author wishes to thank Mike Schwartz and Ismail Sumairah for useful comments.

REFERENCES

Altschuler, A. A. (1970) *Community Control*, New York: Random House.

Hester, R. (1975) *Planning Neighbourhood Space with People*, New York: Van Nostrand Reinhold.

King, S. (1989) *Co-Design: A Process of Design Participation*, New York: Van Nostrand Reinhold.

McDowell, Lynn (1987) 'Community Design', *The Calgary Herald, Sunday Magazine*, 11 October, pp. 19–25.

McClure, W., Byrne, A. and Hurand, F. (1997) 'Visualization techniques for citizen participation', in W. McClure (ed.) *The Rural Town: Designing for Growth and Sustainability*, University of Idaho, Moscow: Center for Business Development and Research, pp. 47–55.

Nelessen, A. C. (1994) *Visions for* a New *American Dream*, Chicago: Planners Press.

Sanoff, H. (1978) *Designing with Community Participation*, New York: McGraw-Hill Book Company.

Sanoff, H. (1990) *Participatory Design: Theory and Techniques*, Raleigh, NC: Bookmasters.

Sanoff, H. (1991) *Visual Research Methods in Design*, New York: Van Nostrand Reinhold.

Towers, G. (1995) *Building Democracy: Community Architecture in the Inner Cities*, London: University College London Press.

Chapter 25

A *praxis* of public participation GIS and visualization[1]

John B. Krygier

25.1 INTRODUCTION

Public Participation GIS (PPGIS) have been conceived as an integrative and inclusive process-based set of methods and technologies amenable to public participation, multiple viewpoints, and diverse forms of information (for a review, see Obermeyer 1998). Public Participation Visualization (PPVis) is an important component of PPGIS. Geographic visualization (GVis) is conceptualized as a predominantly private type of map use involving high human–map interaction wedded to exploratory analyses (MacEachren 1994). Such visual analysis is linked to the analytical component of GIS: maps and other visual representations are not merely the output of GIS analysis, but are part of the analysis itself. GVis Research has focused on highly skilled scientists engaged in scientific research using advanced computing technologies. However, rapid advances in technology are allowing a much broader array of non-scientific users to engage in visualization-type map use. Developments in WWW-based programming languages are making advanced, highly interactive GVis and GIS applications available to anyone with an internet connection. Users can not only access existing geographic information, but also can interactively explore 'what if' scenarios and amend and add information to WWW-based GIS databases. Users can 'make' and 'un-make' information and thus shape and reshape the way they understand their neighbourhood, region, county, and the world. This is an active process of 'sense-making' (Dervin 1999) by diverse people, using geographic information from a variety of sources, represented in maps, images, text, and sound.

A *praxis* or theorized practice of PPVis and PPGIS consists of an explicit awareness of the concepts and theories of information, its representation, of people, social relations, power, and how these shape and are shaped by socially infused technologies such as PPVis and PPGIS. Such awareness must be brought to bear on actual applications that, in turn, will reshape the *praxis*. This chapter reviews a *praxis*-based prototype PPGIS/PPVis WWW site developed for a low-income, inner-city neighbourhood in Buffalo, New

York. This chapter does not prescribe a particular *praxis*, but instead suggests that PPGIS research should proceed within the context of a theorized practice.

25.2 CONCEPTUAL ISSUES IN THE *PRAXIS* OF PPGIS AND PPVis

The Buffalo WWW application has focused and reshaped theoretical and conceptual issues surrounding PPGIS and PPVis. My concern is in developing a theoretically informed practice of PPVis and PPGIS that weds conceptual and theoretical ideas to the actual implementation of a site in a community. Conceptual issues include the geography in PPGIS and PPVis, the medium and site content, non-threatening graphics, and evaluation.

25.2.1 The geography behind PPGIS and PPVis

Traditional maps and GIS provide access to *where* particular phenomena are, but Geographers (and others) have developed more sophisticated methods for analysing and understanding geographic phenomena. For example, many concepts and models and methods for analysing economic data exist and are used by geographers, planners, and regional analysts. The technology for providing such geographic methods of analysis via the WWW exists or will exist soon. While it is important to include these sophisticated methods in PPGIS and PPVis applications, it is also important to consider the potential problems and benefits of the general public having access to such geographic methods and models. The users of such applications need to learn to use and understand such methods, and this implies that an educational component must be central to the development of PPGIS and PPVis applications. This component of PPGIS and PPVis may be guided by existing literature on the design and implementation of educational multimedia and other pedagogic materials (see discussion in Krygier *et al.* 1997a). The importance of geographic education in the context of PPGIS and PPVis cannot be underestimated.

25.2.2 The medium and site content: representation, visual forms and hypermedia

PPGIS and PPVis are not only maps and GIS, but also images, video, text, and sound: an array of *visual forms* (Krygier 1994). The way these interrelated representations are hyperlinked together, the *intellectual design* of PPGIS and PPVis, must be carefully considered (Krygier 1999). This intellectual design is guided by cognitive, social, and geographic theories and may (should?) be open to modification by users of the site. This research focuses on the manner in which current concepts and theories in human geography relate to certain fundamental aspects of visualization and PPVis: the significance of interconnected representational forms (Cosgrove 1984; Krygier 1997b), the spatiality of the map, linked to the development of spatial components in social theory (Sayer 1992; Krygier 1995; 1996), and hypermedia, linked to hypertextual theory (Bolter 1991; Landow 1992; Krygier 1995; 1996). Issues of representation are, then, linked back to the concepts and theories of geography discussed in the previous section.

25.2.3 Public participation and non-threatening graphics

Enhancing public participation with the use of IT consists of more than just making the technology available to people. One can have access to tools that provide a sophisticated geographical analysis of environmental data for an area, but not actually understand the analysis itself. Of particular importance, is the idea of graphics that encourage rather than discourage participation: what can be called 'non-threatening graphics'. Planners involved in engaging public participation in traditional settings (such as public meetings) have noted that participation can be diminished if the graphics used to present information about planning alternatives look too polished, professional, and finished. Sketchy and less-finished looking graphics, however, tend to encourage public participation: the graphics look like the proposal is still in a 'sketchy' and undecided stage. This phenomenon is briefly discussed by MacEachren (1995: 456). The issue of non-threatening graphics is broader than graphics, and includes all aspects of the design of a PPGIS and PPVis application in order to insure effective use by the public. Some possible design strategies for non-threatening graphics in PPGIS/PPVis include:

- use game- and role-playing metaphors,
- allow people to explore issues at home (rather than only in public meetings),
- use intermediaries in public meetings to do what people ask,
- use sketchy (rather than refined and finished) graphics,
- use panoramic views as 'hinge' between situated view and map view,
- use interactive software which moves people through increasing levels of complexity,
- use interactive software to make people critical (different perspectives on same issue), and
- use an on-line encyclopedia of concepts that need to be understood in order to participate.

25.2.4 Evaluation

Evaluation of the impact and consequences of the use of PPVis and PPGIS is a complex and important issue. A broad approach to evaluation described in Krygier *et al.* (1997a) has been adapted to the context of PPVis and PPGIS (Krygier 1999). Evaluation should play a role through an entire project, helping to shape and reshape the design in the process of its development and implementation. Evaluation can be conceived as consisting of four interrelated functions: (1) goal refinement; (2) documentation; (3) formative evaluation; and (4) impact evaluation. *Goal refinement* entails creating a detailed plan of action and set of goals prior to project implementation. *Documentation* is simply documenting what is actually done in the process of creating the application. *Formative evaluation* consists of the systematic collection of information during the process of creating the application to get preliminary feedback on its viability (and to reshape the application in the process of creating it). Finally, *impact evaluation* consists of evaluating the effectiveness of the final application. Each of these evaluation functions can be facilitated with a range of evaluation methods, including interviews, focus groups, questionnaires, observations, ratings assessment, expert review, and achievement tests (a range of both qualitative and quantitative methods). An important approach to impact evaluation for PPVis and PPGIS may be Dervin's sense-making approach (Dervin 1999; Gluck 1998).

For practical purposes, sense-making has well-tested methods and numerous applications in many fields. Sense-making should be particularly viable as a means of understanding and evaluating the complex interactions between users and PPVis applications. A major advantage of sense-making is that it is based on the same conceptual and theoretical ideas that infuse contemporary human geography and social science. Sense-making conceptualizes humans moving through complex time/space contexts, and is similar to Hagerstrand's time geography (Hagerstrand 1982) and Giddens' structuration theory (Giddens 1984). Dervin brings these important theories into the realm of information design by arguing that all information is designed: '...made, confirmed, supported, challenged, resisted, and destroyed' (Dervin 1999: 41). Sense-making provides both theory and methodology which help guide the development of systems which not only deliver information to people, but which allow people to modify, change, and adapt the systems and information. 'Sense making...explicitly privileges the ordinary person as a theorist involved in developing ideas to guide an understanding of not only her personal world but also collective, historical, and social worlds' (Dervin 1999: 46). This is the goal of PPGIS and PPVis, to empower users rather than only provide them with existing information. Sense-making can be a vital element of the praxis of PPVis: an explicit theoretically informed approach to information design which, as Dervin argues, assists 'humans in the making and unmaking of their own informations, their own sense' (Dervin 1999: 43).

25.3 PPVis AND PPGIS IN APPLICATION: THE BUFFALO, NEW YORK CASE STUDY

The conceptual and theoretical issues discussed in the previous section initially shaped ideas about a PPGIS/PPVis application, and were modified by attempting to implement these ideas in an actual community. A grant funded the development of a prototype PPVis/PPGIS website. The project is documented in a Master's Poject and at the WWW site associated with this project (Chang 1997; URL in references). Goal refinement, formative evaluation, and documentation from the project have served as the basis of an evaluation of the software and technology. Issues investigated, and discussed below, include the skills needed to create such applications, available map and GIS functions, necessary hardware, and time involved. The ultimate question is, of course, if the approach taken is viable and worth pursuing beyond the prototype stage, where impact evaluation (such as Dervin's sense-making) can be applied.

25.3.1 Buffalo's Lower West Side community

An inner-city neighbourhood on Buffalo's Lower West Side was chosen as the geographic context for the prototype WWW application. Work began in the summer of 1997 in cooperation with Buffalo's Lower West Side Development Corporation (LWSDC) and its Director, Mark Kubeniec. The Lower West Side Community is diverse, dominated by Hispanics and recent Latin-American immigrants. It is also home to a significant number of Asian-Americans, African-Americans, and Whites. While nearly 50% of the residents have incomes below the poverty level, the eastern edges of the community overlap the fashionable Allentown area, a historic neighbourhood dominated by middle- and upper-income whites and their refurbished, Victorian-era homes.

25.3.2 Choosing an appropriate technology

There are many technologies available for PPGIS and PPVis. Paper maps and coloured pencils are a cheap and relatively effective technology. Digital technologies are diverse and have their own benefits and problems. The primary alternative to WWW-based mapping and GIS is the provision of mapping and GIS functions on microcomputers in community centres (Ghose 1994). However, such physically located resources may be difficult for certain individuals to access. Delivery of mapping and GIS via the WWW can maximize public access to mapping and GIS, and may be the most cost-effective means of providing people (and particularly those in marginalized communities and areas) with analytical tools that would not otherwise be affordable. Familiarity with the interface of web browsers may enhance usability. Users can focus on learning geographic concepts, mapping and GIS functions, rather than struggling with a new GIS software interface. Finally, the WWW provides access to extensive additional on-line information in a multimedia and hypermedia format, which may supplement applications of WWW-based mapping and GIS.

Several methods exist for providing mapping and GIS capabilities on the WWW. A spatial data library can provide access to spatial data and analytical software. The user must perform their own analysis on their own computers after downloading the data and software. Another method is to undertake a GIS analysis and generate maps independent of the WWW, possibly in response to a query from an interested user, and post the results on the WWW. This process can be automated with the use of a map generator. Users set the parameters of a map or GIS analysis on a WWW-based form, which in turn is passed to a map or GIS server, which generates a map or series of maps and posts the results on the WWW page. The US Census Bureau's *Tiger Mapping Service* (http://tiger.census.gov) is a good example of this type of technology. Real-time map browsers, such as ESRI's Map Objects and Internet Map Server provide similar functionality in a package explicitly aimed at component- and WWW-based GIS developers. Early in the research, it was decided to use real-time map browser technology for the Buffalo project, as it provided more sophisticated, real-time GIS and mapping capabilities than spatial data libraries or pre-generated map-analysis approaches. The project maps and databases would reside on a SUNY-Buffalo Geography WWW server, and could be accessed and used by anyone with an Internet connection and a computer.

25.3.3 Developing the prototype WWW site

Discussions with Kubeniec and others from the LWSDC resulted in preliminary foci for the prototype PPVis/PPGIS site. One of the primary goals of the LWSDC is to confront problems caused by absentee landlords in the community, and to subsidize home sales to community members. Thus it was decided that the site should focus on housing issues. To this end, the site needed two map scales with associated databases: a neighbourhood-scale map with streets, lots, and building outlines (see Figure 25.1) and a more generalized city-scale map (see Figure 25.2).

Users of the site can view information about housing in their neighbourhood, then compare their neighbourhood to the city as a whole. Importantly, potential users seemed very comfortable (not threatened!) by the neighbourhood scale maps. Our hope was that the comfort in working at the neighbourhood scale could be used to ease users into using a more abstract, smaller scale map of the city, while enhancing their understanding of their neighbourhood by broadening its spatial context. The city-scale map existed

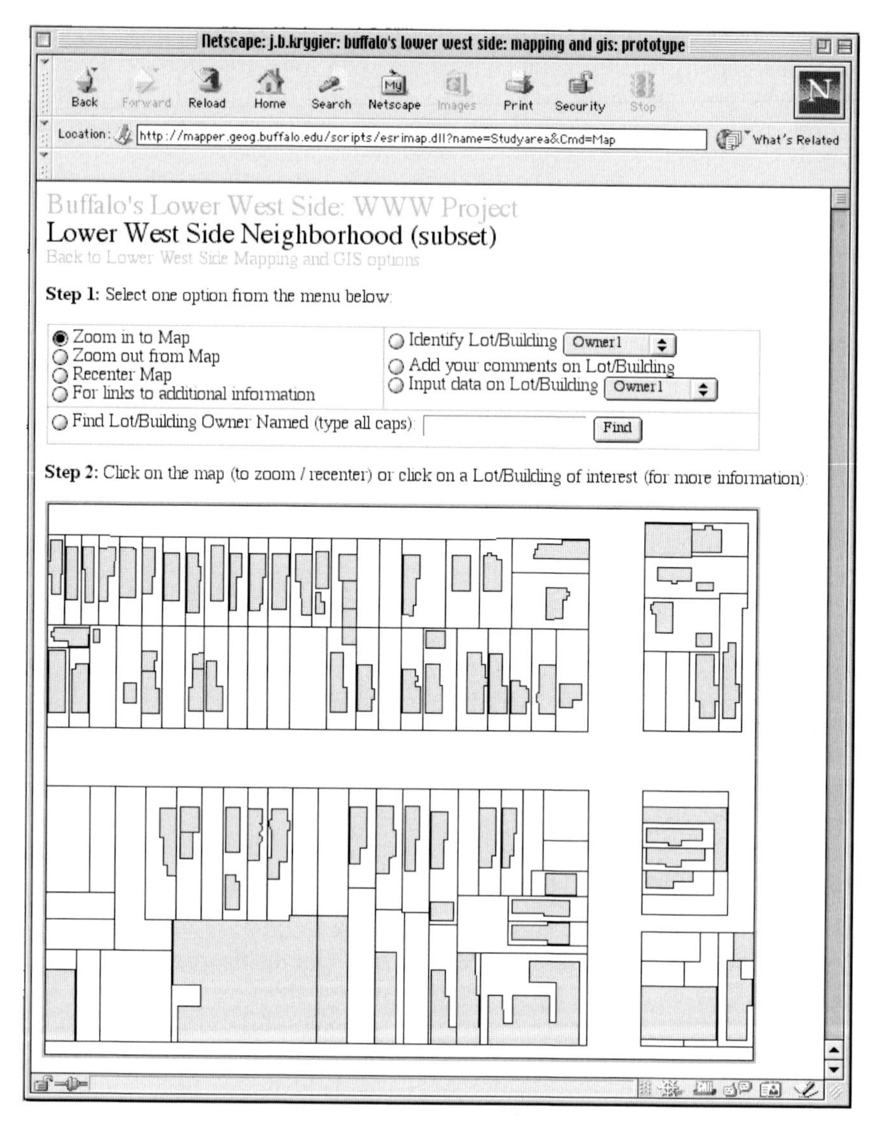

Figure 25.1 Neighbourhood-scale map in Buffalo PPGIS site.

in a compatible (ArcView) format, but the neighbourhood map had to be digitized from paper maps. Multimedia, including images of the neighbourhood, particular homes, and even the use of animation and sound, were seen as a vital part of the site by the LWSDC.

Basic mapping and GIS functions on both neighbourhood and city maps, shown in Figures 25.1 and 25.2, include zoom in to map, zoom out,

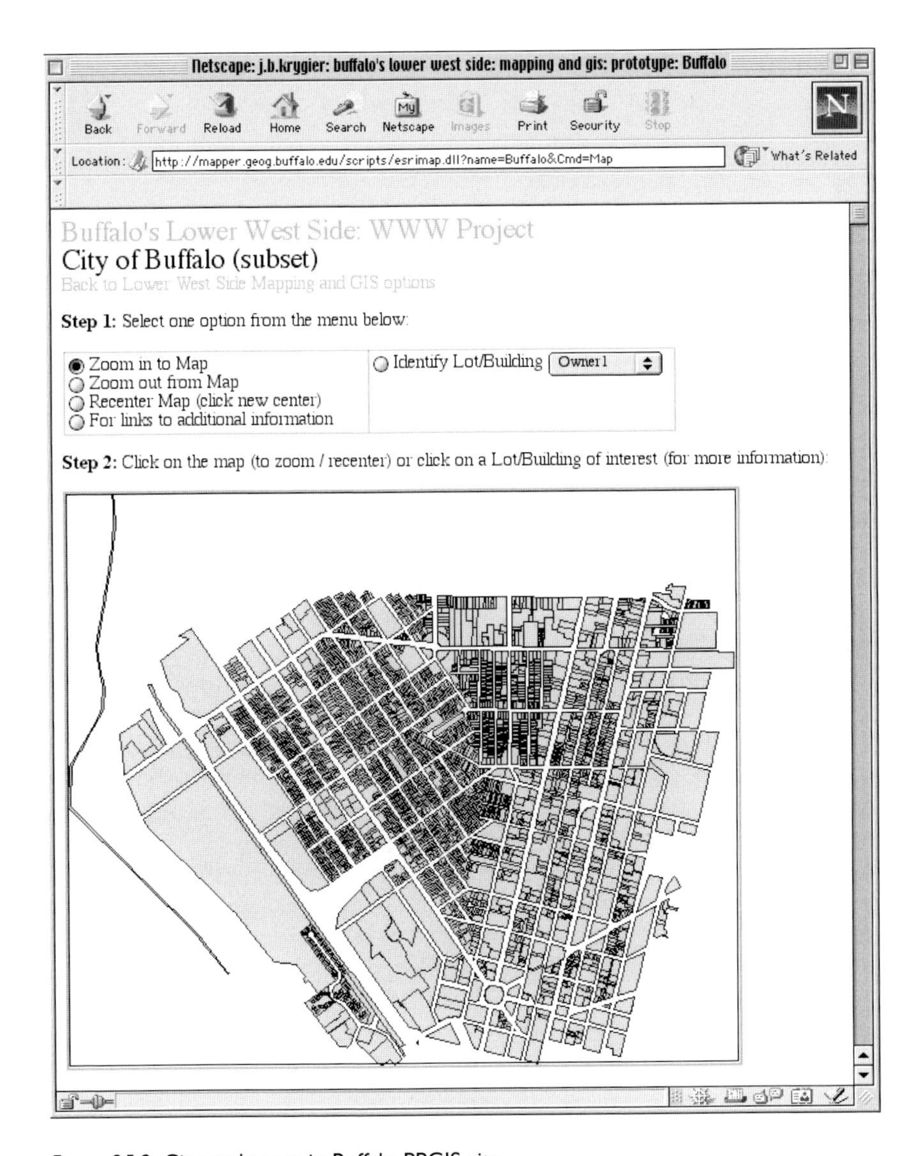

Figure 25.2 City-scale map in Buffalo PPGIS site.

re-centre map, and hyperlink (for example, a click on a city-owned lot links to the Buffalo City WWW pages relevant to that property). Both neighbourhood and city maps also include an identity function (see example in Figure 25.3) which supplies data, such as the name of the lot owner, when a lot is clicked. The neighbourhood map also includes a find function, where the user can locate a lot if the owner's name is known.

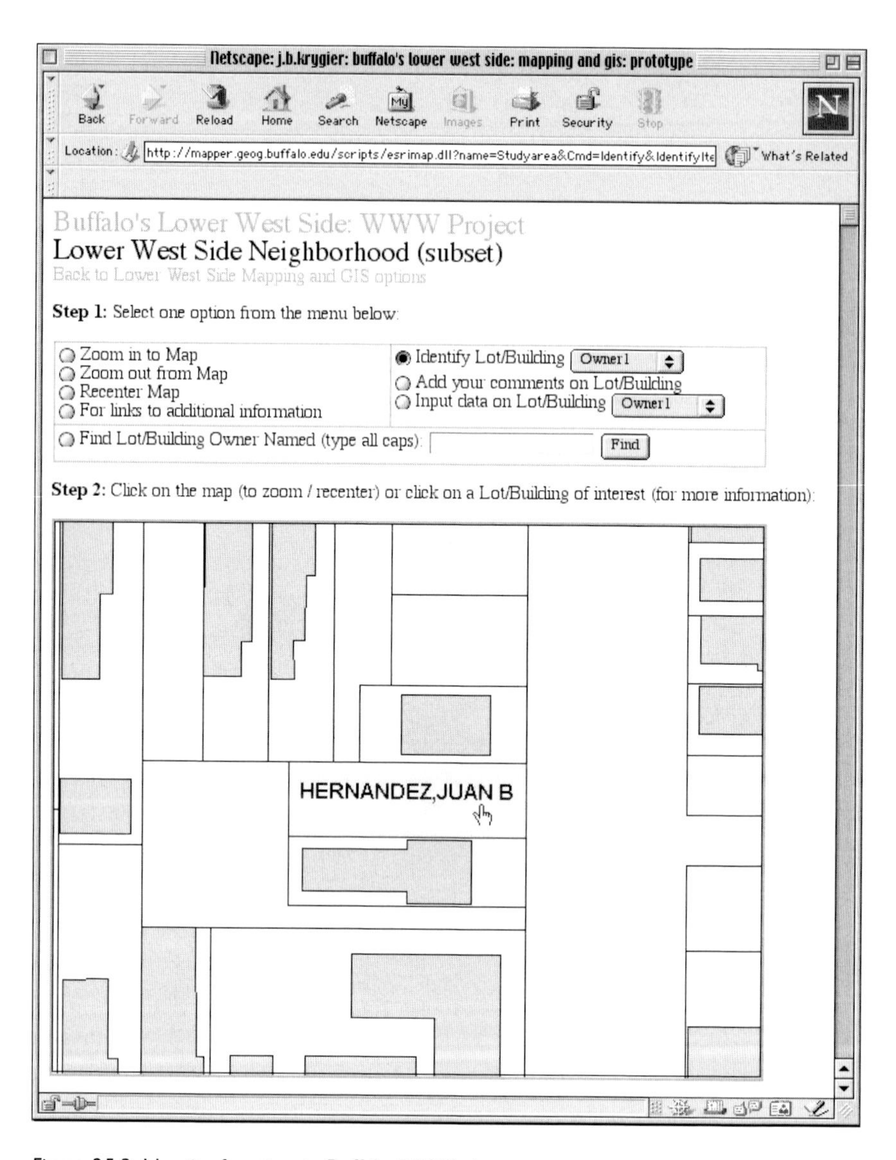

Figure 25.3 Identity function in Buffalo PPGIS site.

Two additional functions were added to the neighbourhood maps, and raise some interesting questions about WWW-based GIS and what can be called 'open databases'. Use of the comment function (see Figure 25.4) takes you to a page which includes a photo of the selected property, basic information (Figure 25.4, left), and a form to submit comments on the

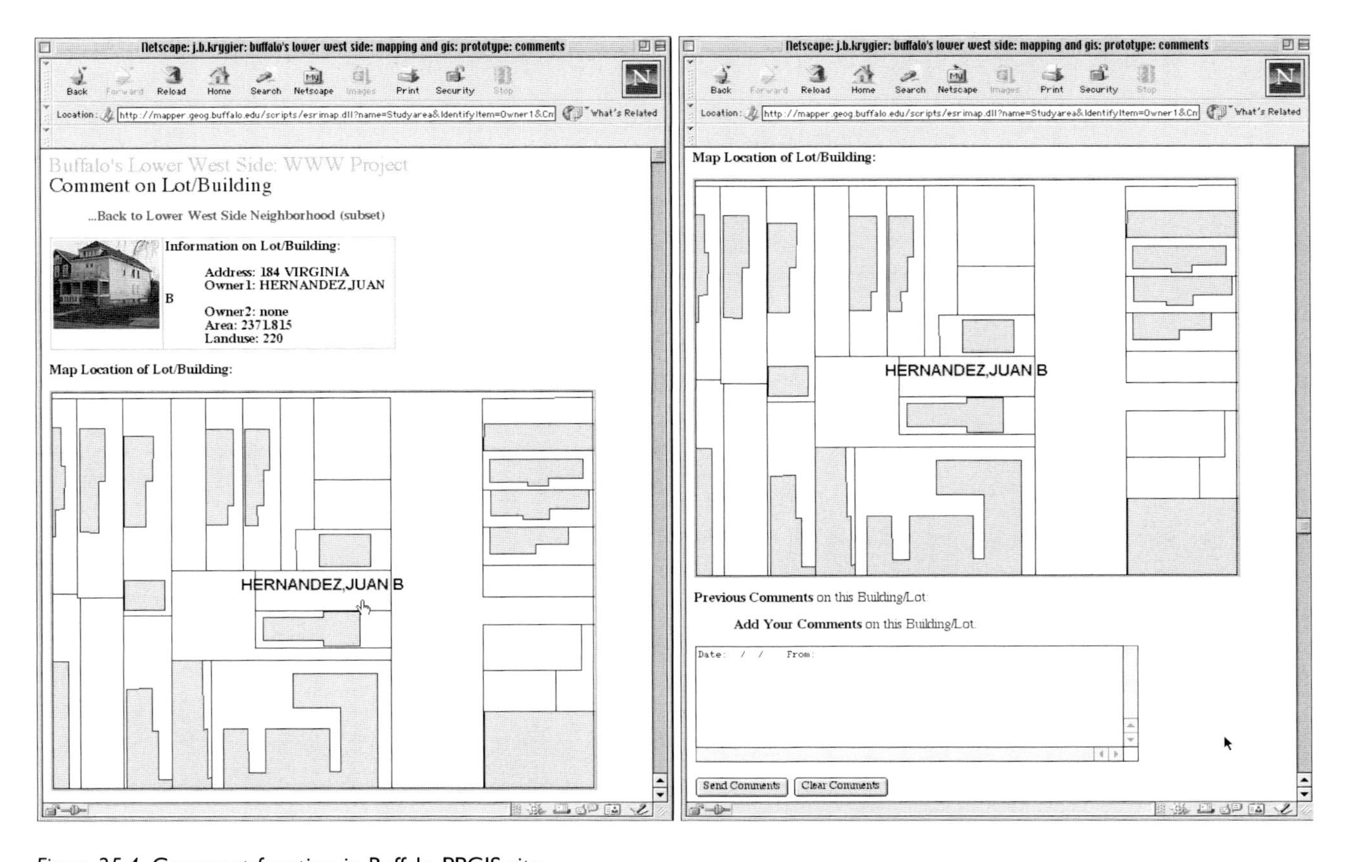

Figure 25.4 Comment function in Buffalo PPGIS site.

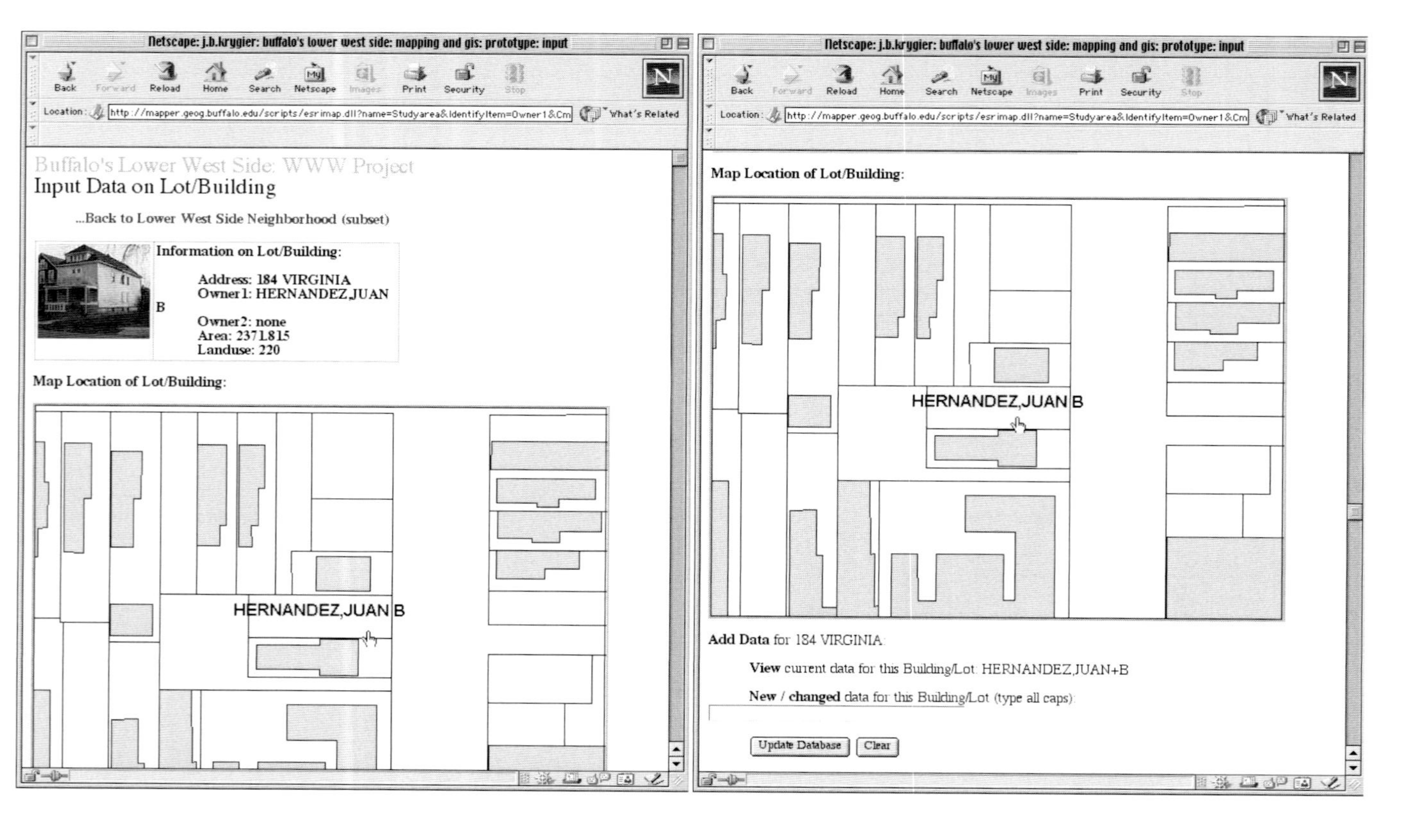

Figure 25.5 Change-database function in Buffalo PPGIS site.

property (Figure 25.4, right). For example, a user may comment that a particular property seems to have been abandoned, or that drug activity was observed. Comments are accumulated on the WWW page corresponding to the property.

An input data function (see Figure 25.5) allows a master user (such as Kubeniec) to change the GIS database associated with the site via the WWW site to, e.g. change ownership status or update information on housing violations on a particular property. Thus a master user in the neighbourhood can change the GIS database using any computer connected to the Internet (and need not have a computer with GIS software installed). These 'open database' functions were incorporated to empower members of the community, allowing them to access, amend, and build information about their neighbourhood. An 'open database' that can be added to and modified via the WWW may be misused, yet it is simply not enough to provide existing information via a PPGIS/PPVis application. The site must allow users to 'make and un-make information' and thus shape and reshape how they understand and represent their neighbourhood. The use of open databases and community-driven WWW-based GIS sites will provide many challenges to the developers of PPGIS/PPVis sites, and to the way we think about GIS databases.

The prototype site was created prior to extensive discussions with all members of the community. The prototype would allow us to assess technology costs and capabilities, and to assess if the technology would be appropriate in the given context. Further, few community members were familiar with GIS or mapping. The prototype, once finished, would give community members a sense of what the technology could do, and would hopefully spur community involvement in developing a more sophisticated site. The danger of this approach is that community members may feel like they are seeing a final version of the site and its capabilities, and may not feel comfortable suggesting other functions to which they would like to have access.

25.3.4 Preliminary evaluation of prototype PPGIS/PPVis site

The project to create the basic prototype PPGIS/PPVis site was, in general, successful. Details are provided in Chang (1997; URL in references), but, given particular software and hardware and a moderate amount of programming time, the chosen development platform (ESRI's Map Objects and Internet Map Server) was able to provide us with the basic set of WWW-based mapping and GIS functions we desired. An organization planning to implement a PPGIS/PPVis site will need access to relatively sophisticated hardware and software, and should expect to invest at least 4 to 5 weeks of a programmer's time in setting up the basic site. The project used a computer with an Intel 486DX 33 MHz chip, and required Microsoft Windows

NT and at least 8 Mb of memory and 10 Mb of disk space (the machine actually had 64 Mb memory and 3 Gb disk space). Software and programming skills required included those necessary to use Map Objects, the Map Object Internet Server, ArcView (primarily for digitizing the neighbourhood map), Visual Basic, CGI, and HTML. The project programmer was familiar with Visual Basic and a limited amount of CGI and HTML, but additional programming skills were learned in the process of creating the prototype. Total time spent on the project, including digitizing and all programming, was 240 hours. This would have been less for a programmer with greater familiarity with Map Objects and the Internet Map Server.

Additional (and not unsubstantial) costs to consider include Krygier's time in guiding the project, Kubeniec's time in helping Krygier and providing maps, data, and guidance from the LWSDC. In addition, upgrades of Map Objects, the Internet server, and Visual Basic in the spring of 1998 required an additional 20 hours of modifications to the prototype. Any PPGIS/PPVis site requires a programmer familiar with Map Objects to maintain and update the system. Indeed, the prototype site occasionally crashes, and it requires someone to check that it is working properly on a daily basis. While volunteers can help with data gathering, planning, and some computer components of a WWW-based GIS project, funds would be required to maintain a programmer on a part time basis: a PPGIS/PPVis application as developed for Buffalo's Lower West Side cannot function without some outside assistance and funding. The map server itself is situated in the SUNY-Buffalo computer network and thus there were no internet access costs for this project. If a private provider of internet access was needed, these costs would have to be figured into the project. Grant monies were paid for computer hardware and software. The largest investment was ESRI's Internet Map Server. The Geography Department at the University of Buffalo has an educational license for the Map Server software; otherwise the costs are substantial. Finally, access and use of the site requires at least some computers in the community with internet access. Some access exists in libraries, schools, and in private homes. Other access would have to be developed, possibly in community centres. In any case, the costs of PPGIS and PPVis delivered via the WWW are substantial. While the benefits seem potentially significant, it will always be difficult to find sufficient resources in communities with limited sources of money and the skills required to undertake and maintain such a project. Universities and academics are certainly an important means of providing such resources and skills to marginalized groups and places (see Leitner *et al.* This volume).

25.4 CONCLUSION

This chapter has discussed both technical and conceptual issues related to PPGIS and visualization. The successful development of a prototype

WWW-based site demonstrates that existing software can provide the kind of basic functionality necessary for PPGIS and PPVis applications. A pressing issue concerning the technology is the cost of hardware, software, and a programmer's time. The Buffalo case study provides concrete data about software, hardware and programming time necessary for a basic PPGIS/PPVis site. Universities and academics have an important role to play in providing such resources to marginalized people and places.

A serious issue with any kind of PPGIS/PPVis project has to do with the complexities of communities and the vagaries of funding for community projects from year to year. In late spring of 1998, the Governor of New York cut state funding for the LWSDC (and other similar agencies in the state). Existing community groups in the Lower West Side are not particularly cohesive (split along income and racial lines) and thus it has been difficult to develop a new home for the Lower West Side PPGIS/PPVis project. Unfortunately, in most cases it will be those communities that are more stable, wealthy, and less vulnerable that can support the development of PPGIS and PPVis sites on the WWW. Such insight, while not particularly surprising, is one of the primary results suggested by this research project.

Most vital are the concepts and ideas which shape the development of actual PPGIS and PPVis applications. This chapter has reviewed issues which must play a role in the future development, implementation, and evaluation of such applications. A theorized practice or *praxis* of PPGIS and PPVis is vital. Included in such a *praxis* are the type of issues confronted in the development of the Buffalo prototype site: selection and implementation of the concepts and theories of geography which underpin the analytical capabilities we provide to PPGIS and PPVis users; the design and construction or the GIS and visualization tools to enhance the user's understanding of, and participation in shaping, the content and analysis available; designing PPGIS and PPVis sites so that they encourage, rather than discourage participation; and finally, evaluating and making sense of the impact of such tools given the complexities of the public who use them. The untimely demise of the LWSDC suspended the development of a full PPGIS/PPVis application for the community, and evaluation of the prototype stopped short of the vital impact evaluation of the project in the community. However, Dervin's (1999) sense-making approach should provide a sophisticated means of evaluating a fully developed PPGIS/PPVis application in an actual community in future applications.

This research suggests that many of the vital issues in GIS today are not technical issues. As GIS plays an expanding role in the way we manage, analyze, and understand spatial phenomena, the societal consequences of GIS come to the forefront of research in GIS. Research on PPGIS is important not only as a means of understanding the impact of existing technologies on society, but in also imagining and engineering new technologies for diverse communities who increasingly have access to GIS and mapping.

NOTE

1. Materials related to the research reported in this chapter can be found at the project WWW site: http://www.owu.edu/~jbkrygie/krygier_html/lws/ppviz.html.

REFERENCES

Bolter, J. (1991) *Writing Space*, Hillsdale NJ: Lawrence Erlbaum.

Chang, K. (1997) 'The design of a web-based geographic information system for community participation', Master's Project, State University of New York at Buffalo, Department of Geography. http://www.owu.edu/~jbkrygie/krygier_html/lws/ppviz.html

Cosgrove, D. (1984) *Social Formation and Symbolic Landscape*, Totawa NJ: Barnes and Noble.

Dervin, B. (1999) 'Chaos, order, and sense-making: a proposed theory for information design', in Jacobsen (ed.) *Information Design*, Cambridge: MIT Press, pp. 35–58.

Ghose, R. (1994) 'Use of information technology for community empowerment: transforming GIS into community information systems', presented at *Association of American Geographers* Annual Meeting, April 1994, Chicago IL.

Giddens, A. (1984) *The Constitution of Society*, Cambridge: Polity Press.

Gluck, M. (1998) 'Interpreting geospatial representations: social semiotics for cartographic analysis', *Cartographic Perspectives* 31(Fall): 4–25.

Hagerstrand, T. (1982) 'Diorama, path, and project', *Tijdschrift voor Economische en Sociale Geografie* 73: 323–356.

Krygier, J. (1994) 'Sound and Cartographic Visualization', in Taylor and MacEachren (eds) *Visualization in Modern Cartography*, Oxford: Pergamon Press, pp. 149–166.

Krygier, J. (1995) 'Visualization, Geography, and Landscape', Ph.D. Dissertation, Pennsylvania State University, Department of Geography.

Krygier, J. (1996) 'Geography and cartographic design', in Wood and Keller (eds) *Cartographic Design: Theoretical and Practical Perspectives*, Chichester: Wiley, pp. 19–34.

Krygier, J., Reeves, C., Dibiase, D. and Cupp, J. (1997a) 'Design, implementation and evaluation of multimedia resources for geography and earth science education', *Journal of Geography in Higher Education* 21: 17–39.

Krygier, J. (1997b) 'Envisioning the American West: maps, the representational barrage of 19th century expedition reports, and the production of scientific knowledge', *Cartography and Geographical Information Systems* 24: 27–50.

Krygier, J. (1999) 'Cartographic multimedia and *praxis* in human geography and the social sciences', in Cartwright, Peterson and Gartner (eds) *Multimedia Cartography*, New York: Springer Verlag, pp. 245–255.

Landow, G. (1992) *Hypertext: The Convergence of Contemporary Critical Theory and Technology*, Baltimore: Johns Hopkins University Press.

MacEachren, A. (1994) 'Visualization in modern cartography: setting the agenda', in Taylor and MacEachren (eds) *Visualization in Modern Cartography*, Oxford: Pergamon, pp. 1–12.

MacEachren, A. (1995) *How Maps Work*, New York: Guilford.
Obermeyer, N. (1998) 'Special content: public participation GIS', *Cartography and Geographic information Systems*, 25: 65–122.
Sayer, A. (1992) *Method in Social Science: A Realist Approach*, London: Routledge.

Chapter 26

A model for evaluating public participation GIS

Michael Barndt

26.1 INTRODUCTION

In this chapter, observations about the role of GIS as a tool supporting urban neighbourhood revitalization are presented and specific criteria are offered for the evaluation of PPGIS. To date, PPGIS projects have been restricted by limited resources, small local organizations with non-professional staffs and boards, a 'distance' between grassroots organizations and government and business sectors, and, fundamental political differences among many players. Opportunities for GIS tools to overcome these limitations are often overstated.

A number of models have emerged over the last decade for community information and research incorporating GIS. The Nonprofit Center of Milwaukee Data Center programme began as an independent entrepreneurial model. Most of the initial funding came from community clients. Much community information experience has been episodic, impacting communities for only short periods of time and in fairly limited ways. Few efforts have been comprehensive. Some programmes have organized detailed information, but played a limited role in assuring that the data are used. Other programmes have stressed service, but have struggled with access and resources to improve information systems. Many programmes have been sector specific and limited to housing, health, education, or environmental concerns.

It is necessary to frame critical evaluation questions and criteria that may be used to critique and refine existing PPGIS programmes. For example, how do we assess the programmes? What expressions of broader objectives for community change might guide the way we use this tool in the service of community? In this chapter, three basic contexts for evaluation are proposed:

1 understanding and the value of PPGIS project results,
2 management of PPGIS projects, and
3 PPGIS and community development principles.

I will review topics within each of these contexts and occasionally identify projects of the Nonprofit Center of Milwaukee Data Center programme to demonstrate relevant successes, limitations and failures.

26.2 THE VALUE OF PPGIS PROJECT RESULTS

26.2.1 Appropriate information

Is the material produced appropriate to the tasks community organizations are addressing? Can the data be organized to match the issues that are to be addressed? Available data is often limited and may not be sharply focused. For example, a coalition of organizations seeking to change the process for renewing alcohol licences could protest the concentration of licences in specific areas of the city, but could not clearly isolate the problem licence sites within this data. They were also not able to link crime data to license site data because the level of resolution of crime data was restricted at the time to census tract aggregates. Data may also miss important elements. For example, many vacant lots in Milwaukee's central city are sold to adjacent property owners obscuring the record of the growing number of vacant lots.

Data may also not make important distinctions. A bank loan reported in the HMDA (Home Mortgage Disclosure Act) may be to a speculator rather than to a houseowner. Data may be biased towards a particular perspective. Evidence that housing stock has been increased within a neighbourhood may avoid the question of whether it is affordable to local residents. Data may cover too much ground. Examination of crime rates without recognition of the different categories of crime may limit insight into prevention strategies. Data may represent an incomplete picture. A Children's Hospital may have a complete database on children treated for asthma, but a substantial number of children may have been served elsewhere.

26.2.2 Action oriented

Can the organization receiving the information use it to support decisions, enhance communication or persuade others? Data should be detailed enough to inform actions. 'Indicators' projects tend to focus upon a small set of numbers, which may be used to persuade someone that action is necessary, but will offer little information to programme planning and implementation. Data also needs to be scaled to the level of action. Even census block data is not enough for an organizer seeking to organize a block. The organizer requires 'face-block' information. Data often requires substantial digestion before it can be used effectively. The HMDA data on the RTK Internet site is such an example.

Data should help to refine quantities of information and provide a focus. A map of the location of incidents of lead poisoning will be enhanced by using statistical analysis to isolate 'hot spots' within the data to permit prioritizing and targeting of programmes. Data that only add to existing evidence of a problem may not be at all useful when an organizational response is not feasible. For example, neighbourhood organizations only tend to be interested in crime information when they are in a position to use that information in a local programme. Data must become information; information must become knowledge. Knowledge must then be translated into decision support for informed choices.

26.2.3 Timely

For organizations to adapt to better use of information, it should be available within the schedule of the organization. Strategic planning may be easily managed as an occasional activity with substantial lead-time. But for GIS to fit the management frame of organizations, material needs to be available quickly. For GIS to be relevant to the work of staff, information may need to be available immediately – at the staff member's desktop. For the power of GIS to be fully realized in group decision-making, probing requests for additional information, scenario development and comparative analysis tools should be available in real time, while participants are exploring options. This final step is embedded in a broader PPGIS vision.

The schedule of a local organization may be different than that of a technical assistance resource. For example, university calendars and perspectives towards time may seriously limit the capacity of universities to contribute to short-term needs of community organizations. Timeliness is also compromised by the delays often encountered accessing information, negotiating for access, waiting for preprocessing work to be done by the organization maintaining the data and by the frequent need to rework the data for new uses.

Community organizations set priorities each year among the many possible subjects that they may focus upon. These priorities should not be distorted by which project has immediate access to data. But frequently, the research programmes of data resources are shaped by the pragmatic recognition that some information is available for work and other material is not.

26.2.4 Accurate

How accurate are PPGIS results? In community settings, information is frequently limited. But a high degree of precision may not be required. However, at a neighbourhood level, accuracy is often more important than at larger settings. If 10 per cent of addresses cannot be geocoded, that may be in part a result of a street name change that could represent a large error from a neighbourhood perspective. Extractions of lists that seem complete enough for a community wide view may be deficient at a local level where individual items on the list are important.

Many barriers to accuracy are built into the data. Administrative data often retain little or no historical information. Data may be collected on an as-needed basis, with substantial variations in the age of the information. Those who collect data may introduce errors. Strategies to address these issues over time are also important. Many times, local knowledge can be used to clear up errors and limitations in large data sets. Little has been done to structure larger data systems to incorporate changes that may be contributed by local level actors. The level of detail can matter at all levels. Often zip code data is too crude to generate accurate results because of the high degree of heterogeneity within the areas. Even block level data may be inappropriate when organizations work with face-block information.

26.2.5 Insightful

As many who work with neighbourhood data do not know the neighbourhood well, data and maps leave the impression that the neighbourhood has been revealed. Data are often a weak reflection of reality. These are often used by those who understand an issue or place to communicate to others some of what the local organizations already know. How often do data inform those with an intimate perspective? To what extent are myths about a community challenged by the facts? Local perceptions may be different than reality. For example, neighbourhoods are often convinced that crime problems are substantial when only small increases have occurred.

The search for patterns in data can suggest unanticipated relationships. But these can be the result of errors in data or models or preparation of materials. Generally face validity is a useful check against the meaningfulness of results.

How important is it that work reveals something new? Much of the time it will not happen. But the results should be valued when they do.

26.2.6 Time perspective

The substantial improvement in data access and mapping tools may mask the fact that historical information may be very limited. Such information may be difficult to access and analyse, but often trends are important. It may also be important to place the experiences in one neighbourhood in perspective. How does it compare to other neighbourhoods in the city and with other cities? In the absence of mechanisms for sharing much more than census information, US neighbourhoods are woefully ignorant of how their community compares with others.

26.2.7 Synergistic

An additional concern is that when data and research are focused upon one sector, an advantage may be lost. From the perspective of neighbourhoods, problems are often interlocking. Data available from only one sector, e.g. housing files, may offer an incomplete perspective. Those organizations or departments that collect data tend to focus only upon the data that they collect. An important synergy can result when information from several sources is brought together. Information clearinghouse objectives address this question by creating a central location to bring data together. Additionally, a clearinghouse may be able to explore relationships across datasets by linking records at a level that is not available to the public.

26.2.8 Combining qualitative and quantitative information

There is often a mismatch between those who work with the quantitative interpretation of data and those who approach an issue from the direct knowledge of persons affected by the issue. Organizations who know the story best often reject the use of quantitative information. They have often learned that political arguments can be won with a personalized approach to the issue. But the quantitative and analytical approach to data can also be an important political tool. Ultimately, 'stories' can be disregarded if the listener does not accept how typical the cases may be. Creative ways should be found to link the two perspectives. This is often done by leaving the final responsibility for reports in the hands of those who know the stories. This requires a substantial investment by all parties to learn to work together.

26.3 MANAGEMENT OF PPGIS PROJECTS

26.3.1 Sustainability

How are PPGIS programmes supported and are they sustainable? Foundation supported programmes are often trapped by the reality that the funding is short-term. This limits the capacity of the system to design for a long-term role. Grant-based budgets may also lead to substantial swings in resources from excess to substantial cutbacks. Programmes expected to raise revenue through fees can also be limited within this arena. Funding options may drive the priorities of a programme toward serving the needs of well-endowed organizations who can afford the service and who already appreciate the value of the work.

One element of a funding model should be to find ways to reduce costs and to deliver basic services for less and less cost. Development efforts should use grant funding to create the procedures that routinize long-term delivery of service. A sustainable programme is enhanced by a capacity to expand and contract resources as required by demand. The costs of a programme can be addressed through a number of innovations such as use of variable and low-cost resources, students and volunteers, involvement of the staff of organizations being served, and procedures to routinize repetitive work. Additionally, the system of resources needs to be changed so that the costs of information are built into grants, programmes, technical assistant pools, and other local budgets.

26.3.2 Replicability

Many GIS and analysis tasks are complex, even tedious, the first time they are done. When the tasks are likely to be repeated, an extra effort may be valuable as a way to reduce long-term efforts. More critical is whether the work can be designed to make the process simpler the next time. Often, universities may do the opposite. Certain basic tasks are viewed as valuable experiences for new students to learn. From this perspective, it is better to start at the beginning each time. Template construction may also be a more difficult task for students to understand when they have just learned basic procedures.

Replication efforts are constrained by the wide variety of needs. Predetermined reports, map series and templates may limit an organization with a perspective and priorities different from previous client organizations.

26.3.3 Efficiency

Given the limited resources and potentially large demand for services, the efficiency of the process can be important. More efficient procedures make repetition more likely. An important objective should be to invest additional energy creating programming procedures that substantially reduce the cost and complexity of common tasks. When 80 per cent of the work can be accomplished this way, resources are available for more difficult work. Some tasks may be very difficult to make efficient. For example, the creation of a complex neighbourhood assets map may require the skills and patience of a cartographer to ensure that it is comprensible to the intended audience.

26.3.4 Integral

Data uses range from long-term research, which may take several years, to strategic planning often requiring several months, to programme planning and to programme management that are of much shorter term. As local organizations build GIS and information into their daily work, data should be rapidly available. Convenient and responsive 'intermediary' structures also increase the likelihood of use. Data will be most useful when those unaccustomed to the resource do not have to wait long periods of time for results.

Ideally, access to information should be immediate. Material should be available to inform the decision-making process as it occurs. In some settings, small groups have met with a computer expert around the computer while reviewing various options. The idealized vision of PPGIS suggests that hardware, software and data resources be brought to the centre of an active decision process with citizens directing the use of tools to support the discussion.

At a more basic level, organizations that can use GIS information on a routine basis in their work would benefit from its integration into the local organization. In these cases, GIS would be only one element in case/client management or information and referral or routine monitoring or investigation activity. In Milwaukee, six small CBOs have collaborated in the development of software for their individual use. The software expands upon traditional MIS designs to incorporate linked information about the entire neighbourhood being served – a Community Information System (CIS). And the concept is being further extended by introducing a GIS module. The challenge in this effort is the need to change the style of work within these organizations to fit the information routine.

26.3.5 System complexity

PPGIS techniques can be complex. Often what is called GIS is merely map making. When a series of addresses are presented as a layer of points, the analysis may have only begun. What is the pattern? What does the pattern correlate with? Are certain locations above or below the expected value given correlated effects? On the other hand, complex techniques can be difficult for laymen to understand. Additional efforts are required to insure that the results are clear. But not every task requires sophisticated techniques. Sometimes the appropriate product is simple.

Reducing complexity may also lead to more approachable data. It has been argued that newsprint and crayons may be the best tools to engage citizens in the use of maps for neighbourhood visioning. The technology is then not a barrier to citizens' understanding. However, when GIS systems are designed to be easy to use, the design may compromise complexity. Or the more complex procedures may be so much more difficult than the more automated ones, that laymen will avoid them even when they are important. Popularizing GIS software in simplistic modules may lead to a generation of misuse or under-utilization of the real potential of GIS. It may be argued, e.g. that the introduction of 'Community 2020' as a GIS solution allows a fast start that ultimately limits the flexibility of community users.

26.4 PPGIS AND COMMUNITY DEVELOPMENT PRINCIPLES

26.4.1 Integrate the components of a working CIS

No episodic efforts to work with community data can substitute for a concerted effort to create within each local community the elements of a working information system. As this objective is so difficult to achieve, it should be a part of early PPGIS initiatives. As this objective requires concerted action, it should be a public part of all communication with others. A consensus should be encouraged toward a community mission to support the information infrastructure. No national or state initiatives can replace the local effort, because critical local data is the responsibility of local organizations.

A working local information system should include:

1 Data sources that have developed efficient, accurate database systems, a recognition of a public responsibility to allow access to the data, a local cost procedure to share the data, and, a set of protocols to ensure the confidentiality of data that should not be public.
2 A clearinghouse operation, likely to be independent of data providers, that can guide the creation of a community information system, acquire data from all sectors, serve as a reliable custodian of confidential data, work through the technical problems of data organization and linkage, archive and consolidate data, generate summaries, trend lines and indicators, co-sponsor cross-sector research and support access by others to public data.
3 A service provider – a technical resource to less technical organizations – providing consultation, education, and product to others on demand.
4 Community research analysts examining the data to identify patterns and suggest policy conclusions that are driven by questions raised by community participants.
5 Community organizations and participants who are accustomed to integrating community data and research into their strategic planning, programme development, administration, evaluation, and impact assessment.
6 Sources of support to meet the costs of these activities. Few organizations have allocated resources to information functions. Many local organizations have few resources to spare.

26.4.2 Rights of information access

Access to data can be a significant problem. When one organization has resolved that problem by negotiating for access, that may not improve the general right of access by others. The barriers to data access may be most often the costs of data. It is important to cover the reasonable costs of data in a way that reduces the costs for others. The Internet is viewed as a critical vehicle to ensure broad access to data. But most of the models for access are limited to either access to individual records or to a crude aggregation tool with little flexibility. Much more powerful on-line analysis tools are required to allow these sites to be of value. And, ideally, selective extractions of the data for local use would be available as well.

When local organizations do not have access to resources that would assist in organizing and interpreting data for their own needs, access to data is not sufficient. Only the strongest organizations tend to benefit when data is only available in undigested formats.

26.4.2 Community priorities and capacity building

As the needs for expertise increases, the capacity of local organizations and coalitions to guide agendas should not be compromised. Many other perspectives may compete to control priorities. Universities are often influenced by research agendas that may be independent of immediate local issues. Foundations tend to set themes that drive the work of community organizations eager to be funded. Local government agencies set priorities that may conflict with local neighbourhoods. But these organizations often control the resources.

Mechanisms are required that allow local organizations to participate in the process from the beginning, the 'Policy Research Action Group' (PRAG) at the University of Loyola, Chicago is one excellent example. Research round-tables involve both academic and community leadership. The Milwaukee Data Center programme also benefits from its independence. The programme is a part of an association of non-profit organizations. The 280 members of the Nonprofit Center of Milwaukee represent almost of the community-based non-profit organizations in the city.

Data, even when packaged in sophisticated ways, may have little effect if organizations are not able to understand how to use it. If organizations are to be transformed by embracing these new tools, then attention must be paid to that transformation. What is the role of education, usually informal, in the relationship between local organizations and those receiving service? Such education may begin with efforts to help organizations become better consumers of information. It may extend to empowering staff of local organizations to conduct much of the analysis work themselves. Education is often complicated by the rapid turnover in small CBOs. Ways need to be found to educate organizations as well as specific persons within them. Today, much of the educational process seems to require endless cycles as personnel changes.

26.4.3 The value of co-production

Joint activities between local organizations and larger research-focused organizations may lead to much richer data. 'Co-production' uses the resources of a community to expand the data gathering process and to involve users in the creation of data. Surveys of housing condition, current retail uses and resident priorities are often best performed by local organizations. Recently, a number of cities have developed 'youth mapping' exercises, involving youth in a neighbourhood in the development of an inventory of assets in their community. In these cases, the outcomes are not merely the gathering of data, but the broader benefits to the involvement of youth.

26.4.4 Increase the capacity of local community system to use the technology

Community development proponents frequently follow the adage: 'Teach a man to fish and he'll eat for a lifetime.' These days options range from 'Teach him how to select the best fish in the marketplace' to 'Find out who owns the pond and buy it.' Given the technical challenges involved, an intermediary role is often important. As the intermediary educates others, how important is the transfer of knowledge and skills?

Generally, technology transfer is important only to the point that consumers are better at utilizing the results of technology. That suggests that a better consumer of services is most important. Too much emphasis upon empowering local organizations to do their own work may divert the energy of organizations from organization and advocacy work best performed at that level. When substantial differences over method and perception are likely, the control over the technical process is more important.

Certain functions may need to be organized on a centralized basis. Data clearinghouse and data development activities would be more complex if fragmented. They require the highest professional skills and the time frame to build capacity over many years. To increase the power of less powerful neighbourhoods and organizations, priorities need to be established to identify organizations that can most benefit from such support. It is often much more convenient to work with organizations who have the strength from the beginning to assimilate this technology.

26.4.5 Integrate into a broader community development process

The use of data is incidental, of course, to broader community development objectives. It is important that data initiatives be linked to these other processes. Use of PPGIS to serve single programmes may be less valuable than service to coalitions and neighbourhood wide strategic planning. The tools can be more meaningful when they assist with decisions that allocate resources rather than set speculative priorities or demonstrate needs that will not likely be met.

Neighbourhood based organizations should be empowered to work with information on a casual basis and not just programme management data, but neighbourhood wide data. Organizations and local leadership need to develop the capacity to respond to the challenges facing their community on an independent, locally controlled basis.

26.5 CONCLUDING COMMENTS

The questions raised in this discussion are not meant to suggest mandates for the ideal CIS model. A number of contrasting models appear to be viable ones. Often, instead of suggesting a better way of operating, the questions point out a dilemma. How far should models go to transfer capacity to local organizations? How complex or how centralized an activity is appropriate?

The issues raised should be used less as a scoring system and more as an assessment tool. It may not be appropriate to compare apples to oranges, but as advocates for more effective use of information and GIS tools we should be seriously critiquing the whole fruit basket of alternatives as they proliferate.

Chapter 27

Public participation, technological discourses and the scale of GIS

Stuart C. Aitken

27.1 INTRODUCTION

If, as some researchers suggest, local struggles are characterized as scale dependent, and if the works of community activists are 'spatially fixed' at the local level, then it is likely that they will continue as relatively unsupported endeavours because they fail to gain recognition and respect from larger political constituencies (cf. Herod 1991; Smith 1992; Delaney and Leitner 1997). The case studies in this volume suggest that, at its best, PPGIS offers the possibility of respect and credibility for residents, activists, and concerned citizens involved in planning, development, and environmental management. Is it possible that PPGIS enables a breakthrough of local practices and community concerns from what John Agnew (1993: 252) calls 'hidden geographies' of scale?

The purpose of this chapter is to raise questions about the kind of participation that is afforded by 'user-friendly' PPGIS and the potential for enabling certain local issues to 'jump scale' (Smith 1993) and forge a larger political constituency. The first part of the paper discusses what constitutes public participation, and draws on contemporary critiques of Habermas' notion of the 'public sphere'. The second locates some of the work on PPGIS in this debate by assessing the ways in which it may politicize issues and overcome hidden geographies of scale. Concerns are raised about some forms of PPGIS that may perpetuate instrumental discourses as barriers to democracy and communication in the public sphere.

27.2 RE-THINKING PUBLIC PARTICIPATION

It may be argued that the acceptance of GISs as spatial data platforms and analytic resources upon which informed decisions can be made in many ways legitimizes certain local issues as larger public concerns. The increasingly user-friendly status of this technology, and the development of GIS research and applications within public service institutions, such as universities, combine to make public participation and community focus inevitable. For example, as one of the more cited uses of PPGIS, ongoing work in Minneapolis by a team of researchers at the University of Minnesota seeks to ingratiate the capabilities of GIS and MapInfo to community groups so that they may access publicly available information on local toxic hazards through Toxic Release Inventory (TRI), Petrofund and Superfund sites, and also resource databases on schools, community centres, senior care, daycare centres and local parks (McMaster *et al.* 1997; Leitner *et al.* this volume). The point is that researchers can share their knowledge in a participatory setting that might enable appropriate and ethical kinds of collaboration with community groups. In addition, web-based GIS technology is now relatively accessible to the extent that some argue that criticism of GIS's elitism is no longer valid (Kingston, this volume) and virtual GISs are appropriate conduits for participatory planning in low-income neighbourhoods (Krygier, this volume).

If the costs and hierarchical constraints to access are eroding, then so too are geographic limitations as GIS becomes a valuable tool to highlight local problems in more remote parts of the globe (cf. Laituri, this volume; Jordan, this volume). Indeed, some evidence suggests that PPGIS is not only valuable but is increasingly appreciated by previously skeptical locals (Kyem, this volume). Optimistic rhetoric surrounds much of this research, with phrases like 'empowerment of marginalized people' joining with notions of 'public participation' and 'community involvement'. So, perhaps we have come a long way from John Pickles' (1995) cautioning about technological elitism, but I want to argue that there is still concern about how PPGIS is situated in larger discourses of planning and policy-making. And so I'd like to step back a little from the optimism to consider what precisely is meant by public participation, and what is enabling about GIS technology.

In some ways I am revisiting concerns that Suzanne Michel and I raised about how GIS modelling in the global north is situated in instrumental notions of planning that obfuscated the face-to-face communications of practical day-to-day planning and policy making (Aitken and Michel 1995). Globalization processes expand to most countries in the world the arguments we made about GIS modelling incorporating inappropriately mechanistic and instrumental forms of planning. As Trevor Harris and Dan Weiner point out, current GIS developments in the global south are located within a modernist 'development' paradigm, which is top-down, technicist, and elitist. As a result, Western definitions of knowledge and meaning are perpetuated globally as technical data and spatially integrated decision-support systems (Harris *et al.* 1995; Weiner *et al.* 1995; Harris and Weiner 1996, and this volume). Proponents of PPGIS argue that alternative forms of GIS production are possible and the varied case studies in this volume suggest that these can be context specific rather than general, and they can be communicative rather than instrumental. It might be argued then, that some forms of PPGIS go a long way toward resolving the criticisms of general modelling and instrumentality in GIS that Suzanne and I raised. But public participation carries with it a host of connotations that require careful consideration.

27.2.1 The public and private status of actions

When talking about participation, there is sometimes confusion over the public and private status of *actions* such as environmental activism, 'cleaning up' neighbourhoods, community participation and local planning endeavours. Some feminists voice concern that these actions tend to mirror women's domestic concerns (child-care, housing safety, the environment) without any obvious impact upon larger political and civic cultures (Wilson 1991; Garber 1995; Staeheli 1996). Put simply, there is concern that because some activities amount to 'public housekeeping' they are easily dismissed at the scale of cities or regions by the same 'city fathers' who shrug-off responsibilities over social and local welfare in the first place. This raises not only the issue of the content of local activism and planning, but also the scale at which it is practised and the notion of hidden, and enervating, geographies of scale. The inability of some local activist groups to make headway against city, state, and federal jurisdictions and the dismissal of local groups' concerns for environmental, domestic and child-rearing issues speak eloquently of the persistence of a public political culture that denies access to certain groups. The question that PPGIS raises relates to the access and, by extension, the legitimacy that is offered by technological approaches to the analysis of spatial data and their attendant visualization techniques (see Krygier, this volume). Setting aside the well-worn arguments about the impediments of cost, knowledge about, and access to the technology, does GIS garner legitimacy for local housekeeping issues in a largely patriarchal society? Are GIS-savvy arguments sufficient to enable community-based constituencies to jump scale from the local to larger public political cultures?

Much of contemporary academic understanding of public political culture comes from Jürgen Habermas's critique of a modern lifeworld colonized by the logic of instrumental rationality and strategic management that denies the need for face-to-face contact. Habermas was particularly interested in the conditions that allowed the public sphere to be established, how it was materially transformed over time, and what that transformation meant for the possibility of a progressive formal democracy (Calhoun 1992). He argued that progressive democracy is offset by a contemporary public realm that is alienating, and calls for a 'paradigm shift' from a philosophy of *consciousness* and *self* to a philosophy of *language* and *communication* embedded in his theory of reasoned action (Habermas 1984; 1989). In this formulation, *space* and *action* not only convey information, but also transmit collective political and moral meaning and, consequently, notions of justice. In an important sense, issues of justice come down to *who* gets heard, *how* they get heard, and *where* they are constructed in relation to the public sphere. It seems to me that this is where PPGIS may offer some legitimate public engagement that transcends and transforms notions of local activism out of the arena of public housekeeping.

Lynn Staeheli (1996) points out that the public and private status of *actions* is often equated with the *spaces* in which they occur such as homes, community centres, planning departments, or council chambers. Public policy-makers and analysts tend to equate public actions with public spaces, and private actions with private spaces. Local actions, then, become part of a community politics that loses power through a rigid and static conceptualization of scale with the home being the lowest point of entry and the state the highest. The introduction to this book argues that the potential of web-based PPGIS enables a 'public participation ladder' that begins with a simple 'right to know' but ends with full participation in decision-making, arguing that PPGIS and the web break down potential barriers to participation at each level. Traditional public participation has been limited to 'the right to know', 'informing the public', and 'the public's right to object' (the first three levels of participation). The web, Kingston (this volume) argues, enables higher levels of participation. The ability to define interests, determine agendas, assess risks, recommend solutions and participate in decision-making is enhanced by a web-based platform because, among other things, 'certain psychological elements which the public face when expressing their points of view at public meetings' are erased. This is an important point to the extent that the web muddies up the private/public divide because it establishes a public arena that people can access from the privacy of their homes. But the seemingly magical ability to surf around a virtual council meeting not only hides the technologies, platforms and capital that makes this possible, but it also hides the ways that technologies, platforms and capital create scale. The boundaries, borders and processes of access in the web and in GIS technology combine to create complex stories that include financiers, computer programmers, software and hardware developers, as well as the users of the technologies. By focusing on these seemingly innocuous constructions of scale, it is sometimes possible to uncover manipulations of the web that foster quite profound social and political ramifications.

27.3 SPATIAL STORIES AND SCALE DEPENDENCIES

Stories about the construction of PPGIS are of some interest with regard to how scale relations are created. Moreover, the mechanics of decision-support systems, websites, search engines and issues of who controls access relate not only to the creation of scale relations but also to the use of technologies and graphic interfaces to represent scale. Well-grounded local PPGISs are replete with examples of cartographically based hierarchical scale relations that are created ostensibly to help the user sift through a flood of information by, e.g. zooming up and down from the macro to the micro, but which also represent forms of steering that create boundaries and hide commercial and political influences. In a recent paper, I use several examples of GIS-based websites to make the point that the apparent ease with which scale relations are visualized by the technology point to very complex stories of how sites are linked, how public and private coalitions are created, and how free access is determined and vested with commercial interests (Aitken 1999). Some proponents of PPGIS technologies embrace the virtual environments that they create as a mirror on reality or as an appropriate alternative reality with little consideration of the political and cultural implications of how these environments produce space and scale. The contrivance of scale and the seemingly natural delineation of regions are particularly susceptible to the vagrancies of these new technologies. Space and scale are social constructions and the very notion of 'fixing' them so that we may travel through or up and down them with ease presupposes a particular way of thinking about the world that is based on Cartesian logic and forged out of instrumental and strategic reasoning. This is the 'god-trick' (Haraway 1991), the idea that powerful people are able to take on positions as disembodied master subjects. It derives from a Cartesian objectivity that determines a view of everywhere from no particular location, 'a view from nowhere' (Nagel 1986).

The 'view from nowhere' is facilitated further when Cartesian logic (sometimes in the form of maps, plans and fly throughs) is embellished with technical and instrumental discourses that do not necessarily serve local needs. Sarah Elwood notes how the introduction of technology at the local planning level actually changes the way some residents think about the planning process. PPGIS is empowering in some ways, but she notes that it may also disenfranchise certain sectors of a community (Elwood and Leitner 1998; Elwood 2000, and this volume). Elwood's work points specifically to discourse and the use of language. A longitudinal study of local activists in Powderhorn, a multi-ethnic inner-city neighbourhood in Minneapolis, revealed important changes in the ways some participants used certain words and phrases to actualize their agenda. It was quite clear that those who adopted technical GIS and planning jargon felt more empowered and respected as legitimate partners in the planning process (Elwood 2000). A consequent shift in the goals formulated by some residents was evident, such as the adoption of a 'Variance Matrix' to assist with neighbourhood decisions regarding housing and land-use with particular emphasis on requests for code variances. Elwood argues that the variance matrix shifted local emphasis to a strongly instrumental approach to neighbourhood space, focusing on a quantitative GIS data base and standardized decision-making using the Matrix. Those who were suspicious of the technology or grounded their discourse in everyday language felt alienated, that their voices were not being heard or, worse, that those who learned the technical language were capitulating to an Orwellian system of 'double-speak'.

A different example of scale sensitive research that is concerned about the complex constituencies of the public and the private is Mei-Po Kwan's (1999a,b; 2000) use of GIS technology to help unravel the day-to-day constraints on women's activities. Her concerns about the public and the private focuses on actions inside and out of the home, and how these are contextualized as 'fixity constraints' that emanate from larger urban scale accessibility patterns. She proposes that GISs be used to help interpret constraints on women's daily rounds. Whereas Elwood's study embraces the complexity of the interface between technology and users, and is advancing the way urban planning contexts (and democracy) are theorized, Kwan is concerned specifically about scale and public and private spaces. Unfortunately, the kind of work that Elwood and Kwan engage comprises only a very small part of what constitutes GIS research.

27.3.1 Transforming the public sphere

The implication of what I am saying here is that the maps and discourses that surround PPGIS, planning and environmental management may be the primary means through which boundaries are established and spatial differentiation takes place. This, of course, simplifies an extremely complex set of processes but signals a need to look more carefully at what Gregson and Lowe (1995: 224–225) call the over-identification of geographers (and planners) with the instrumental logic and language of capitalist production of time-space when they really need to focus on 'the full range of geographic scales' including the day-to-day contexts of lived experience that are not ensconced in standardized codes. Staeheli suggests that concerns over scale may help us focus on the *relationship* between activity and space in which the questions that develop are about the *transgression* of certain socially coded spaces and activities. In other words, it is the *constitution* and *transformation* of public space – and hence public and private spheres, and planning at both local and regional scales – that is of crucial importance (Staeheli *et al.* forthcoming). The ways that PPGIS potentially transforms public space is through a reconstitution of scale dependencies, but not if they become embroiled in specialized and potential debilitating technical and instrumental discourses. This begs the question of whether it is possible to jump scale using PPGIS while engaging discourses that are communicative and do not obfuscate, what happens to those who are 'planned for' in the planning process? The question returns me to Habermas and feminist agendas that arise from the critique of his work.

27.4 STRONG PUBLICS AND THE POLITICIZATION OF LOCAL CONCERNS

Nancy Fraser (1997: 70) criticizes any conceptualization of Habermas's public sphere that suggests it is constituted as *anything* outside of the private sphere. A public sphere used in this way conflates scale relations between the nation, the community, and the economy of paid employment on the one hand, and arenas of public discourse on the other. Fraser contends that we need to focus on multiple public spheres, a point that Habermas neglects with his focus only on the emergence of the bourgeois public sphere. *Counterpublics* (as Fraser calls them) contest the exclusionary norms and scale dependency of Habermas's singular bourgeois public. Habermas's public sphere is not only problematically singular, but also presupposes the desirability of separation between civil society and the state and, thus, it distinguishes practice in speech communities from ideological space (i.e. the state). Fraser argues that this distinction promotes *weak publics* where action consists exclusively of opinion formation through communities of speech and does not encompass decision-making which is left to the state. *Strong publics* encompass both opinion formation and decision-making which is authoritative to the extent that strong publics are able to set the terms of the debate for weaker publics. Strong publics help construct any given 'common sense' of the day and they usually figure strongly in defining, however vaguely, what is political in the discourse sense (Fraser 1989: 167).

Perhaps the most poignant example of PPGIS raising a strong public that enables local issues to jump scale revolves around the activism and research that outlines concerns of *environmental racism*. In this research and activism, GIS is used to visualize and conceptualize a form of racism whereby waste and pollution facilities are located disproportionately in poor and minority urban neighbourhoods. Sui (1994) cites the groundbreaking work of Burke (1993) who uses socio-economic data from the Census's TIGER files and TRI data to determine where toxic release facilities are located in Los Angeles County. Burke's statistical analysis suggested a strong association between low income, minority status, and the location of toxic release sites. In general, the poorer the area and the higher its minority percentage, the greater the number of toxic waste facilities in the area. Laura Pulido and her colleagues (Pulido 1996; Pulido *et al.* 1996) caution that works that focus solely on problematic census variables (like race) often miss the importance of evaluating social processes, including class formation and local conceptualizations of racism. As a consequence, Burke's work was followed by other studies on urban environmental health issues that used qualitative methods and ethnographies as well as quantitative assessments (Cole and Eyles 1997; McMaster *et al.* 1997; Jerrett *et al.* 1998). The important point about this work for what I want to say here is that the GIS community enabled a strong public in helping to define 'environmental racism' as part of a larger political culture and, at the local level, qualitative data and quantitative spatial analyses empowered community decision-making.

Fraser (1989: 167) notes that 'it is the relative power of various publics that determines the outcome of struggles over the boundaries of the political.' Her writings deny a homogenized, mechanistic and institutional space for democracy and justice wherein communication and consensus may evolve because such a space denies the practical implications of a social and hierarchical construction of scale that makes access from one scale to another or, alternatively, from a weak to a strong public, difficult. The issue that Fraser broaches in highlighting these kinds of scale relations pivots on how local concerns become politically charged at other scales so that they cannot be dismissed as public housekeeping. Scale as conceptualized here is political, not Cartesian. As suggested in the opening sentence of this chapter, Cartesian logic suggests the myth of actions bound by scale. But as William Bunge (1977: 65) noted many years ago, 'geography recognizes that people operate at various scales simultaneously' and, in this sense, it offers a sophisticated understanding of scale that is missing from other sciences. Disregarding how social relations are scaled to create difference misses a significant potential of PPGIS. Assuming that some unseen democratic process simply propels political action along a clearly delimited trajectory from the local to the urban to the national evokes the metaphor that cream rises to the top or, in reverse, that the trickle-down effects of large-scale economic policies actually help inner-city neighbourhoods. Worse still is the assumption that scale arises simply out of some simplistic notion of cartographic hierarchy and a representation of space in this way enables political struggle, progressive or reactionary, to shape political discourse. Clearly this is not the case, but PPGIS can be part of creating strong multiple publics that augment democracy. They do so by enabling people to become involved at a level that does not obfuscate their daily lives through maps and language drawn from instrumental, strategic logic. Rather, to be effective, the maps and language of PPGIS must communicate spatial stories that clarify and ultimately politicize the issues about which local people feel concern.

REFERENCES

Agnew, J. (1993) 'Representing space: space, scale and culture in social science', in James Duncan and David Ley (eds) *Place/Culture/Representation*, London and New York: Routledge, pp. 251–271.

Aitken, S. C. (1999) 'Scaling the light fantastic: geographies of scale and the web', *Journal of Geography* 98: 118–127.

Aitken, S. C. and Michel, S. (1995) 'Who contrives the "real" in GIS?: geographic information, planning and critical theory', *Cartography and Geographic Information Systems* 22(1): 17–29.

Bunge, W. (1977) 'The point of reproduction: a second front', *Antipode* 9(2): 60–76.

Burke, L. M. (1993) 'Race and environmental equity: a geographical analysis in Los Angeles', *Geo Info Systems* 9: 44–50.

Calhoun, C. (ed.) (1992) *Habermas and the Public Sphere*, Cambridge, MA: MIT Press.

Cole, D. C. and Eyles, J. (1997) 'Environments and human health and well-being in local community studies', *Toxicology and Industrial Health* 13 (2/3): 259–265.

Delaney, D. and Leitner, H. (1997) 'The political construction of scale', *Political Geography* 16(2): 93–97.

Elwood, S. (2000) 'Restructuring participation and power: geographic information technologies and the politics of community-based planning', paper presented at the annual meetings of the Association of American Geographers.

Elwood, S. and Leitner, H. (1998) 'GIS and community-based planning: exploring the diversity of neighbourhood perspectives and needs', *Cartography and Geographic Information Systems* 25(2): 77–88.

Fraser, N. (1989) *Unruly Practices: Power, Discourse and Gender in Contemporary Social Theory*, Minneapolis: University of Minnesota Press.

Fraser, N. (1997) *Justice Interruptus: Critical Reflections on the 'Postsocialist' Condition*, New York and London: Routledge.

Garber, J. A. (1995) 'Defining feminist community: place, choice, and the urban politics of difference', in J. A. Garber and R. S. Turner (eds), *Gender in Urban Research*, Thousand Oaks, CA: Sage Publications, pp. 24–43.

Gregson, N. and Lowe, M. (1995) 'Home-making: on the spatiality of daily social reproduction in contemporary middle-class Britain', *Transactions of the Institute of British Geographers* 20: 224–35.

Habermas, J. (1984) *The Theory of Communicative Action: Reason and the Rationalization of Society, Vol. 1*, Boston: Beacon Press.

Habermas, J. (1989) *The Structural Transformation of the Public Sphere*, Cambridge, MA: MIT Press.

Haraway, D. (1991) *Simians, Cyborgs and Women: The Reinvention of Nature*, London: Routledge.

Harris, T. M. and Weiner, D. (1996) 'GIS and Society: The Social Implications of How People, Space and Environment are Represented in GIS', NCGIA Technical Report 96–97, Scientific Report for Initiative 19 Specialist Meeting, South Haven, MN, 2–5 March 1996.

Harris, T. M., Weiner, D., Warner, T. and Levin, R. (1995) 'Pursuing social goals through participatory GIS: redressing South Africa's historical political ecology', in J. Pickles (ed.) *Ground Truth: The social implications of geographic information systems*, New York: Guilford, pp. 196–222.

Herod, A. (1991) 'The production of scale in the United States labour relations', *Area* 23 (1): 82–88.

Jerret, M., Eyles, J. and Cole, D. (1998) 'Socioeconomic and environmental covariates of premature mortality in Ontario', *Social science & medicine* 47(1): 33–49.

Kwan, Mei-Po (1999a) 'Gender, the home-work link, and space-time patterns of non-employment activities', *Economic Geography* 75(4): 370–394.

Kwan, Mei-Po (1999b) 'Gender and individual access to urban opportunities: a study using space-time measures', *The Professional Geographer* 51(2): 210–227.

Kwan, Mei-Po (2000) 'Gender differences in space-time constraints', *Area* 32(2): 145–156.

McMaster, Robert, Helga Leitner, and Eric Shepherd (1997) 'GIS-based environmental equity and risk assessment: methodological problems and prospects', *Cartography and Geographic Information Systems* 24(3): 172–189.

Nagel, T. (1986) *A View from Nowhere*, New York: Oxford University Press.

Pickles, J. (1995) 'Representations in an electronic age: geography, GIS and democracy', in J. Pickles (ed.) *Ground Truth: The Social Implications of GIS*, New York: Guilford Press, pp. 1–30.

Pulido, L. (1996) 'A critical review of the methodology of environmental racism research', *Antipode* 28(2): 142–159.

Pulido, L., Sidawi, S. and Vos, R. A. (1996) 'An archaeology of environmental racism in Los Angeles', *Urban Geography* 17(5): 419–439.

Smith, N. (1992) 'Geography, difference and the politics of scale', in J. Doherty, E. Graham and Mo Malek (eds) *Postmodernism and the Social Sciences*, London: MacMillan, pp. 57–79.

Smith, N. (1993) 'Homeless/global: scaling places', in J. Bird, B. Curtis, T. Putnam, G. Robertson and L. Tickner (eds), *Mapping the Futures: Local Cultures, Global Change*, London: Routledge, pp. 89–119.

Staeheli, L. A. (1996) 'Publicity, privacy, and women's political action', *Environment and Planning D: Society and Space* 14: 601–627.

Staeheli, L., Mitchell, D. and Aitken, S. C. (forthcoming) 'Urban geography', in G. Gail and C. Wilmott (eds) *Geography in America: Towards the Twenty-First Century*, Oxford University Press.

Sui, D. (1994) 'GIS and urban studies: positivism, post-positivism, and beyond', *Urban Geography* 15(3): 258–278.

Weiner, D., Warner, T. A., Harris, T. M. and Levin, R. M. (1995) 'Apartheid representations in a digital landscape: GIS, remote sensing and local knowledge in Kiepersol, South Africa', *Cartography and Geographic Information Systems* 22(1): 58–69.

Wilson, E. (1991) *The Sphinx and the City: Urban Life, the Control of Disorder, and Women*, Los Angeles: University of California Press.

Chapter 28

Conclusion

William J. Craig, Trevor M. Harris and Daniel Weiner

> Working in Africa, I've found cockroaches vie with power surges as computer killers. It is really a serious issue. Putting mesh over the holes certainly helps, but then you risk overheating the machine – all the more as the weather is hot anyway. Current computer casing seems designed for use in cool climates with low insect densities. It's about time some smart manufacturers jumped into this market gap.
>
> (Rob Denny of One World International; Digitaldivide e-mail list, 26, January 2001)

28.1 INTRODUCTION

There is a spontaneous coming together of community participation with geographic information systems and technologies, and this event is taking place in a diversity of social, political, and geographic contexts. Computer-killing cockroaches in Africa are a stark reminder that PPGIS are indeed context dependent, and this important reality is demonstrated by the case studies in this book.

Within the broad umbrella of what has become known as PPGIS, applications range from Internet-dependent spatial multimedia systems to conventional field-based participatory development methods with a modest GIS/GIT component. These diverse PPGIS case studies have in common the application of GIS to address concerns articulated by community participants and the blending of local knowledge with 'expert' information. As a result, data products and the scale of analysis must be appropriate for the needs of the participating community, and community data access must be assumed. Establishing and maintaining community trust is also essential for successful PPGIS production and implementation. These are critical ingredients for any participatory research and development project, and they indicate the centrality of the *nature of participation* in understanding PPGIS.

There has been a tendency in the past to focus on the technical challenges of community GIS. The case studies in this book suggest, however, that the political complexities inherent in community participation may be larger obstacles for system implementation, and that technical challenges may be overestimated. PPGIS is purposefully value-laden and redefines the meaning of 'accuracy'. Its objective is to include 'peoples' maps and narratives to more fully understand complex socio-economic, cultural and political landscapes. This is why positivist truth statements are used with discretion. The ability of a PPGIS project to influence spatial decision making is, therefore, of central importance in evaluating the potential impact of community GIS initiatives. The digital countermapping of PPGIS tells the spatial stories of marginalized people and communities. Whether this can be translated into real power and political influence remains to be seen. However, the potential for PPGIS to augment place-specific political struggles is intriguing. Stuart C. Aitken (Chapter 27) asks whether 'PPGIS can be part of creating strong multiple publics that augment democracy by enabling people to become involved at a level that does not obfuscate their daily lives through maps and language drawn from instrumental, strategic logic'. This possibility of 'jumping scale' with PPGIS is an important example of how new ITs can impact the terrain of political struggle. All technologies are contradictory, however, and GIS is no exception, for PPGIS simultaneously empowers and marginalizes people and communities.

PPGIS is also a platform for integrating qualitative and quantitative information. This is significant for social scientists because of the historic dualism between researchers who employ qualitative methods and those who employ quantitative methods, and because of the unfortunate difficulties in merging the two. In this way, PPGIS highlights *place*, and in ways that conventional GIS systems normally do not. Such unanticipated benefits of PPGIS are important for geographers and other social scientists who (once again) have discovered the importance of place for scientific enquiry and development projects.

28.2 PPGIS IN PRACTICE

GIS are being integrated in communities to serve many purposes, and with various degrees of effectiveness. The contributions in this book provide a broad view of the current state of PPGIS practice in the United States and around the world. As outlined by Leitner *et al.*, community groups are accessing GIS and data in a wide variety of ways. Some communities use PPGIS to administer and manage territory under their control (e.g. Elwood; Walker *et al.*; Kyem; Jordan; Bond) and to make informed input into local planning processes (Sieber; Parker and Pascual; Ventura *et al.*; Kingston; Bosworth *et al.*). There are also cases where PPGIS has helped communities to develop their own spatial strategies and policies (e.g. Sawicki and Burke; Tulloch; McNab; Laituri; Harris and Weiner). Bosworth *et al.* show the multiple ways a government can make data available to communities, while Kingston, Ventura *et al.* and others demonstrate how PPGIS is rapidly merging with the Internet. Dangermond describes Community 2020 and the Geography Network as examples of growing access to data and analytical services available online.

Sawicki and Peterman document the diversity of institutional arrangements for PPGIS production and implementation. Most PPGIS are not produced and sustained within participant communities. An interesting exception to this is Powderhorn Park (Minneapolis), an inner-city neighbourhood organization that created its own in-house capability to support local day-to-day housing efforts (Elwood). There are many potential paths for developing in-house GIS capability. In Australia, Walker *et al.* collaborated with a group of organizations to create a centre that serves their spatial information needs, needs that could not be met by individual organizations in the area. In New Jersey, NGOs developed GIS with help from the state environmental agency (Tulloch), and the Intertribal GIS Council provides a support base for its Native Americans constituents. But not every organization should, or can, have in-house GIS capability (Sieber). Stonich's coalition, working to resist industrial shrimp farming, does not possess the resources to acquire or maintain an in-house PPGIS.

Many community information needs can be met by conventional maps and reports delivered by a government service center on compact disc or over the Internet. Casey and Pederson call this 'public records GIS', and many cities and counties now provide this type of public data inventory. Such an approach does not, however, fulfill the needs of what they call 'community-based GIS'. A community-based GIS provides relevant local data and is capable of performing spatial analysis for participating communities. For example, the Data and Policy Analysis Group of the Atlanta Project provides sophisticated maps to assist local committees in understanding the nature of prioritized community issues, and to help them develop policy recommendations (Sawicki and Burke).

One of the greatest difficulties with implementing community-based GIS is incorporating complex and socially differentiated information. Harris and Weiner overcome this difficulty with the production of socially differentiated mental maps with particular themes, and then incorporate that information into a spatial multimedia database. Al-Kodmany employs an innovative graphic design method to extend GIS to incorporate block-specific community views. But community organizations do not necessarily represent the views of a majority of community members. Kyem's case study in Ghana identifies the common contradictions inherent in practices of community participation. For example, women are excluded, some people are intimidated by the technology, clans have a difficult time working together, and the existing power structure is often disinterested in empowering citizens. Laituri talks about the unwillingness of indigenous people to contribute data they consider sensitive for fear of being exploited. Elwood discusses how aligning a community group with the culture of municipal government has transformed the internal politics within the participating community. Bosworth *et al.* use a communication pyramid to show that most people choose not to get involved in community activity, but clearly some aspects of organizations and technology tend to systematically exclude some individuals.

A final point about PPGIS practice is concerned with viewing PPGIS as a process. Walker *et al.* demonstrate that communities working together to create a GIS centre helped resolve many conflicts among the participating groups. Process was also a central theme of Jordan's case study in Nepal and the study by Meridith *et al.* in Canada and Mexico. The latter identified 'second order cybernetics' whereby people working together become more aware of their situation, and thus make personal adaptations to accommodate community needs and desires.

28.3 PPGIS FUTURES

The contributing chapters in this book provide many perspectives on how community participation is being linked with GIS and GIT. For the first time it is possible to observe specific instances of what PPGIS is and how it might evolve in the future. PPGIS is presently both academic research and community development planning. Despite the underlying theme of community participation and GIS, the chapters demonstrate that many different variants of PPGIS exist. In drawing upon these chapters, we wish to identify six core themes that both summarize current trends and point toward the future.

28.3.1 PPGIS and socio-geographic context

PPGIS in urban and industrialized regions are increasingly Internet-based. Elsewhere PPGIS combines conventional participatory field methods with a GIS/GIT component. In the future, it is likely that the Internet, with associated spatial multimedia, will become the dominant PPGIS platform. Nevertheless, context and place will inevitably remain important and will influence specific PPGIS production and implementation. As such, there is no universal PPGIS model, and place-based methodologies that navigate local politics and production relations should predominate.

28.3.2 Defining communities and the nature of participation

Community participation is the cornerstone of PPGIS. This volume demonstrates that participation is practised in a diversity of ways. There is a tendeny to homogenize communities, and this is problematic. In the future, community GIS projects must explicitly recognize the complex social differentiation within participant communities. Internet-based PPGIS will further complicate the definition of a community and practises of participation. Virtual communities present significant opportunities and challenges as participation is broadened, but becomes less place-based. Community participation from the home computer will ultimately transform PPGIS in ways that we do not yet understand.

28.3.3 Appropriate technologies and data

PPGIS produces information that is desired by communities, and employs accessible technologies that are not limited to GIS. It is thus possible to question the role of GIS in PPGIS futures. At present, PPGIS uses very limited GIS functionality, and mostly involves digital cartography that links local (qualitative) and expert (quantitative) knowledge. It is questionable to what extent the Internet-based spatial multimedia configurations of the future will rely on the advanced spatial analytical capabilities of GIS. Evolving community spatial decision support systems will likely draw upon a variety of technologies and software interfaces. The role of GIS in this mix is thus ambiguous, and might even bring the term PPGIS into question.

28.3.4 How empowerment and disempowerment occur

PPGIS can empower communities when digital countermaps communicate spatial stories that are integrated into local decision-making. Success stories to date include crime prevention, housing condemnation and renovation, smart growth and land-use planning, natural resource management, and the preservation of indigenous territories. Disempowerment has been observed through the reconfiguration of established community groups and threats of existing elites in response to the introduction of new technologies. Changes in the planning discourse associated with PPGIS have altered existing community power relationships. Disempowerment can take place when government agencies limit data access to community groups that are deemed to be too radical. Unequal access to the Internet also empowers and disempowers simultaneously. To date we have seen only glimpses of this empowerment/disempowerment nexus. As a result, the specific mechanisms by which PPGIS empowers and disempowers people and communities remain fundamental areas for research.

28.3.5 PPGIS as research methodology

PPGIS research contributes to geographic information science and interdisciplinary studies of place. One perhaps unintended consequence of PPGIS for the discipline of geography is a more contextual GIS-based analysis of place. Future PPGIS academic research can thus contribute significantly to geography and to the social sciences in general. As with any participatory research, however, it is imperative that community participants fully understand why they are participating before a project is initiated. The chapters include a number PPGIS case studies that do not directly support community-based spatial decision-making.

28.3.6 Democratizing spatial decision-making

Perhaps the greatest challenge for PPGIS is to contribute to more inclusive spatial decision making. Although the chapters do provide some anecdotal insight as to how this might take place, there has been little systematic long-term evaluation of the contribution of PPGIS to local and regional spatial decision-making. This is understandable given that PPGIS is in its infancy and is only now penetrating the administrative and bureaucratic structures of planning agencies, development organizations, universities, NGOs, and the private sector. The monitoring and evaluation of PPGIS projects over a longer time span will provide insight into the effectiveness of such implementations. Most, if not all, PPGIS projects intend to support community involvement in some type of spatial planning process. The effective transition from PPGIS product to implementation in the context of the local and regional landscape of economics and politics must be a central focus of future PPGIS work.

Index